Manual of Offshore Surveying for Geoscientists and Engineers

Manual of Offshore Surveying for Geoscientists and Engineers

R.P. LOWETH

CHAPMAN & HALL

London · Weinheim · New York · Tokyo · Melbourne · Madras

Published by Chapman & Hall, 2–6 Boundary Row, London SE1 8HN, UK

Chapman & Hall, 2–6 Boundary Row, London SE1 8HN, UK

Chapman & Hall GmbH, Pappelallee 3, 69469 Weinheim, Germany

Chapman & Hall USA, 115 Fifth Avenue, New York, NY 10003, USA

Chapman & Hall Japan, ITP-Japan, Kyowa Building, 3F, 2-2-1 Hirakawacho, Chiyoda-ku, Tokyo 102, Japan

Chapman & Hall Australia, 102 Dodds Street, South Melbourne, Victoria 3205, Australia

Chapman & Hall India, R. Seshadri, 32 Second Main Road, CIT East, Madras 600 035, India

First edition 1997

Printed in the United Kingdom at the University Press, Cambridge

ISBN 0 412 80550 2

A catalogue record for this book is available from the British Library

Library of Congress Catalog Card Number: 96 – 72156

Printed on permanent acid-free text paper, manufactured in accordance with ANSI/NISO Z39.48-1992 and ANSI/NISO Z39.48-1984 (Permanence of Paper).

Contents

List of figures

Front cover photograph supplied ccourtesy of Elf Caledonia Limited.

List of tables

Foreword

The *Offshore Surveying Manual for Geoscientists and Engineers* has been published to satisfy the needs of the following groups of people:

Hydrographic surveyors

The author has some 20 years' experience in hydrographic surveying, but at no time during that 20 years has there been a book available containing the algorithms, equations, techniques and practical information in the detail required by the profession. This manual attempts to fill that gap, providing the hydrographic surveyor with all the necessary tools.

Seismic navigators

Writing seismic navigation processing and quality control software for Unix systems and riding seismic vessels as a supervisor and quality controller have been two of the author's main occupations for the past few years. Much of the experience gained in both of these occupations has been translated into this manual. The work of seismic navigators is actually quite distinct from that of hydrographic surveyors, so chapters explaining the techniques of binning, Kalman filtering and streamer/source positioning systems have been included.

Operations geophysicists

Operations geophysicists are, in general, responsible for the conduct of 2D and 3D marine seismic surveys, and this manual will provide them with a wealth of information to help with planning, specifying and conducting those surveys. In particular, the appendices covering the UKOOA formats and typical navigation specifications will be found invaluable, as will the sections on planning, quality control and processing.

Tertiary education

The author has taught at degree and post-graduate levels in hydrographic surveying and navigation, and during that time longed for a book which could be used as a practical guide for students in offshore surveying. This manual has been organized into discrete, complete chapters which will be used as building-blocks during training and as references during working life.

Programmers

There are many algorithms detailed in this manual. Wherever possible they have been laid out in a manner suitable for computer programming. The chapters on least squares and Kalman filters show the algorithms necessary to code up sophisticated solutions for navigational programs.

This manual assumes that the reader has a reasonable knowledge of mathematics, in particular matrices and their manipulation. Inevitably there will be some readers whose knowledge surpasses the information in this book, and some who are unable to 'see the light' after reading it. Therefore we will be pleased to accept any ideas and suggestions aimed at improving the manual. Such suggestions should be forwarded to:

The QSEA Company

22 Bournemouth Parade

Trigg

WA 6029

Australia

Tel: +61 9 341 8603

Fax: +61 9 245 1424

e-mail eccles@iinet.net.au

R.P.Loweth

21 January 1997

1 Introduction

1.1 Sextants to satellites

The timespan between the exclusive use of sextants, celestial navigation and dead reckoning for positioning surveys to the almost complete elimination of these tools in favour of electronic positioning is only about 40 years. In terms of offshore positioning, the achievable accuracy has increased from about a kilometre to approximately 3 m in that time.

The power of the sextant should not be underestimated. Used to measure vertical angles between the visible horizon and the stars and planets, and relying on the knowledge that the apparent motion of the heavenly bodies can be accurately predicted, an inexpensive mechanical tool made from brass and glass can be used to position a vessel to about a kilometre in the open ocean. It does not have any power requirements other than a steady hand, does not require continuous manning and relies only on clear weather and the user's knowledge of celestial navigation for successful operation.

When two or more sextants are used together to observe horizontal angles between shore marks within a few kilometres of the vessel, the position can be accurately hand plotted to within 3-5 metres. Furthermore, there is no ambiguity with a sextant (qualified by 'circular' fixes).

Since the advent of electronics and radio, the surveyor is now confronted with a huge selection of positioning and navigational systems. These can be listed as follows:

- Optical systems — the sextant as mentioned above.
- Terrestrial radio systems — depending upon frequency these can be used from a few metres to a few thousand kilometres offshore. Their accuracy varies greatly from system to system.
- Acoustic systems — used mostly for relative positioning within a spread but can also be used for absolute positioning in the long baseline mode.
- Sonar Doppler — used as bottom-tracking device to measure course and speed.
- Inertial navigation systems — based on gyros and accelerometers and used in submarines.
- Satellite systems — now in very common use worldwide and accurate to a few metres when used in the differential mode.
- Laser — used for short-range relative positioning.
- Digital magnetic compasses — used for streamer shaping.
- Gyro compasses — used to obtain absolute direction.

1.2 2D and 3D marine seismic surveying

The timespan from the evolution of 3D seismic surveying to the present day is only a few years, particularly with regard to in-spread navigation techniques. Prior to that, seismic surveying consisted of acquiring a line of common mid-points (CMPs), and processing that data to produce a two-dimensional, vertical section through part of the earth's crust — hence the term 2D seismic. The evolution of 3D techniques has seen the basic operation expand into recording data over an area, rather than along a line. The area to be surveyed is divided into rectangles or bins, and the requirement is to steam through the area and populate each bin with a certain number of CMP data points. Once processed, the data forms a three-dimensional cube or volume which can be viewed and analysed from any direction. The aim is to ensure that there are enough data points in each bin that when the seismic data associated with those data points is stacked the resulting section is complete, noise-free and well defined. Since it is now common to tow several streamers each 4 km long astern of the vessel, and to fire a shot every 10 seconds or so using a 'flip-flop' source, and because the accuracy required for each CMP is often less than 10m (absolute) the navigation requirements for 3D surveys have become critical. Sometimes there is a tendency to overload the spread with observations, in the hope that since there are so many observations, the adjustment is bound to succeed. Conversely, sometimes the geometry is very weak and the resultant CMP positions are known with little reliability.

3D seismic was originally conceived as a means of looking at problematic development and appraisal projects, but it is gaining increasing importance as an exploration tool. For many offshore operators, 3D seismic can occupy up to 90% of the seismic budget. Needless to say, the operation is very expensive and the amount of exploration carried out is driven largely by the expected success rate and the current oil and gas prices. Further development of the technique over the past few years has led to ocean bottom cable surveys, where the streamers are laid on the seabed and only the source is mobile. This technique has significant advantages in shallow-water transition zone areas.

Although 3D acquisition has led to multi-streamer, multi-source arrays designed to populate more than one line of bins at a time, the seismic data aspect of the acquisition is not much different from 2D. But the navigation requirement has changed considerably. From simply positioning the vessel and estimating the streamer angle from radar, sophisticated techniques have had to be developed to cater for the most demanding positioning requirement of 3D surveys, which is to position the CMPs to a predetermined tolerance. This tolerance varies with the bin size, but later in the manual it is shown that a tolerance of about 5 m for the vessel and about 8 m for every CMP, given a bin size of 25 m, is realistic. Even the word "tolerance" itself needs a concise definition, and this will also be dealt with.

The level of sophistication of navigation systems has increased dramatically over the past five years, and so has the power of computers. Unfortunately, the link between the two, i.e. the software, has not advanced so quickly, and the industry is currently in a transitional stage as systems move from the old mini-computer style of environment towards the new scalable processor (SPARC) open architecture systems. It has been said that the power and speed of computers is doubling every 18 months, and the choice of available machines is quite staggering. For this reason a chapter has been dedicated to computer technology, in which not only hardware but the techniques and present limitations of real-time interfacing are reviewed.

Whilst the prime target of 3D positioning is to get all the CMPs in the right bins relative to each other, there is always the possibility of there being a systematic error in the whole dataset; considerable effort has been made in this manual to demonstrate practically how this can be prevented.

2 Datums and projections

2.1 Geodetic nomenclature

Geodesy is the determination of the shape of the earth. The subject is complex, and this chapter gives a general understanding combined with many of the basic formulae required for the UTM projection. The shape of the earth is very difficult to map mathematically, so many different models have been created to represent a 'best fit' solution. Most of these models use a primitive shape, such as a sphere or ellipsoid to approximate the shape of the earth.

An appreciation of the various terms used in geodesy is given in the Glossary.

2.1.1 The earth's surface

The surface of the earth is quite irregular and almost impossible to define mathematically. It is roughly ellipsoidal in shape, and the maximum departures from an ellipsoid are about 8km (these departures being the highest mountain ranges and the deepest ocean trenches). However, the earth's surface is very important, as it is the surface on which most observations are taken and on which most points of interest lie. The fact that it is nearly impossible to define mathematically means that some other surface must be used to approximate its shape.

2.1.2 The geoid

The geoid is the equipotential surface of the earth's attraction and rotation. This surface coincides (for the most part) with mean sea level in the open ocean, and also is nearly coincident with the imaginary surface produced by assuming the sea was free to flow in frictionless channels under the land masses. An excellent treatment of the meanings of gravity, potential and equipotential surfaces is given in Bomford (1985) and Cross, Hollwey and Small (1981).

The geoid can be quite well approximated by an ellipsoid; the maximum departure from a best-fitting ellipsoid is only about 110m. The geoid is of great importance in offshore surveying because many observations are directly or indirectly related to it — for example the reduction of soundings to mean sea level, and the reduction to mean sea level of all measured ranges.

2.1.3 The ellipsoid

There is only one surface of the earth, and only one geoid; unfortunately there are very many ellipsoids. Ellipsoids are ellipses rotated about their minor axes and they represent the nearest simple mathematical shape to use as a datum for observations. Many ellipsoids have evolved over the past 300 years, and it is not sufficient to express a position purely in terms of latitude, longitude and height; such a set of coordinates cannot be related to a position on or near the earth's surface without also specifying the ellipsoid to which those coordinates are related.

Figure 2.1 shows the relationship between the surface of the earth, the geoid and an ellipsoid.

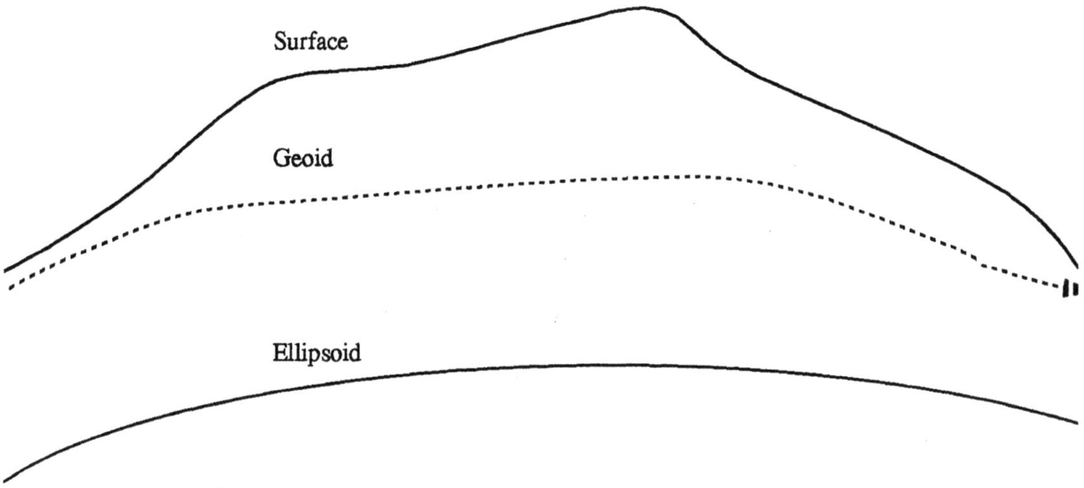

Figure 2.1 Relationship of surface, geoid and ellipsoid.

There are some potential problems. For example global positioning system (GPS) receivers refer to heights based on an ellipsoid while a spirit levelled point refers to the geoid: Figure 2.2 shows the differences.

N is called the geoid-ellipsoid separation. h is the orthometric height derived by spirit levelling and H is the ellipsoidal height. The three are connected thus:

$$H = N + h \tag{2.1}$$

This is actually an approximation since heights are always measured normal to their respective datum and the geoid and ellipsoid may not be parallel at the point of measurement. This non-parallelism is always less than 60" of arc, so for practical purposes it can be disregarded.

What cannot be disregarded is the possibility of measuring height on different datums. This happens all too often in offshore surveying and it is a potent source of gross error. The rule is to ensure that all quoted heights are referred to the same datum.

2.2 Reference systems

There are a number of different reference systems that can be used for offshore surveying. This section concentrates on describing the models available; a discussion on their practical use is given in the Chapter 4.

A geodetic datum is defined by the size of a reference ellipsoid and by its position and orientation with respect to the earth. There are eight parameters needed to define a geodetic datum:

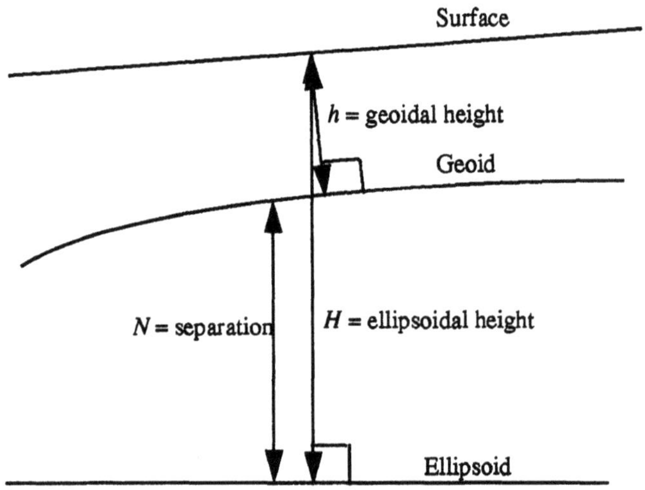

Figure 2.2 Geoid-spheroid separation.

Firstly any ellipsoid can be described in terms of two parameters, a and b.

a is called the semi-axis major and b the semi-axis minor. Their relationship is:

$$\frac{x^2}{a^2} + \frac{y^2}{b^2} = 1 \tag{2.2}$$

The flattening f is given by:

$$f = \frac{(a - b)}{a} \tag{2.3}$$

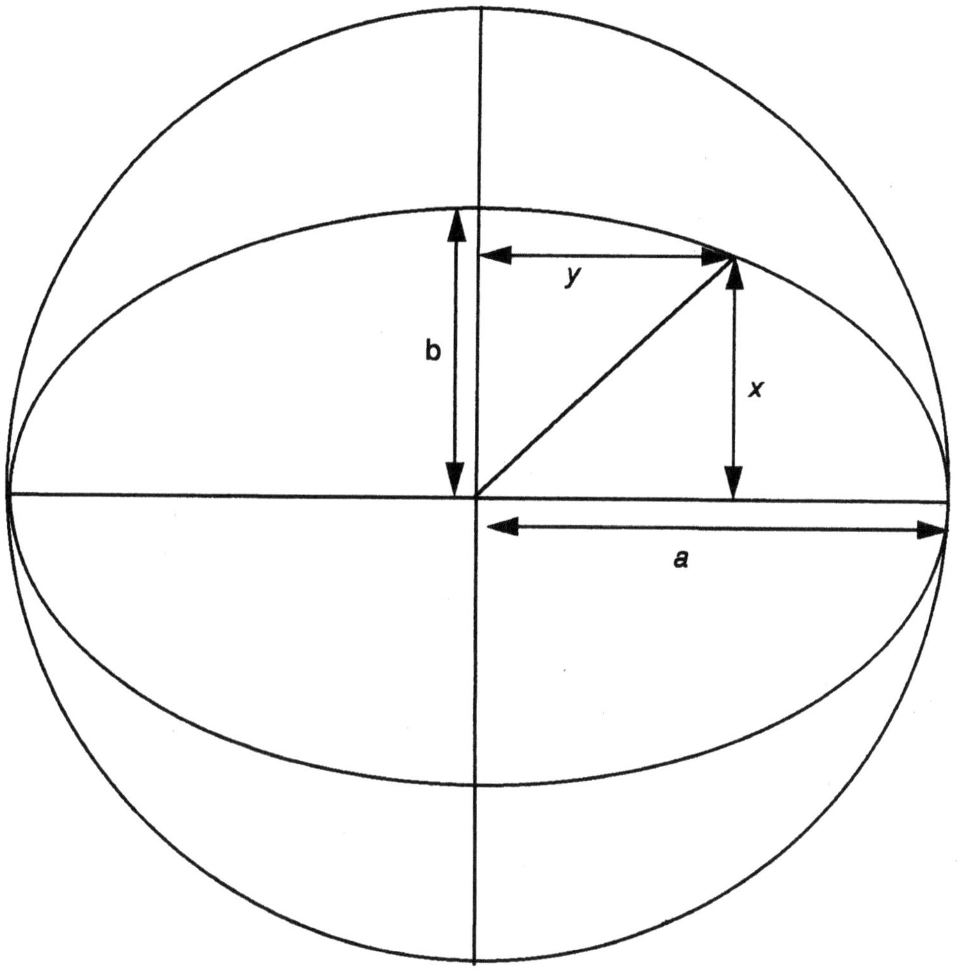

Figure 2.3 Ellipsoid parameters.

and the eccentricity e by:

$$e^2 = \frac{(a^2 - b^2)}{a^2} \tag{2.4}$$

a and *b*, or *a* and *f*, or *a* and *e* are the first two parameters of a geodetic datum.

Three parameters are now required to fix the datum's centre with respect to an earth-based fixed point. It would be a very neat solution if there were only one ellipsoid in use and if it were centred on the earth's centre of mass. Ellipsoids centred in this way are called geocentric, and a good example is WGS84, which is used as the datum for the GPS coordinate system and which is gaining in popularity throughout the world. But there are a large number of other ellipsoids in use, most of which are based on a locally fixed datum. AGD84 (currently used in Australia) is a good example — it is based on the latitude, longitude and height of a survey station in the centre of the continent.

Geocentric ellipsoids will have a reasonable fit worldwide, whereas locally fixed ellipsoids such as AGD84 are unlikely to be a good fit except in the area for which they were designed.

The last three parameters needed to fix the ellipsoid as a datum are the rotations in three mutually perpendicular axes to orient the ellipsoid. This orientation is usually such that the minor axis is parallel to the axis of rotation of the earth. This itself is a sophisticated subject since the spin axis of the earth changes with time, and the ellipsoid has to be oriented either with the spin axis direction at a certain time or with the mean direction over a period of time.

2.3 Projections

A projection is a method for transferring a point on one surface to another surface.

The overhead projector is a projection system. It projects a flat surface on to a flat surface. In doing so it also decreases the scale and increases the size of the image on the projected surface.

There are many projection devices, ranging from movie projectors to microscopes; all of them modify the original image to produce a new image.

The surveying-related use of projections is more complex. There are two problems:

- to project a curved surface on to a flat surface
- to project a flat surface on to a curved surface.

The problem in dealing with a curved surface can be described with a thick skinned navel orange. It is very difficult to peel the skin off an orange and then flatten the skin. No matter what is done the skin will not be truly flat. There are several approaches that could be tried. One would be to cut the skin into a series of slivers and lay them next to each other. The result would be a very ugly configuration such as in Figure 2.4. Quite obviously, this approach would not be suitable for map-making, purely from an aesthetic standpoint. Interestingly enough, if the slivers are made narrow enough we can preserve shape and scale.

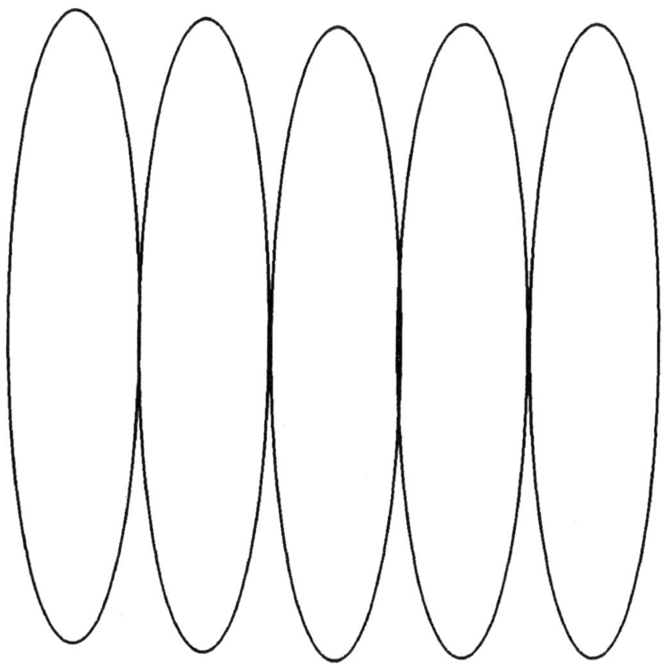

Figure 2.4 Cut orange.

The skin can be cut up into a grid, but the result is still poor as in Figure 2.5. Note, however that these small areas are almost flat. This is useful. If the area of the survey is small (as in a site survey) then the curvature of the earth can be ignored, within limits, and the assumption can be made that distances and angles measured can be plotted directly. There is another potential use here; if the origin of the 'projection' can be thought of as the navigational reference point of a vessel towing seismic equipment, then the entire seismic spread (i.e. streamers, sources, tailbuoys, sleds) can be thought of as flat. The secret here is to continuously move the centre of the 'projection' with time.

The problem has produced many interesting approaches with no one solution being the answer for all applications. The solution to mapping a curved surface on to a flat surface again relies upon making some assumptions and approximations.

We can take a shape that will lay out flat when only one cut has been made. Make it the same size as the earth. Project the points on to it. Cut it and lay it flat. This is the basis for nearly all projections used. There are some projections that use multi-dimensional shapes (4D,5D etc.) but these are not really of interest to the hydrographic surveyor.

Figure 2.5 Grid cut.

Figure 2.6 shows how cone and cylinder projections are laid out.

There are some problems that arise out of these projections:

- SCALE — only at the points where the cone or cylinder touches the surface of the earth is the scale unity. For the cylinder the scale changes constantly but evenly further away from the intersection line. The cone changes scale away from the intersection line but at a non-linear rate.
- ANGLES — depending on how the projection was made angles can change.
- SHAPE — the shape of objects will vary.
- AREA — the areas of closed figures on the Earth will change depending on the projection.

No one projection will preserve all of the above. Most projections are a trade-off between all elements. The variety of projections has occurred for two reasons. To enhance the scale ratios it is possible to intersect the shape in two lines. Figure 2.7 shows the methods for intersecting and the resultant deformation circles.

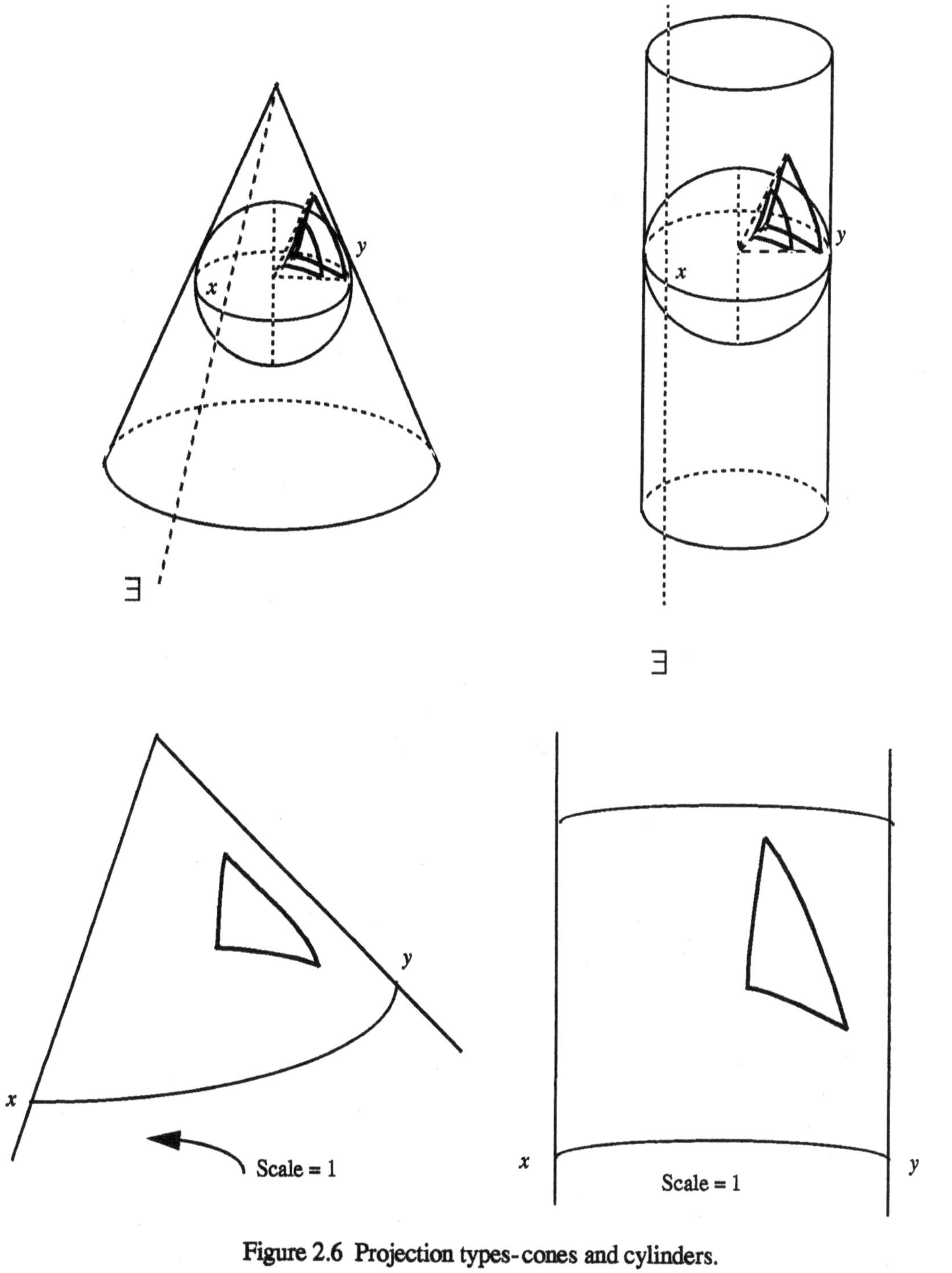

Figure 2.6 Projection types-cones and cylinders.

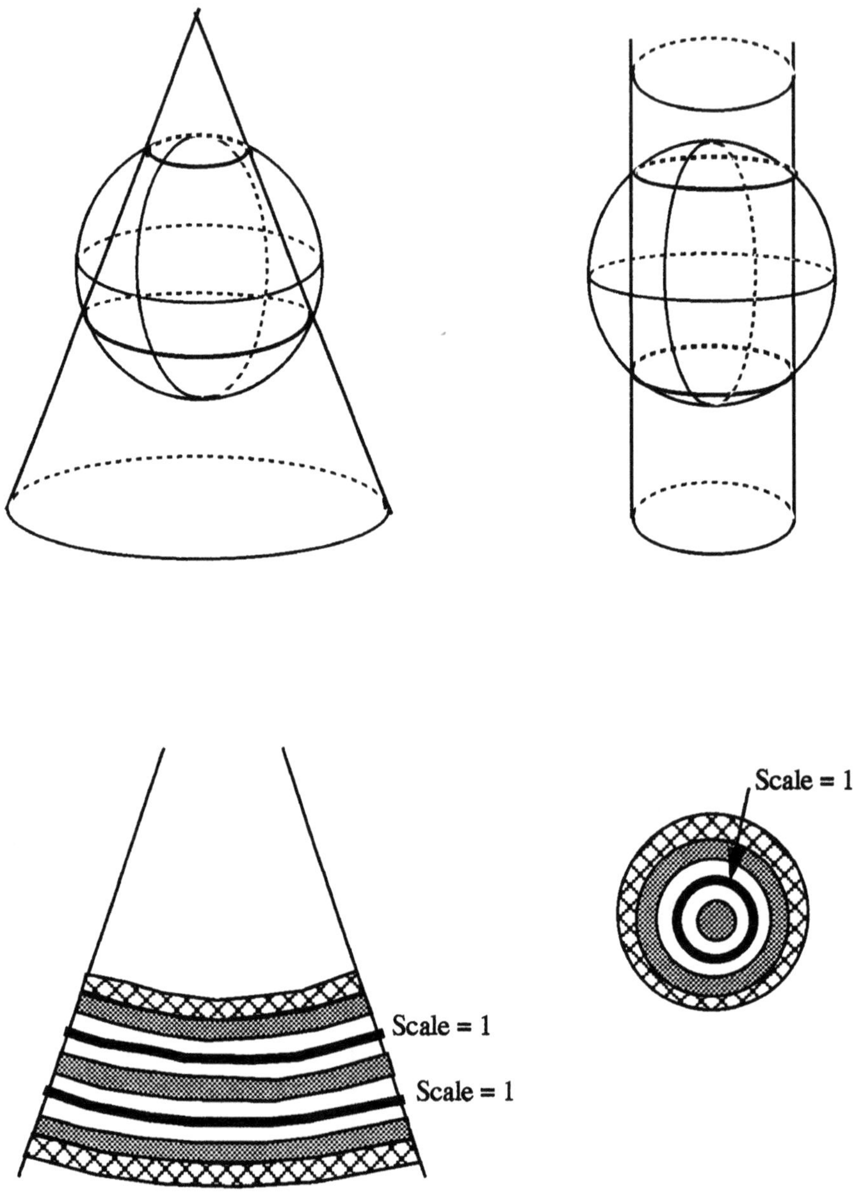

Figure 2.7 Methods of intersection.

There are many methods of projection. Figure 2.8 shows some of the different ways projections can be made onto cones and Figure 2.9 on to cylinders.

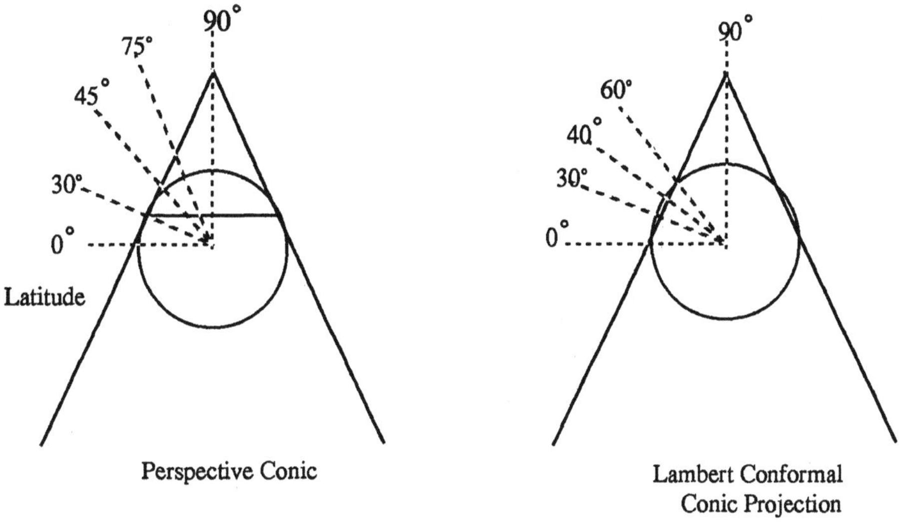

Figure 2.8 Conic projections.

The differences in the method used to project result in differing attributes for the projection.

The Universal Transverse Mercator Projection (UTM) is slightly different from the rest. It is an intersecting cylinder projection. The area swept out between the intersection lines is six degrees. This is the only area used by the UTM. To cover the earth the cylinder is moved six degrees and the next section drawn. Therefore this requires 60 'strips' to cover the earth.

Figure 2.10 shows one of these six-degree zones and details the scale changes that occur.

The scale changes only occur across the zone (east—west) but are constant along the zone (north — south). UTM is the basis for the Australian Map Grid (AMG) and for most projections used around the world.

2.4 The Australian Map Grid (AMG)

The AMG is a plane rectangular coordinate system adopted in Australia on 23 April 1965. It is intended to cover the mainland, Tasmania, and close to the extents of the continental shelf. The AMG was readjusted in 1982; these results were adopted in 1984 and are referred to as GMA82 superseding AMG66.

The AMG is a transverse Mercator projection with zones of six degrees and overlap belts of 80 kilometres on each grid boundary.

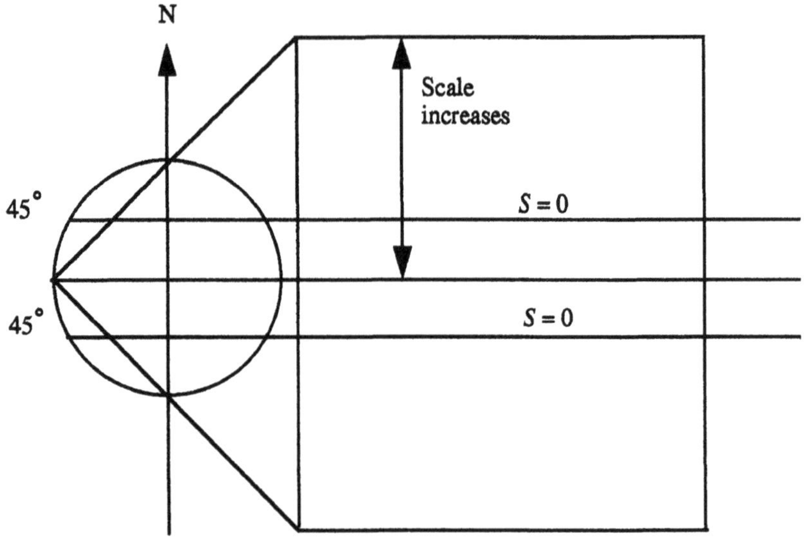

Galls Stereographic Cylindrical Projection
(cylinder fitted at 45° N/S.)

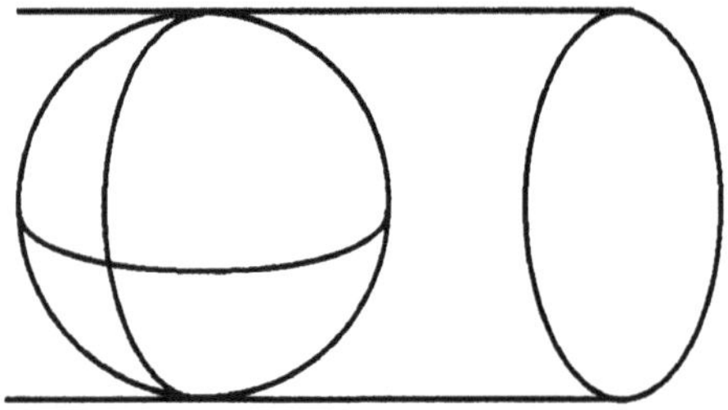

Transverse Mercator
Projection

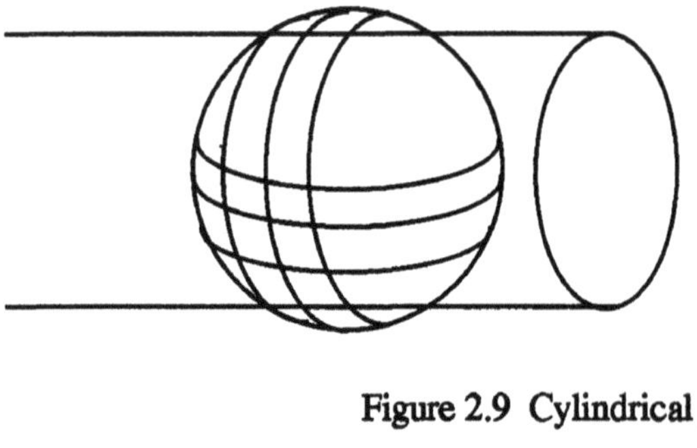

Universal Transverse
Mercator Projection

Figure 2.9 Cylindrical projections.

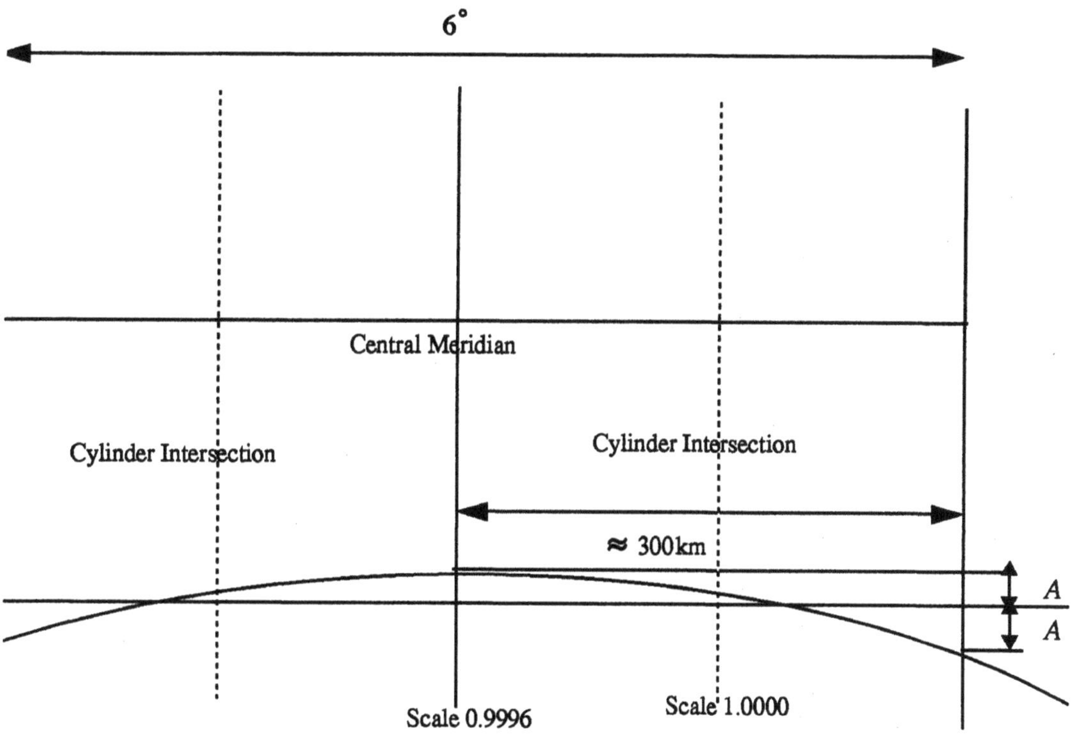

Figure 2.10 UTM zone.

The AMG uses a different spheroid from the WGS84 coordinate system, the difference vector being approximately 195 m.

The AMG zones which refer to each six-degree belt of the projection are numbered from 49 to 57, west to east. The central meridian has its origin at the intersection of the meridian with the equator. This is then translated into a false origin by adding 10000000m to the northings and adding 500000m to the eastings. It is the false origin that is used as the start point for eastings and northings.

There is a relationship between zone number and the central meridian:

$$Zone = \frac{(long + 180)}{6} \tag{2.5}$$

Figure 2.11 summarizes these points.

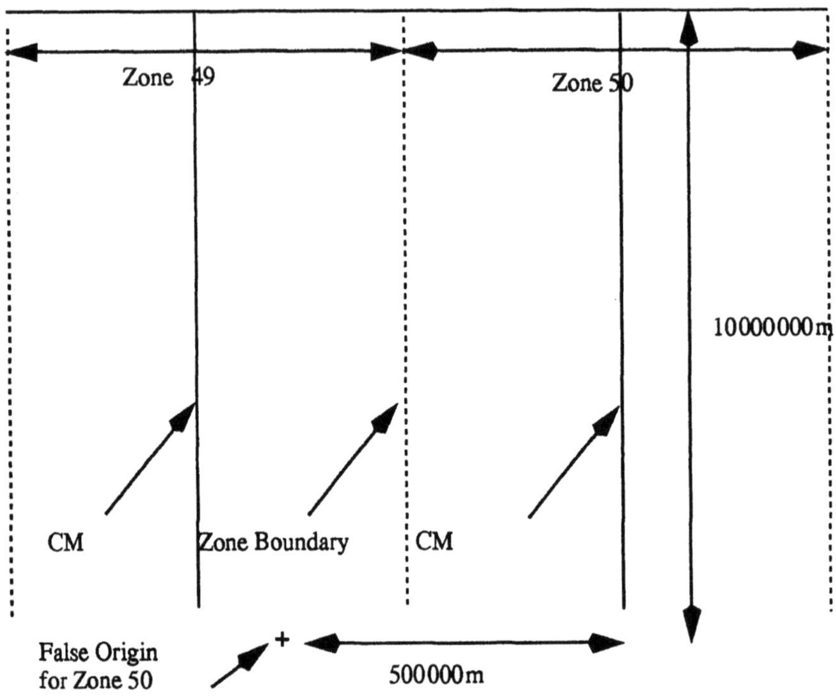

Figure 2.11 Summary of AMG.

It is not possible to quote a set of AMG coordinates as northing, easting and height. This is a coordinate common to all eight AMG zones. Therefore we must quote either a zone number or central meridian to give the location of one point in the AMG system.The AMG is defined as:

$a = 6\,378\,160$ m

$f = 1/298.25$ exactly

with the datum being Johnston Geodetic Station:

Geodetic latitude 25° 56' 54".551 5 S

Geodetic longitude 133° 12' 30".077 1 E

Spheroidal height 571.2m

The cylinder is not round in section but elliptical.The purpose of the AMG is to represent large areas of the earth's surface on a map. To allow ease of computation the map coordinates are plane rectangular. This means that simple trigonometry holds true. Distances and bearings can be scaled off the map and areas calculated. Field measurements are also relatively simple.

The relationship between what is recorded and what is mapped is more difficult. All distances, angles and bearings must be converted from the field to the map. This also means that the setting out of AMG coordinates is difficult as the corrections must be made for every point set out.

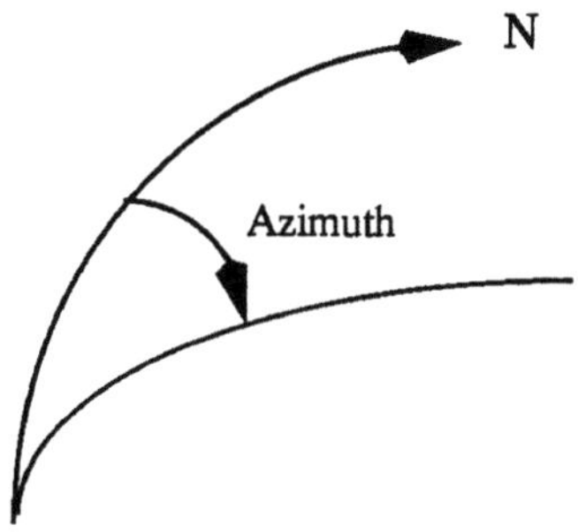

Figure 2.12 Azimuth.

- Azimuth — a horizontal angle measured from the spheroidal meridian clockwise from north through 360 degrees.

- Distance — there are several which can be used. The spheroidal distance is the shortest distance between two points on the spheroid. If this is projected on to the grid it appears as a curve. The plane distance is the length of a straight line on the grid. The plane distance is some times incorrectly called the grid distance, but for our purposes the difference is minor.

- Scale factor — a result of the projection. It is 0.9996 at the central meridian and at 3 degrees either side is 1.0003. As it is not linear any line drawn has a variable scale along it, so a variable scale factor must be applied. Only lines running north—south will have a constant scale factor. Scale factor makes reasonably large differences to the computation of lengths so it must always be applied.

- Grid convergence — the factor added to an azimuth to obtain a grid bearing. As will be shown later, grid convergence carries the opposite sign in the southern hemisphere from that in the northern hemisphere, and is a common source of gross error.

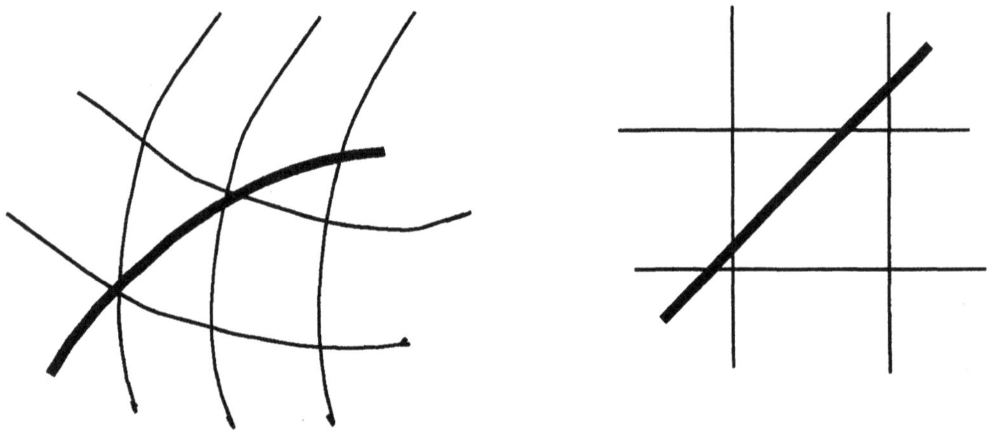

Spheroidal Distance Plane Distance

Figure 2.13 Spheroidal and plane distances.

- Grid bearings — bearings obtained from the reduction of AMG coordinates. Arc-to-chord correction is required to convert a grid bearing to a plane bearing.

Figure 2.14 is a summary of some of the major terms commonly used in conjunction with AMG.

The AMG has several drawbacks for most exploration work. The northings and eastings are particularly long. If millimetres are expressed, then the northing is ten places long. The formulae used to calculate on the grid often use power series expansions resulting in a minimum of 20 significant figures being required to maintain the accuracy of the calculation. The storage requirements in both hard disk and RAM memory are increased significantly and the speed of the computer is slowed considerably as it has to maintain all of these significant figures.

There is also the possibility of incorrectly applying corrections such as arc to chord, grid convergence and meridian convergence. This is due to the algebraic sign reversal that occurs when moving from the northern hemisphere to the southern hemisphere. This has presented itself several times recently with contractors based in the northern hemisphere working around the southern hemisphere. The best solution is to have a QC system developed in the southern hemisphere on board for all surveying operations.

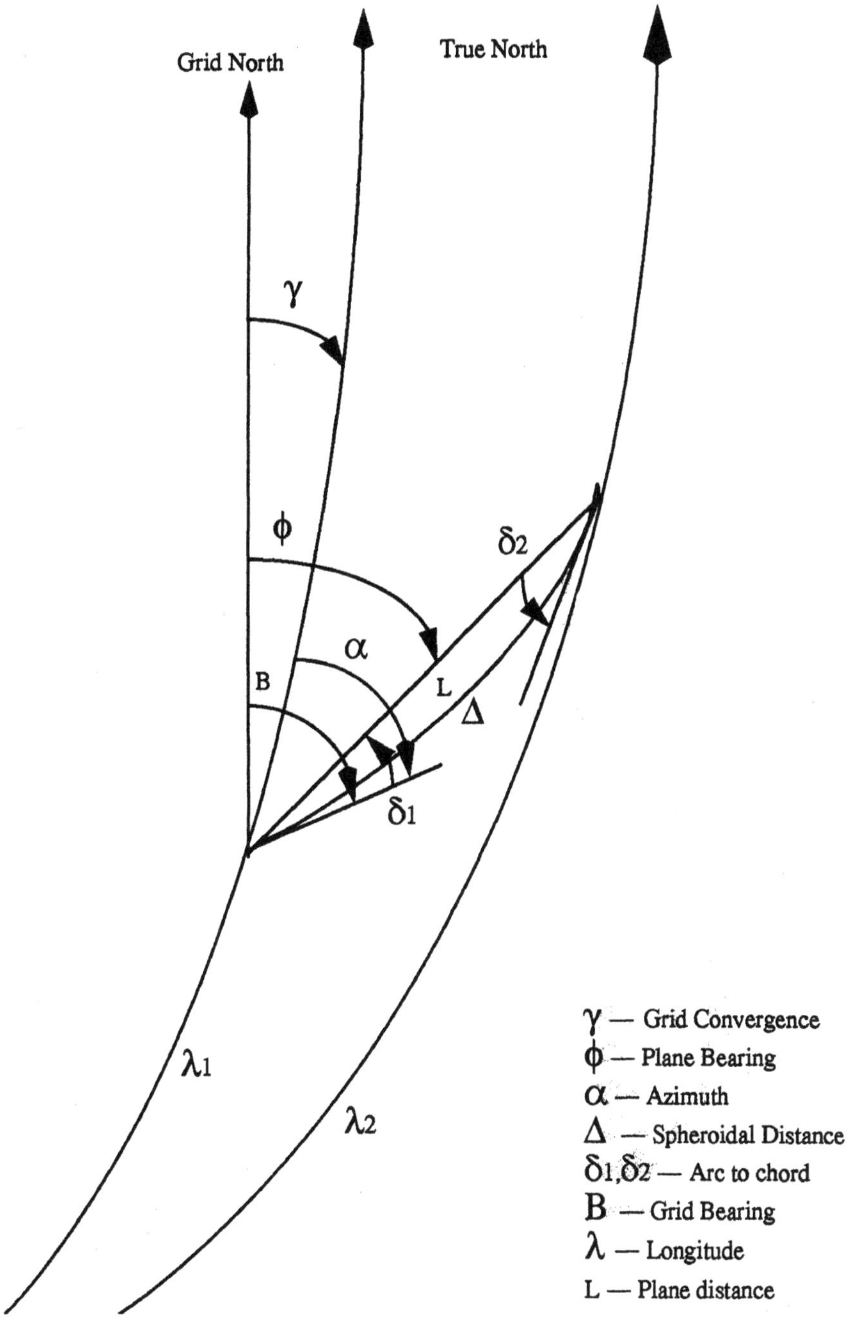

Figure 2.14 Summary of AMG terminology.

Another solution is to use a local coordinate system for the operation and calculation of the survey and then if necessary convert the points to AMG after the survey is completed.

The generalized formulae given at the end of this chapter will work on all TM projections, including AMG.

2.5 World Geodetic System 84 (WGS84)

The World Geodetic System 84 has been adopted by the US government for all positioning activities around the world and to reduce the many datums and coordinate systems that were used in the US to one.

WGS84 is the system employed by GPS satellites. There are some differences between the AMG and the WGS system. Most of the formulae used are the same but the datum is very different. WGS84 is a geocentric datum. It uses the gravity mass point centre of the earth as its zero origin with the axis defining north being in a different location to north as described by the AMG. The mass point observations are highly accurate and are derived from sources such as the motion of satellites, very long baseline interferometry, lunar laser rangings and precise geodetic observations.

The separation between AMG and WGS is a translation. If a map in each system is overlaid, then the coordinates are shifted by a 195m vector. This is a common error to find in surveys, but one that can always be avoided by careful calibration and checking.

The WGS84 has only minor limits to the extents of its use. It can be used on, above or below the Earth. It is used to refer to the positions of satellites replacing the old right ascension and declination references. The WGS84 should not be used North of latitude 80 and South of latitude $79^{\circ}30'$. These areas provide too many anomalies and should be represented with a polar stereographic projection.

2.5.1 Comments on the WGS 84 EGM

The official WGS84 Earth Gravitational Model (EGM) is complete through degree (n)=180, and order (m) = 180 and is comprised of 32755 geopotential coefficients. While such a model is needed for certain applications on the earth's surface and in near-earth space, it is not required and is impractical to apply for many applications. Official DMA guidelines are not complete regarding when to apply the complete WGS84 EGM and when the use of various truncation levels is appropriated. However, the following points can be made.

The WGS84 EGM through $n = m = 180$ is to be used when calculating:

- WGS84 geoid heights via a spherical harmonic expansion
- WGS84 gravity disturbance components (or deflection of the vertical components) via a spherical harmonic expansion
- WGS 84 $1^{\circ}\text{x}1^{\circ}$ mean gravity anomalies via a spherical harmonic expansion

Expansions to this degree and order ($n = m = 180$), and higher, are needed to accurately model variations in the earth's gravitational field on or near the earth's surface.

The WGS 84 EGM through $n = m = 41$, which reflects the contribution of ground-based satellite tracking data (as does the complete $n = m = 180$ model) , is more appropriate for satellite orbit calculation and prediction purposes. The use of higher degree and order models for such applications is not recommended at this time. However, for each satellite orbit class, the Department of Defense [DOD] users will need to conduct orbital analyses and ascertain the EGM truncation level that is best' for the satellite involved, document results, coordinate recommendations as necessary, and then use this particular truncated EGM for the satellite being analysed.

The WGS84 EGM through $n = m = 12$ is the UNCLASSIFIED portion of the complete $n = m = 180$ model. It is authorized for use only in two instances:

1. *When a situation or application requires the use of an UNCLASSIFIED EGM, thereby precluding use of the CLASSIFIED WGS 84 EGM;*
2. *When satellite orbital analyses reveal that the WGS84 EGM truncated at n - m = 12 (or lower) provides results of acceptable accuracy.*

The WGS84 EGM coefficients through $n = m = 41$ were obtained from a least squares solution of a normal equation matrix developed by combining individual normal equation matrices formed from Doppler satellite tracking data, satellite laser ranging data, surface gravity data, oceanic geoid heights deduced from satellite radar altimeter data, NAVSTAR Global Positioning System (GPS) data, and 'lumped coefficients'. The effect (contribution) of coefficients through $n = m = 41$ was removed from a worldwide 1° x 1° mean gravity anomaly field h, leaving a worldwide residual 1° x 1° mean gravity anomaly field. An independent determination of EGM coefficients through $n = m = 180$ was then made using the residual field. The coefficients through n=m=41 from the least wquares solution and the coefficients above $n = m = 41$ from the independent determination comprise the $n = m = 180$ WGS84 EGM.

2.5.2 Spherical harmonic expansion of the geopotential

GM	= *Earth's gravitional constant*
a	= *Semi-axis major (of the ellipsoid)*
r	= *Radial (geocentric) distance*
n,m	= *Degree and order*
ϕ, λ	= *Geocentric latitude, geocentric longitude*

$$U = \frac{GM}{r}\left|1 + \sum_{n=2}^{\infty}\sum_{m=0}^{n}\left(\frac{a}{r}\right)^n (\bar{C}_{n,m}\cos m\lambda + \bar{S}_{n,m}\sin m\lambda)\,\bar{P}_{n,m}\sin\phi'\right. \quad (2.6)$$

$$\bar{P}_{n,m}(\sin\phi') = \text{Normalized associated legendre function} \quad (2.7)$$

$$\bar{C}_{n,m} and \bar{S}_{n,m} = \text{Normalized geopotential coefficients} \quad (2.8)$$

2.5.3 Conclusion

There are many more projections and datums that can be used. The majority will exhibit similar characteristics to those described in this discussion, but will have minor although often significant impacts on the survey.

When problems arise with projections they are often linked to the incorrect application of algebraic signs or to the use of the wrong datum. The wrong datums should manifest as translations of the coordinates. Incorrectly applied signs will be much more difficult to find. They are, as we have seen, non linear in nature and can result in small differences. In the worst cases, such as the incorrect application of grid convergence or scale factor, the errors will render the work useless, until they are corrected.

2.6 Transverse projection formulae

The formulae presented here do not carry proofs. They have been written down for ease of coding into computer programs. The formulae are not comprehensive, but they do cover the most commonly encountered computations.

2.6.1 Symbols

ϕ	geodetic latitude, negative south of equator
λ	geodetic longitude, positive eastwards
λ_0	geodetic longitude of a central meridian
ω	geodetic longitude measured from a central meridian, positive eastwards
E'	easting measured from a central meridian, positive eastwards
N'	northing measured from the equator, negative southwards
E	E' + 500000m
N	N' (+1000 000m in the southern hemisphere)

ρ	radius of curvature in meridian
ν	radius of curvature in prime vertical
α	azimuth, clockwise through 360deg from true north
β	grid bearing, clockwise through 360deg from grid north
θ	plane bearing, clockwise through 360deg from grid north
γ	grid convergence, positive when grid north is west of true north in southern hemisphere: $\beta = \alpha + \gamma$
δ	arc to chord correction with sign defined by
	$\theta = \beta + \delta = \alpha + \gamma + \delta$
s	spheroidal distance
s'	geoidal distance
S	grid distance
L	plane distance
m	meridian distance, true distance from the equator, negative southwards
G	mean length of an arc of one degree of the meridian
σ	m/G
a,b	major and minor semi-axes of the spheroid
f	flattening
e^2	eccentricity squared
k_0	central scale factor = 0.9996
k	point scale factor
K	line scale factor
t	$\tan \phi$
ψ	ν/ρ
ϕ'	footpoint latitude = latitude at which $m = N'/k_0$
R^2	$\rho\nu$
r^2	$R^2 k_0^2 = \rho\nu\, k_0^2$
h	spheroidal height
H	height of a point above the geoid (or orthometric height)

N $h - H$ = geoid — spheroid separation

2.6.2 Reduction of distance to spheroid

Both physical and geometic corrections must be applied to a measured wave path to reduce it to a spheroidal distance. The physical corrections include delays and velocity corrections and are dealt with elsewhere in this book. All geometric corrections can be combined into a single equation (which is rigorous) as follows:

$$s = 2R\alpha \text{as in}\left(\frac{d}{2R\alpha}\right) \tag{2.9}$$

where

$$d = \left[\frac{x^2 - \langle h_A - h_B \rangle^2}{\left(1 + \frac{h_A}{R_\alpha}\right)\left(1 + \frac{h_B}{R_\alpha}\right)}\right] \tag{2.10}$$

and x is the wave path chord.

2.6.3 Meridian distance - series method

$$m = a\,(A_0\phi - A_2 \sin 2\phi + A_4 \sin 4\phi - A_6 \sin 6\phi) \tag{2.11}$$

where

$$A_0 = 1 - \frac{e^2}{4} - 3\frac{e^4}{64} - \frac{5e^6}{256} \tag{2.12}$$

$$A_2 = \left(\frac{3}{8}\right)\left(e^2 + \frac{e^4}{64} + \frac{15e^6}{128}\right) \tag{2.13}$$

$$A_4 = \left(\frac{15}{256}\right)\left(e^4 + \frac{3e^6}{4}\right) \tag{2.14}$$

$$A_6 = \frac{35e^6}{3072} \tag{2.15}$$

2.6.4 Footpoint latitude

The footpoint latitude is the latitude for which $m = N'/k_0$. This can be calculated directly, but three other values n, G and σ must be calculated first.

$$n = \frac{a-b}{a+b} = \frac{f}{(2-f)} \tag{2.16}$$

$$G = a(1-n)(1-n^2)\left(1+\frac{9}{4}n^2+\frac{225}{64}n^4\right)\frac{\pi}{180} \tag{2.17}$$

$$\sigma = \frac{m\pi}{180G} \tag{2.18}$$

Then

$$\begin{aligned}\phi' &= \sigma + \left(\frac{3n}{2} - \frac{27n^3}{32}\right)\sin 2\sigma \\ &+ \left(\frac{21n^2}{16} - \frac{55n^4}{32}\right)\sin 4\sigma \\ &+ \left(\frac{151n^3}{96}\right)\sin 6\sigma \\ &+ \left(\frac{1097n^4}{512}\right)\sin 8\sigma\end{aligned} \tag{2.19}$$

2.6.5 Spheroid to grid - Redfearn's formula

EASTING

$$E' = k_0 \{ \nu\omega\cos\phi \tag{2.20}$$

$$+ \nu\frac{\omega^3}{6}(\cos\phi)^3(\psi - t^2)$$

$$+ \nu\frac{\omega^5}{120}(\cos\phi)^5[4\psi^3(1 - 6t^2) + \psi^2(1 + 8t^2) - \psi(2t^2) + t^4]$$

$$+ \nu\frac{\omega^7}{5040}(\cos\phi)^7(61 - 479t^2 + 179t^4 - t^6)\}$$

NORTHING

$$N' = k_0\{m + \nu\sin\phi\frac{\omega^2}{2}\cos\phi \tag{2.21}$$

$$+ \nu\sin\phi\frac{\omega^4}{24}(\cos\phi)^3(4\psi^2 + \psi - t^2)$$

$$+ \nu\sin\phi\frac{\omega^6}{720}(\cos\phi)^5[8\psi^4(11 - 24t^2) - 28\psi^3(1 - 6t^2)$$

$$+ \psi^2(1 - 32t^2) - \psi(2t^2) + t^4]$$

$$+ \nu\sin\phi\frac{\omega^8}{40320}(\cos\phi)^7(1385 - 3111t^2 + 543t^4 - t^6)\}$$

GRID CONVERGENCE (radians)

$$\gamma = -\sin\phi\omega \tag{2.22}$$

$$- \sin\phi\frac{\omega^3}{3}(\cos\phi)^2(2\psi^2 - \psi)$$

$$- \sin\phi\frac{\omega^5}{15}(\cos\phi)^4[\psi^4(11 - 24t^2) - \psi^3(11 - 36t^2) + 2\psi^2(1 - 7t^2) + \psi t^2]$$

$$- \sin\phi\frac{\omega^7}{315}(\cos\phi)^6(17 - 26t^2 - 2t^4)$$

POINT SCALE FACTOR

$$k = k_0 \{ 1 + \frac{\omega^2}{2} (\cos\phi)^2 \psi \tag{2.23}$$

$$+ \frac{\omega^4}{24} (\cos\phi)^4 [4\psi^3 (1 - 6t^2) + \psi^2 (1 + 24t^2) - 4\psi t^2]$$

$$+ \frac{\omega^6}{720} (\cos\phi)^6 (61 - 148t^2 + 16t^4) \}$$

2.6.6 Grid to spheroid

LATITUDE

Let $x = E'/k_0\nu'$

$$\phi = \left(\phi' - \left(\frac{t'}{k_0\rho'} \right) x \frac{E'}{2} \right. \tag{2.24}$$

$$+ \left(\frac{t'}{k_0\rho'} \right) \left(x^3 \frac{E'}{24} \right) [-4\psi'^2 + 9\psi' (1 - t'^2) + 12t'^2]$$

$$- \left(\frac{t'}{k_0\rho'} \right) \left(x^5 \frac{E'}{720} \right) [8\psi'^4 (11 - 24t'^2) - 12\psi'^3 (21 - 71t'^2)$$

$$+ 15\psi'^2 (15 - 98t'^2 + 15t'^4) + 180\psi' (5t'^2 - 3t'^4) + 360t'^4]$$

$$\left. + \left(\frac{t'}{k_0\rho'} \right) \left(x^7 \frac{E'}{40320} \right) (1385 + 3633t'^2 + 4095t'^4 + 1575t'^6) \right)$$

LONGITUDE

Let $x = E'/k_0\nu'$

$$\omega = \sec\phi' x \tag{2.25}$$

$$- \sec\phi' \frac{x^3}{6} (\psi' + 2t'^2)$$

$$+ \sec\phi' \frac{x^5}{120} [-4\psi'^3 (1 - 6t'^2) + \psi'^2 (9 - 68t'^2) + 72\psi' t'^2 + 24t'^4]$$

$$- \sec\phi' \frac{x^7}{5040} (61 + 662t'^2 + 1320t'^4 + 720t'^6)$$

GRID CONVERGENCE (radians)

Let $x = E'/k_0\nu\Pi$

$$\gamma = -t'x$$ (2.26)

$$+ t'\frac{x^3}{3}(-2\psi'^2 + 3\psi' + t'^2)$$

$$- t'\frac{x^5}{15}[\psi'^4(11 - 24t'^2) - 3\psi'^2(8 - 23t'^2)$$

$$+ 5\psi'(3 - 14t'^2) + 30\psi' t'^2 + 3t'^4]$$

$$+ t'\frac{x^7}{315}(17 + 77t'^2 + 105t'^4 + 45t'^6)$$

POINT SCALE FACTOR

Let $x = E'/k_0\nu'\rho'$

$$k = k_0\left\{1 + \frac{x}{2} + \frac{x^2}{24}\left[4\psi'(1 - 6t'^2) - 3(1 - 16t'^2) - \frac{24t'^2}{\psi'}\right] + \frac{x^3}{720}\right\}$$ (2.27)

2.6.7 Simplified grid formulae

POINT SCALE FACTOR

$$k = 0.9996 + 1.23E'^2 10^{-14}$$ (2.28)

LINE SCALE FACTOR

$$K = k_0\left[1 + \frac{(E'^2_1 + E'_1E'_2 + E'^2_2)}{6r^2_m}\right]$$ (2.29)

3 Accuracy

3.1 The measurement process

3.1.1 Measurement

The measurement of a quantity simply means the assignment of a numerical value to represent that quantity. Seismic navigation is mostly concerned with the measurement of ranges, angles and azimuths, and the first stage in the measurement process is the determination of the **units** to be used.

Common units used in navigation systems are:

Table 3.1 Units of measurement for navigation systems

System	Type	Unit
Syledis	Range	Metres
Argo	Range	Lanes
Micro-Fix	Range	Metres
Mars Laser	Range	Metres
	Azimuth	Degrees
Compasses	Azimuth	Degrees
DGPS	Latitude	Degrees
	Longitude	Degrees
Acoustics	Range	Metres or Milliseconds

In Table 1, the output units are shown, not the measurement units. For instance, in the case of Micro-Fix the internal measurement units are time, not distance.

Having defined the units of measurement, the next stage in the process is to define a **model** which approximates but simplifies the physical reality. For example, in the case of a measured radio range we use a mathematical model which alters the measured range according to transmitter and receiver errors, the velocity of propagation, the frequency of the carrier wave, and so on.

The third and final stage in the measurement process is to define a **procedure**. Using this procedure, **observations** are made and transformed using the mathematical model to produce the **values** required.

3.1.2 Errors

The error of a value is defined as

$$E = M - T \tag{3.1}$$

where E is the error

M is the measured quantity

T is the true value.

At this early stage it must be realized that the true value is never known! Nevertheless, the above definition allows us to determine the **accuracy** of an observation, which is the **difference between** M **and** T. Unfortunately, since we never know the true value, we can also never know the accuracy of an observation. What we need is a method of estimating the accuracy.

As an example, suppose we measure the distance between a shore station and an antenna on the vessel. The raw measured range is (say) 12510.3m. We then modify this range to take account of known errors at the transmitter (the station ashore) and the receiver (the antenna on the vessel) and the assumed velocity of propagation for the (known) frequency of the transmitted signal. How do we know that the reduced range is now accurate? The truth is, we don't. All we can do is take the utmost care with both the measurement process and the reduction process. One of the best precautions we can take is to ensure that we measure several ranges from different stations to lessen our chances of making a large mistake.

Now we need to introduce another term — **precision**. Many people use the terms accuracy and precision indiscriminately, but there is a fundamental difference between the two terms. Accuracy is the difference between the measured value and the true value; precision is a measure of the **reliability** of an observation. We can make a very precise measurement of a range between a shore station and the vessel, but if the shore station coordinates are incorrectly defined the range will be very inaccurate. Precision will be defined in statistical terms in a later section.

There are three types of errors:

(a) Systematic errors

Systematic errors arise from an incorrect choice of mathematical model. Consider the following example:

In Figure 3.1 the arrowed lines represent measured laser azimuths from the stern of the vessel to targets on the paravanes supporting the two sources and streamers.

We can model the azimuths as follows

$$X_T = X_M + E \tag{3.2}$$

where X_T is the true value

X_M is the measured value

E is the alignment error.

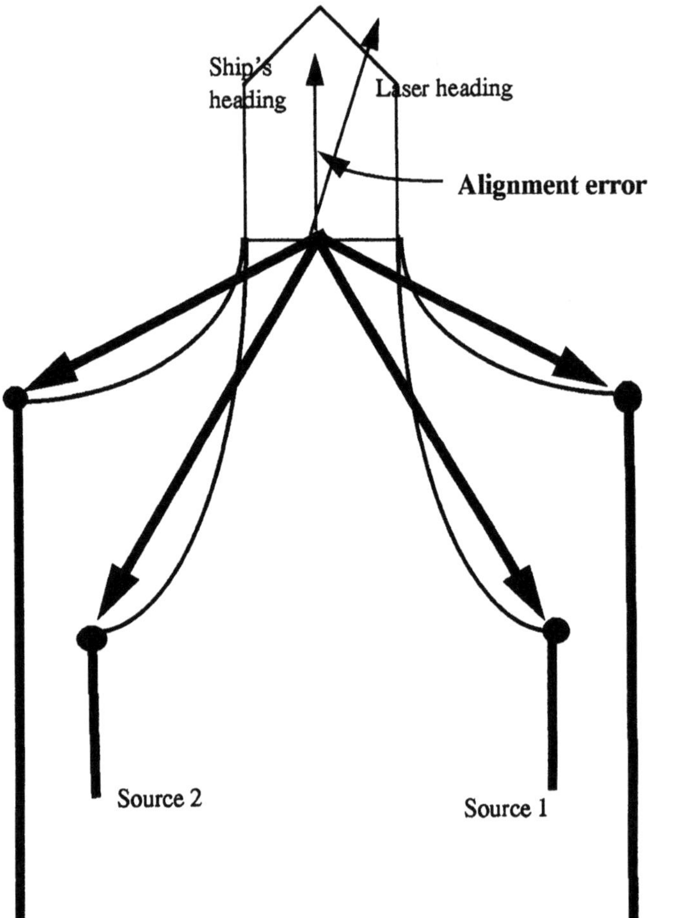

Figure 3.1 Laser azimuths.

If we do not calibrate the laser, the alignment error is unknown and equation (3.2) can never be satisfied, so we have created a systematic error because all measurements will be in error by a constant amount.

(b) Random errors

If two acoustic transducers were deployed from booms either side of a vessel travelling at constant velocity and we then measured the acoustic range between them at 2 second intervals, the measurements would vary slightly even if we observed in ideal conditions of no wind, tide, etc. The slight variations in the measured range will be due to variations in the temperature, salinity and pressure of the water, variations in the acoustic path and the inability of the equipment to resolve an exact range. The variation in the readings is a measure of their **precision** — the lower the variation, the more precise the readings. The variations themselves are **random errors.**

(c) Blunders

Blunders are third in the list of error types because they are the least common - systematic and random errors are almost always present, which is not the case for blunders. But blunders often have the most disastrous effects.

Typical blunders experienced in our work are using the wrong algebraic signs for transforming one datum to another — for instance substituting minus signs for plus signs in the transformation from WGS84 to AGD84 would leave the AGD84 coordinates about 400m from their real values! Another good example is using the incorrect offsets when computing receiver group positions down a streamer.

The catastrophic effects of blunders are the main reasons why surveyors are taught to check their work by independent methods. This applies from the most basic of trigonometric computations to the most complex of networks. It also applies to seismic surveys; **each step of the measurement process should be independently checked.**

3.2 Statistical terms

3.2.1 Random variables

A random variable is one whose value depends upon chance. The acoustic ranges measured in section 3.1.2(b) constitute a **sample** of the acoustic range — the acoustic range is the **random variable.** The fact that we are using the word sample implies that we have only looked at part of the whole. The whole is called the **population**, which is what we would end up with if we continued our measurements an infinite number of times.

As a rule we only take small samples of each random variable in positioning; only in cases such as the final GPS phase measurements to locate the drillstem of a rig would we take a large number of observations. So we are confronted with the problem of having to deduce characteristics of the population from a very small sample. Furthermore, we have to use these deduced parameters to estimate the reliability of the measurements and the most likely value of the random variable.

3.2.2 Probability density function

The probability density function, $\phi(x)$, of a random variable x is the function whose integral gives the probability $P(a, b)$ of x lying between the values a and b. Expressed as an equation this is:

$$P(a,b) = \int_a^b \phi(x)\, dx \tag{3.3}$$

This is equivalent to the shaded area in Figure 3.2 :

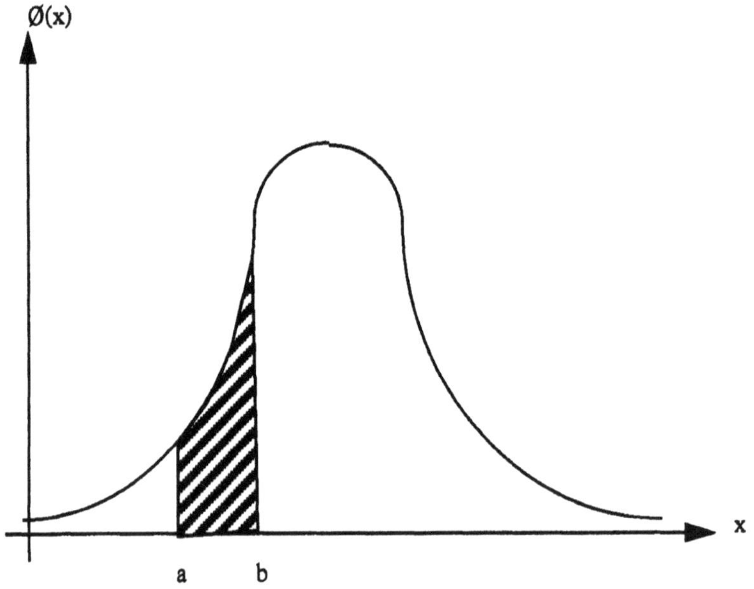

Figure 3.2 Probability density function.

Now, a probability of 1 is taken to mean that an event is certain to occur, and we know that x must lie between - infinity and infinity, so we can write:

$$\int_{-\infty}^{\infty} \phi(x)\, dx = 1 \tag{3.4}$$

3.2.3 Expectation

The **expectation** of a random variable is defined as the **mean value** taken over the population, which is:

$$E(x) = \int_{-\infty}^{\infty} x\phi(x)\,dx \tag{3.5}$$

The important point arising here is that if we have an estimate of x which is equal to the expectation of x, then the estimate is said to be an **unbiased estimate**.

3.2.4 Variance and standard deviation

If a random variable x has expectation ζ, its **variance** σ_x^2 is defined as:

$$\sigma_x^2 = E\{(x-\zeta)^2\} \tag{3.6}$$

The **standard deviation** of x is the positive square root of the variance.

Now, the above definitions are in terms of expectations, and we have said in Section 3.2.3 that the expectation of a random variable is its mean value taken over the population. We have also said that we only take a sample of the population, so we now define the mean, variance and standard deviation in terms of the sample.

The **mean** of the sample is

$$\bar{x} = \frac{1}{n}\sum_{i=1}^{n} x_i \tag{3.7}$$

which is simply the sum of all the observed values divided by the number of values.

The **variance** of the sample is

$$\sigma_x^2 = \frac{\sum_{i=1}^{n}(x_i-\bar{x})^2}{(n-1)} \tag{3.8}$$

and the **standard deviation** of the sample is

$$\sigma_x = \sqrt{\frac{\sum_{i=1}^{n} (x_i - \bar{x})^2}{(n-1)}} \tag{3.9}$$

3.2.5 Covariance

The last of the statistical terms we need to look at is **covariance.** If two random variables x and y have expectations ξ and ψ respectively, then the covariance of x and y is defined as

$$\sigma_{xy} = E\{(x - \xi)(y - \psi)\} \tag{3.10}$$

Translating this from expectations to samples, we get

$$\sigma_{xy} = \frac{\sum_{i=1}^{n} (x_i - \bar{x})(y_i - \bar{y})}{(n-1)} \tag{3.11}$$

Covariance is a term used extensively in least squares and Kalman filtering, and we shall investigate its uses later.

3.3 Absolute accuracy

3.3.1 General discussion

Marine seismic survey positioning specifications are very often issued as a disjointed series of figures such as:

- Argo +/– 10 m
- DGPS +/– 5 m
- Feather < 10 deg
- Shot interval 25 m +/– 1.5 m
- Compasses < 1 degree from mean of all compasses

- Acoustics range +/– 2m

What we really should be doing is specifying the survey in terms of the quantities we require. If we position the vessel to better than 10 m and the compasses are less than a degree from their mean value, what does that do to our receiver group positions? If the acoustic ranges are better than 2m what does that mean in terms of our source position? To say that we require the Argo tolerance to be better than 10m is really quite nonsensical. What we have to do is define a foolproof method so that we can ensure that, providing our specifications are met, our common mid point positions will have the required accuracy.

Consider Figure 3.3 :

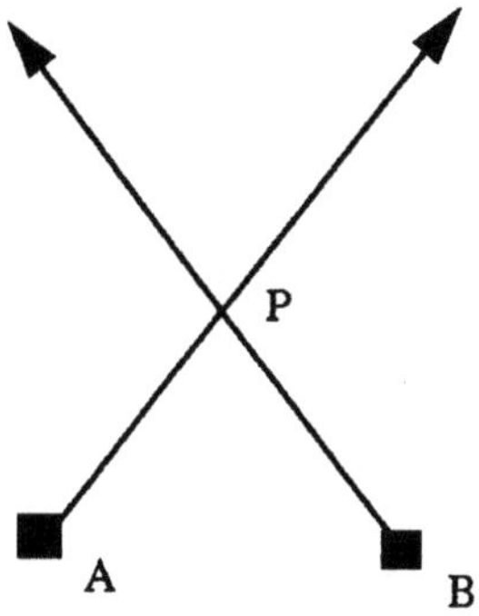

Figure 3.3 Two-bearing fix.

The arrowed lines represent two bearings from stations A and B. They intersect at point P, which is the position of the vessel. We therefore need just two lines of position (LOPs) to define a two-coordinate position in terms of eastings and northings or latitude and longitude or x and y. But nobody would accept that position because if one of the bearings is incorrect, then so too is the position. Now we add a third bearing from station C, as in Figure 3.4 :

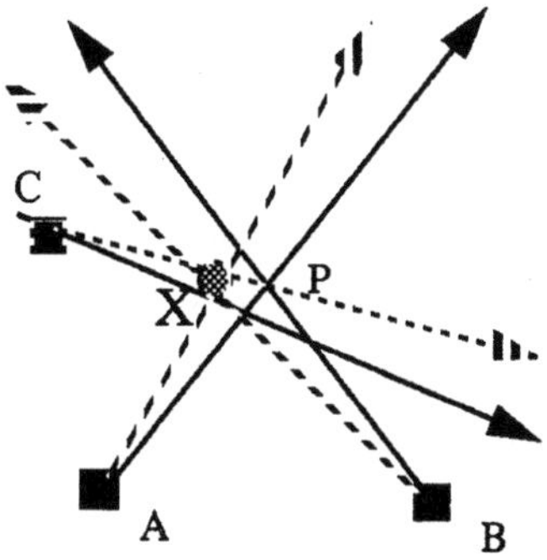

Figure 3.4 Three-bearing fix.

Now the point P is not so easily defined; the obvious choice is to place it inside the triangle created by the three LOPs, but suppose the three bearings contained a systematic error which, if compensated for, shifted all three LOPs to the left? The result is that the position of the vessel is shifted to X, a point which is outside the original triangle. The answer here would be to model for the unknown systematic error — something which is done quite regularly for radio positioning ranges containing an unknown systematic scale error due to propagation.

We can summarize the above by saying that although the position of the vessel is not known so precisely when we increase the number of bearings, there is a very good chance that if one or more of the bearings are in error the effect of the vessel's position will be less catastrophic, thereby increasing the accuracy.

The rule is that to solve for ***n*** unknowns we need at least ***n*** observations. If we have more than ***n*** observations, we have **redundancy**.

We can propagate this rule throughout the spread, ensuring that there is redundancy in finding all the parameters we want to know about, including the vessel position, the source positions, the streamer shapes, the tailbuoy positions and the receiver group positions.

- Specification 1 — ensure that there are redundant observations for every facet of the entire spread.

We must also be careful not to go to the other extreme and provide over-redundancy in the system, because the computing time required to adjust a particular network increases as the square of the number of observations: in other words if we look at the vessel position on its own, the difference in computing time between using ten ranges and using three ranges is 100/9 or 11 times as long. Not to mention the increase in cost, which probably rises at the same proportional rate.

3.3.2 Bin size and accuracy

The only value in positioning the ship accurately is so that we can propagate that accuracy through our various in-spread systems. If we could somehow find absolute positioning points on the sources and streamers, the position of the vessel would become irrelevant. This is becoming a reality with the advent of shock-mounted DGPS and RGPS receivers on the sources and streamer heads.

Let us suppose that we are conducting a 3D survey in which the bin size is to be 50 m long and 25 m wide. How should we determine the accuracy of the common mid-points so that we can be confident of them falling in the right bins? We need to go back to our theory of standard deviations.

Figure 3.5 shows the probability density function. The curve is a **normal distribution**, and it can be shown that for any population this curve will result. Its equation is

$$\phi_x = \frac{n}{\sigma\sqrt{(2\pi)}}\exp\left(\frac{-x^2}{2\sigma^2}\right) \tag{3.12}$$

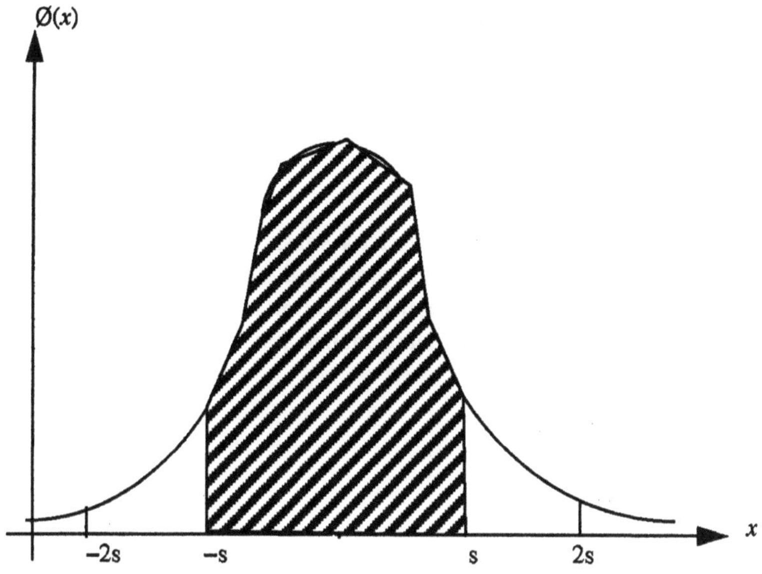

Figure 3.5 Probability density function.

where σ is the standard deviation and n is the total number of observations. The shaded area under the curve in Figure 3.5 is the area lying between $-\sigma$ and $+\sigma$. We can use equation (3.12) to evaluate the area under the curve for various limits - see Table 3.2:

Table 3.2 Area under the normal curve for various standard deviations

Limits	Area %
$-\sigma$ to $+\sigma$	68.3
-2σ to $+2\sigma$	95.4
-3σ to $+3\sigma$	99.7

Now the percentage area under the curve given in column 2 of Table 3.2 is the percentage probability that any deviation is between the corresponding limits.

Going back to our problem of how to specify the accuracy required for a bin width of 25m, if we set the required standard deviation of a common mid point (CMP) to be 6.25m we will be 68.3% sure that all CMP's fall within ±6.25m, 95.4% fall within ±12.5m, and 99.7% fall within ±18.75m. Only 0.3% will fall outside the limits of ±18.75m. We specified a bin width of 25m, so we can be 95.4% certain of CMPs falling in the right bin. This can be generalized into another specification:

- Specification 2 — the standard deviation of all common mid points should be 0.25 × nominal bin width.

A basic assumption here is that the bin width is the overriding parameter, since the receiver groups cannot move so much in the inline direction. In the case of a bottom cable survey both bin width and bin length must be considered equally.

Theoretically it should then be the contractor's task to demonstrate that he is achieving the desired standard deviation for the CMPs, but in practice this is all but impossible given the current state of available software. We shall show later in the book that when we compute a solution for a position we end up with a variance — covariance matrix that looks like this:

$$\begin{bmatrix} \sigma_E^2 & \sigma_{EN} \\ \sigma_{NE} & \sigma_N^2 \end{bmatrix} \tag{3.13}$$

where $\sigma_E{}^2$ is the variance in the Eastings direction,

$\sigma_N{}^2$ is the variance in the Northings direction,

and $\sigma_{EN} = \sigma_{NE}$ is the covariance of both parameters.

The matrix in equation (3.13) gives us the standard deviation in the eastings and northings axes, but we can use the matrix further to give us the maximum and minimum axes of the resultant error ellipse.

Note that we may wish to show the ellipse with axes of $2.5\sigma_{max}$ and $2.5\sigma_{min}$ to guarantee to 95% that the true position was within the ellipse.

When surveyors refer to error ellipses they mean these ellipses derived from the variance-covariance matrix associated with the resultant position. We must be careful to note what scale factor (if any) has been applied to the ellipses.

The navigation software on board the seismic vessel will be capable of giving an error ellipse for the vessel position, and very often for the individual nodes of the in-spread network, but seldom for the sources, receiver groups and CMPs themselves. This is because the acoustic, laser, compass and tailbuoy elements are individually computed, and the task of propagating errors through the system to the sources and groups is quite complex and cumbersome. If the software gives an integrated solution then the source, group and midpoint ellipses can be derived very easily.

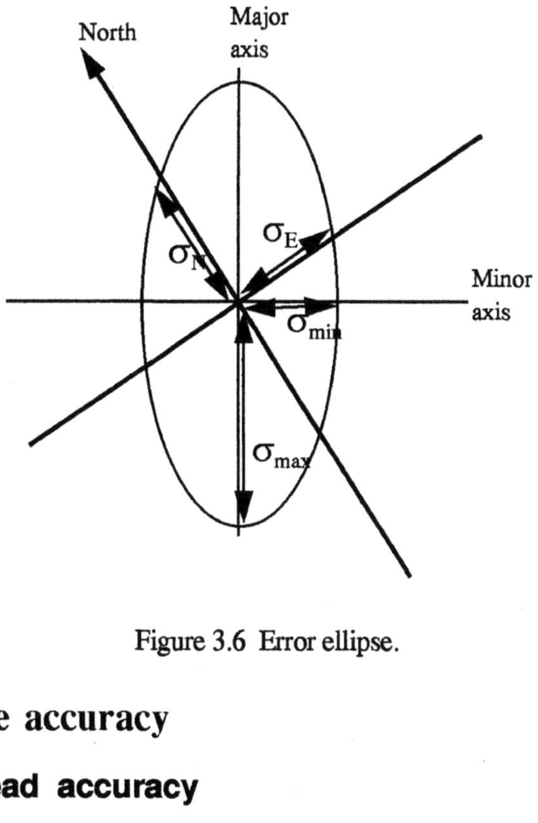

Figure 3.6 Error ellipse.

3.4 Relative accuracy

3.4.1 In-spread accuracy

In Section 3.3.2 we concluded that rarely would the navigation system be capable of producing absolute error ellipses for the CMPs so that we could ascertain the likelihood of any CMP falling in the correct bin.

It is now well accepted that for a 3D survey the conventional use of layback and offset measurements to determine the source and front receiver group positions is inadequate; such positions are determined using a network of acoustic and/or laser range and bearing measurements.

A simple 3D spread may look like this

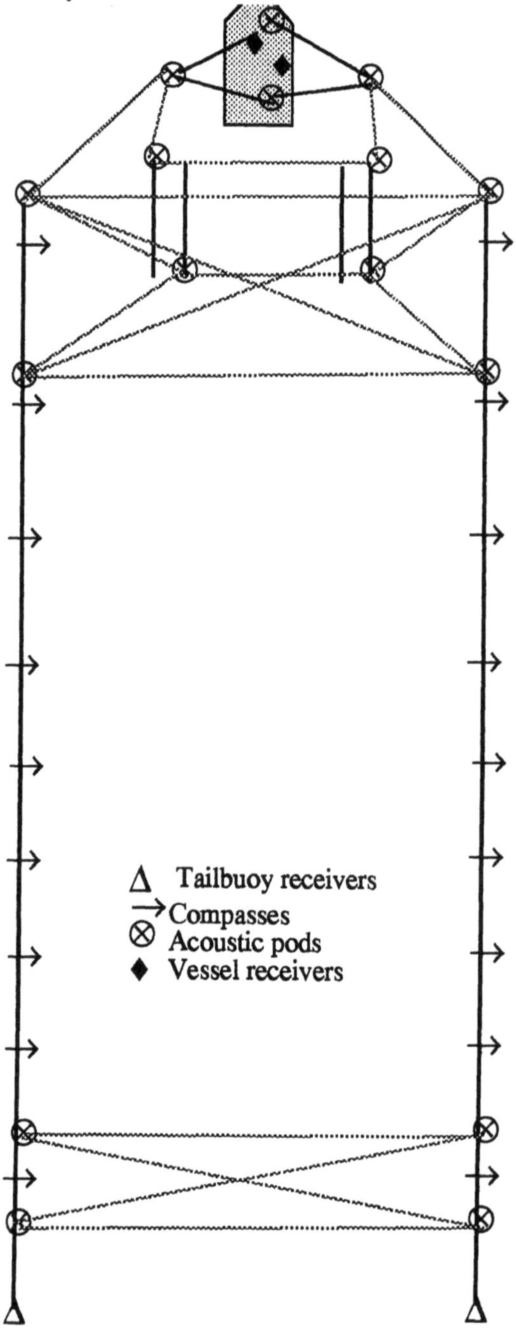

Figure 3.7 Simple 3D spread.

In Figure 3.7 we have two acoustic transponders towed from booms either side of the hull. They are themselves positioned acoustically from transducers mounted on the hull, and form a baseline of about 50m in length. The front ends of both streamers, and the source positions, are then determined by acoustic ranges from the boom nodes. Now, as Figure 3.7 is drawn, the network formed by the acoustic ranges at the front end looks very solid; all the internal angles in the network are quite large, and it could be assumed at first glance that we will derive positions for the sources and front receivers of high precision and high accuracy, also assuming that we use an acoustic system capable of measuring ranges with a precision of better than 1m that has been properly calibrated. What could possibly go wrong?

The first point is that the whole network must be orientated into a grid of some sort. Exactly which grid is used doesn't matter at this stage; it may be based on the course made good, a local Cartesian grid, the nominal sail line, the projection's grid, and so on. Whichever grid is used, the orientation of the network is critical. It can be ascertained using the vessel gyro, the front streamer compasses, or a combination of both. It is most common for the boom baseline to be computed first and then to be orientated using the vessel's gyro, because the hull transducers are fixed with a known offset. The boom baseline is then also held fixed and the front end network is computed using that inviolate baseline. So the gyro is absolutely critical to the computation.

The second point is that Figure 3.7 has been deliberately drawn at a distorted scale, because that is how the front-end network is usually displayed in order that the individual nodes and ranges can clearly be shown. With an inline offset from navigation reference point (NRP) to sources of 200m, NRP to front groups of 350m, and a boom length of 50m, the diagram drawn to scale looks as in Figure 3.8 .

Now we can see that the internal angles of the network have become very small, and will also recognize that small errors in the acoustic ranges will cause quite large variations in the source and group positions. Consider the following arguments:

- An error of 1.5 deg in the gyro heading will cause the front receiver groups to be misplaced by 9.2m.
- An error of 2m in one range from the booms to the front receivers, and a missing cross range at the front receivers, will introduce an error of no less than 45m in the front receiver group position.

Herein lies the third important specification:

- Specification 3 — the inline offset from the NRP to sources and front receiver groups should be as short as possible, to increase the accuracy of the front-end network.

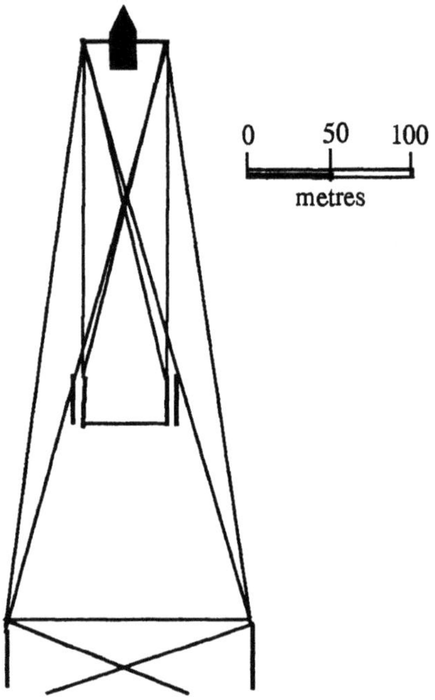

Figure 3.8 Front-end drawn to scale.

If the inline offset has to be large because of problems with vessel noise, towing arrangements, etc., we can minimize the effects by using an integrated approach to positioning and by using laser as well as acoustics. Both of these subjects are covered later in the book.

Assuming that the front-end acoustic network is computed as a separate entity, which is the case with most software packages currently available, we will end up with a variance — covariance matrix which expresses the reliability of the nodes of that network in relative terms; that is, the standard deviations and error ellipses of each node will be known relative to all the other network nodes. The problem we will look at in the next section is how to relate these variances to those of the vessel position and in turn to the CMP positions.

3.5 Propagation of variance

In Section 3.2.4 equation (3.9) gave us the standard deviation of a sample. When this standard deviation is an unbiased estimate it is commonly called the **standard error** of a measurement, s_x.

In general, if

$$z = \phi(x_1, x_2, x_3, \ldots, x_n) \tag{3.14}$$

where z is a parameter related to measurements x by the function $\emptyset$, and the standard error of x_1 is s_1, it can be shown in matrix notation that

$$C_z = \emptyset C_x \emptyset^T \tag{3.15}$$

where C_x and C_z are the covariance matrices for x and z respectively.

This is Gauss's propagation error law for linear equations, and can best be illustrated by an example:

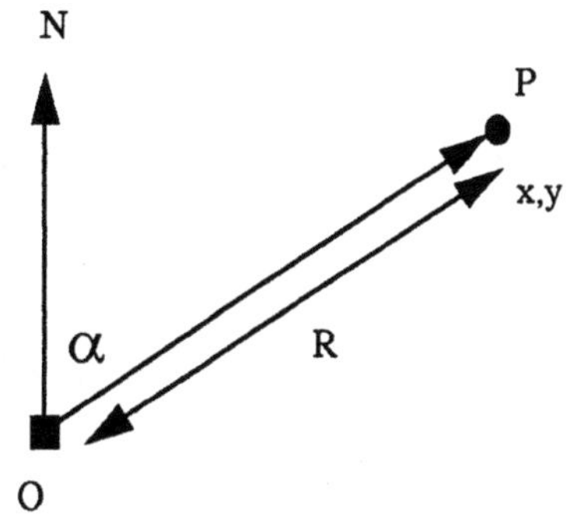

Figure 3.9 Range and bearing.

In Figure 3.9 the point P has been positioned from point O by the measured range r and bearing α. The coordinates of P are then given by

$$x = r\sin\alpha \tag{3.16}$$

$$y = r\cos\alpha \tag{3.17}$$

The measurements r and α are made and their standard errors estimated as follows:

r = 250.00m

α = 30.00 deg

σ_r = 0.05m

σ_a = 0.0055 deg (9.7 E −5 rad)

We want to find values for x and y, the coordinates of P, and their standard deviations.

Unfortunately, equation (3.16) and equation (3.17) are non-linear, and we first need to linearize them so that we can use Gauss's law. This is done by taking partial differentials as follows:

$$\frac{\partial x}{\partial r} = \sin\alpha \tag{3.18}$$

$$\frac{\partial x}{\partial \alpha} = r\cos\alpha \tag{3.19}$$

$$\frac{\partial y}{\partial r} = \cos\alpha \tag{3.20}$$

$$\frac{\partial y}{\partial \alpha} = -r\sin\alpha \tag{3.21}$$

Now we can form the matrix J from the partial derivatives:

$$J = \begin{bmatrix} \frac{\partial x}{\partial r} & \frac{\partial x}{\partial \alpha} \\ \frac{\partial y}{\partial r} & \frac{\partial y}{\partial \alpha} \end{bmatrix} = \begin{bmatrix} \sin\alpha & r\cos\alpha \\ \cos\alpha & -r\sin\alpha \end{bmatrix} \tag{3.22}$$

The covariance matrix $C_{r\alpha}$ is:

$$\begin{bmatrix} \sigma_r^2 & \sigma_{r\alpha} \\ \sigma_{r\alpha} & \sigma_\alpha^2 \end{bmatrix} = \begin{bmatrix} 2.5\times10^{-3} & 0 \\ 0 & 9.4\times10^{-9} \end{bmatrix} \tag{3.23}$$

and we can now use equation (3.15) to propagate the variances:

$$C_{xy} = JC_{r\alpha}J^T = \begin{bmatrix} 1.07\times10^{-3} & 8.31\times10^{-4} \\ 8.31\times10^{-4} & 2.03\times10^{-3} \end{bmatrix} \tag{3.24}$$

Therefore we have solved the problem:

x = 125.00m

y = 216.51m

σ_x = 0.03m

σ_y = 0.04m

Note that, although equation (3.23) has zeros for the covariance of r and α because the two measurements are assumed to be uncorrelated, the values x and y **are correlated** and so have values in equation (3.24).

Also note that the variance — covariance matrices are symmetrical — this is always the case and has important implications when programming computations in terms of speed savings.

The above example is a very simple one. Without going into the mathematics of the problem, we can quickly see the complexity of trying to propagate the variances and covariances of an absolute position on the vessel through a front-end network of acoustic nodes, thence into receiver groups and source positions. This computation and derivation of the associated error ellipses is quite complicated and takes an appreciable amount of computing time. However, if we use an integrated system that can tie all our absolute and relative observations together into a single network, the variances and covariances are already in place as part of the main computation. This is one of the key advantages of using an integrated Kalman filter, and is explored in detail later in the book.

3.6 Summary

We have defined all the main statistical variables used in seismic navigation, and have noted the difference between accuracy and precision. We know the three different types of error and we know how to propagate variances and covariances through a network.

We have elicited three valuable points to look for when writing specifications:

- Specification 1 — ensure that there are redundant observations for every facet of the entire spread.
- Specification 2 — the standard deviation of all common mid points should be 0.25 × nominal bin width.
- Specification 3 — the inline offset from the NRP to sources and front receiver groups should be as short as possible, to increase the accuracy of the front end network

There are other factors we need to address, such as streamer shaping, the use of active tailbuoys and so on. These will be dealt with later.

4 Introduction to computations

4.1 Introduction

This chapter deals with the basic concepts of positioning at sea; we start with an introduction to least squares, on which all modern positioning computations are based, and then develop the various formulae used in the computations.

We are going to put into a single chapter the information that is disseminated to undergraduate surveyors in about a year of study, so some of the detailed explanations and proofs will necessarily be shortened.

Before going straight into least squares, we will briefly revise the coordinate systems available to us in the context of computations.

4.2 Coordinate systems

4.2.1 The ellipsoid

The ellipsoid is the mathematical figure which approximates most closely the true shape of the earth. Unfortunately, many people have tried to establish the best-fit ellipsoid for the earth, and many of the ellipsoids they calculated are in use. Life would be very much easier if there were only one ellipsoid (or spheroid).

In Australia we generally use the Australian Geodetic Datum as a datum for our offshore surveys. Even this is somewhat complicated by the following facts:

- There are two Australian datums in use — AGD66 and AGD84.
- Neither of the two datums is geocentric.
- Australia intends to move to a geocentric datum in 2000.

The AGD66 datum has the following definition:

Semi-axis major: 6378160.0m

Flattening: 1/298.25 exactly.

The minor axis of the spheroid was defined in 1966 to be parallel to the earth's mean axis of rotation in 1962 (this was later changed in 1970), and the meridian of zero longitude was defined as being parallel to the Bureau International de l'Heure (BIH) meridian plane near Greenwich. The centre of the spheroid was defined by the coordinates of Johnston Geodetic Station, a station in the centre of Australia. At that time it was assumed that the spheroid — geoid separation was zero at Johnston, and also zero at all the other geodetic stations listed in the 1966 adjustment.

Since 1966 a huge amount of information on the shape of the geoid has become available, particularly through satellite observations, and it was realized that the 1966 adjustment was no longer accurate. In 1982 all the information then available was put into a new least squares network adjustment called GMA82 (Geodetic Model of Australia 1982). This resulted in a new Adopted Coordinate Set in 1984, which we refer to as AGD84 (geographical coordinates) and AMG84 (grid coordinates). The GMA82 adjustment is a truly spheroidal adjustment, so any observations used in conjunction with AGD84 or AMG84 must first be reduced to the spheroid. In Western Australia the difference in eastings and northings between AGD66 and AGD84 is of the order of 3 — 4m. Unfortunately, but not surprisingly, the geodetic network stopped when it reached the coast, so we can only extrapolate AGD66 positions offshore. Furthermore, the AGD66 adjustment was not a truly spheroidal adjustment, so there is no single exact shift that will move coordinates from AGD66 to AGD84.

At the turn of the millennium Australia intends to change its datum yet again, this time to a geocentric datum. The new datum will be called GDA92 and is very close to WGS84, the earth-centred datum used by GPS. The hope is that the entire world will eventually adopt WGS84s or a similar datum. Because AGD84 is a true spheroidal datum, we can derive a seven-parameter shift to get from AGD84 to WGS84 and vice versa. The seven-parameter shift is listed in Table 4.1:

Table 4.1 Seven-parameter shift from AGD84 — WGS84

Parameter	WGS84-AGD84	AGD84-WGS84
dX (m)	116.00	-116.00
dY (m)	50.47	-50.47
dZ (m)	-141.69	141.69
rX (")	0.23	-0.23
rY (")	0.39	-0.39
rZ (")	0.344	-0.344
S (ppm)	-0.0983	0.0983

Because vessels conducting seismic work often sail great circle lines, it would make sense to compute on the surface of the spheroid. But to obtain latitude and longitude from spheroidal distance and azimuth, or to obtain spheroidal distance and azimuth from latitude and longitude we need to use Robbins' formulae, which involve a lot of sine and cosine functions and are therefore computer intensive. The formulae themselves require ten-figure trigonometrical functions and could never be performed manually with ease. We also need to convert the results of our computations to a projection so that we can view them on a screen or chart.

4.2.2 The projection

As has been said above, the formulae for computation on the ellipsoid are not simple. Purely from a convenience point of view it has become usual to compute on the projection on a plane rectangular coordinate system. The problem now is that systems such as DGPS, Geoloc, etc., output in terms of latitude and longitude, so at every shot we need to convert to the projection. If this conversion were simply a point conversion, e.g. for the vessel position, there would be no problem other than the inconvenience of using Redfearn's formulae to convert between spheroid and grid. However, these formulae are well known and accurate to 1 mm in any zone of the AMG. They are actually the opening terms of an infinite series but can be regarded as exact for our purposes. The problems start to arise when we compute in terms of ranges and bearings on the grid. In the case of a (spheroidal) range we must convert to a grid distance using the line scale factor:

$$k = k_0\left(1 + \frac{(E'^2_1 + E'_1E'_2 + E'^2_2)}{6r^2_m}\right) \tag{4.1}$$

where E'_1 and E'_2 are the eastings of points 1 and 2 measured from the central meridian,

k_0 is the scale factor on the central meridian and

r is the earth's radius of curvature.

In the case of azimuths the problem is more acute because we must convert to a grid bearing using the grid convergence.

The grid convergence is the angle that must be algebraically added to the true azimuth to give the grid azimuth. In Figure 4.1 ,

$$\beta = \alpha + \gamma \tag{4.2}$$

Note that when we are to the east of the central meridian, grid convergence is positive; if we are west, it is negative. Also notice that grid convergence increases with distance from the central meridian, and that it applies no matter which direction the lines are run. Lastly, the grid convergence is zero on the equator and has the **opposite sign in the northern hemisphere!**

If it were not for the problems generated by scale factors and convergence, it would be advantageous to compute on the grid. But there is a better way.

4.2.3 Cartesian coordinates in three dimensions

Consider Figure 4.2 .

The Z axis is parallel to the earth's mean spin axis (called the Conventional International Origin or CIO), X is towards zero longitude and Y towards 90 deg E. The coordinates of point P are then simply the x, y, z distances shown. Now consider another point P', whose Cartesian coordinates are x', y', z'.

The straight line distance *PP'* is then

$$d = \sqrt{(x' - x)^2 + (y' - y)^2 + (z' - z)^2} \tag{4.3}$$

The straight-line distance can be corrected for curvature as follows

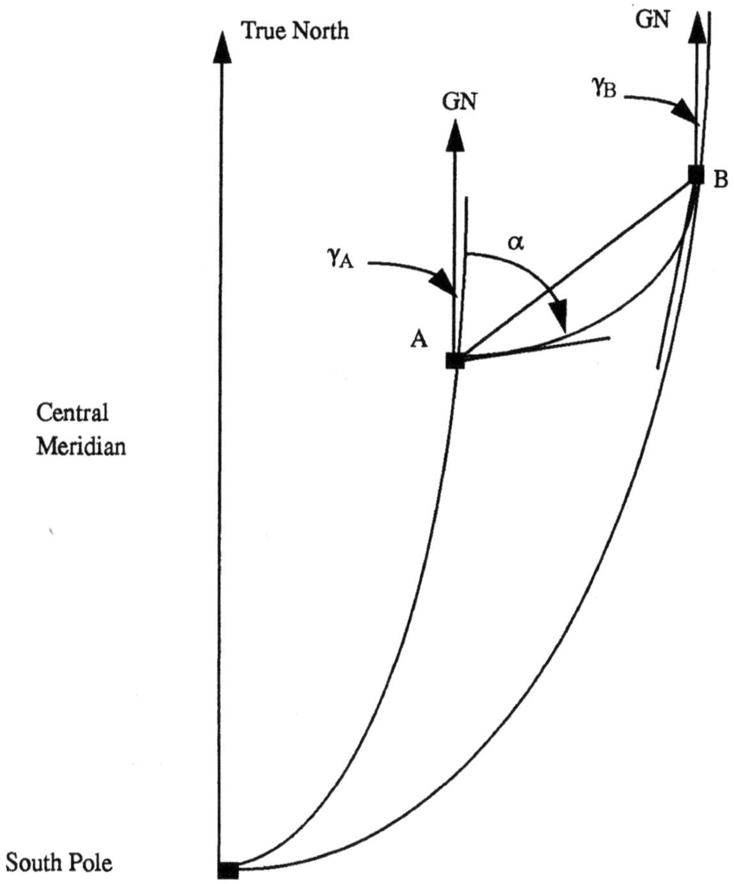

Figure 4.1 Azimuth relationships.

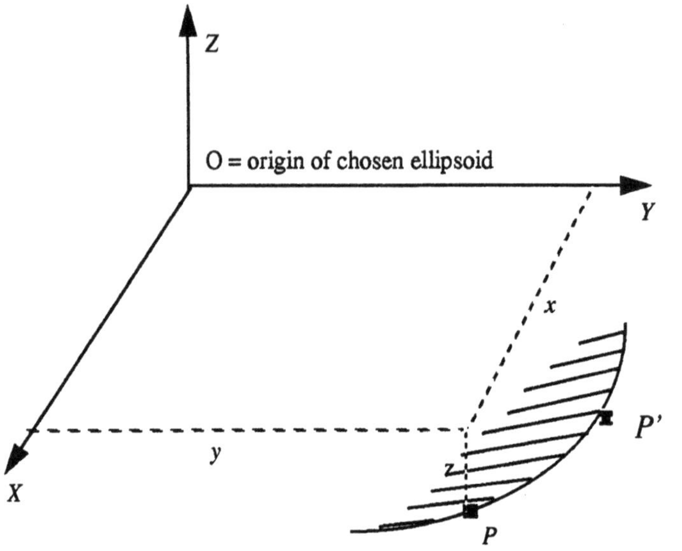

Figure 4.2 3D Cartesian system.

$$D = d + \frac{d^3}{24R^2} \tag{4.4}$$

where R is the earth's radius.

The azimuth from P to P' is

$$\cot A_{PP'} = \frac{[(z' - z)\cos\phi - (x' - x)\sin\phi]}{y'} \tag{4.5}$$

where ϕ is the latitude and assuming the axis of x is taken to lie in the meridian of P so that $y = 0$.

These formulae are obviously very simple; the trouble is that the (x, y, z) coordinates of P are not very meaningful; they will often be large six- or seven-figure numbers, and may often be negative.

One way of resolving this problem is to rotate and translate the global Cartesian system into a system whose origin is a point on the surface of the chosen spheroid, and such that the Y axis points true north, the X axis points 90 deg East and the Z axis is the normal at the point of origin, positive upwards. Now, within a radius of 10km or so from the point of origin, we can define true **distances within the spread** as

$$D = \sqrt{(x'-x)^2 + (y'-y)^2 + (z'-z)^2} \tag{4.6}$$

and we can define **true azimuths within the spread** as

$$A_{PP'} = \text{atan2}\left(\left(x'-x\right), \left(y'-y\right)\right) \tag{4.7}$$

These are the simplest equations of all to use, and they involve no scale factors or convergence!

We therefore propose that the best method to use is as follows:

- Compute the vessel position in terms of the global 3D Cartesian system.
- Transfer the global 3D position to the local 3D system, using an origin which moves from shot to shot and which is located at the vessel's navigation reference point.
- Compute the in-spread data (i.e. sources and streamers) on the local system.
- Transfer the output back to the spheroid and/or projection as required.

Note that the computation from one system to another only ever involves point computations, not lines; therefore scale factors and convergence never enter into the computation. We will use X and Y coordinates throughout the computations rather than E and N, to emphasize that we are working in a local 3D Cartesian system.

4.3 Least squares

4.3.1 Why least squares?

The person responsible for postulating the least squares process was Legendre, in 1806. He proposed that, given a set of n equally reliable measured values $(x_1, x_2, \ldots, x_n)$ of a quantity, the most probable value (MPV) x of that quantity is that which makes the sum of the squares of the residuals a minimum. A residual (v_i) is defined as

$$v_i = (x - x_i) \tag{4.8}$$

It follows from this proposal that the arithmetic mean of a series of equally reliable observations is the MPV:

$$\sum_{i=1}^{n} (x - x_i)^2 = \sum_{i=1}^{n} v_i^2 \tag{4.9}$$

According to the least squares principle,

$$\left(\frac{d}{dx}\right) \sum_{i=1}^{n} (x - x_i)^2 = 0 \tag{4.10}$$

If we differentiate equation (4.10) we get

$$2nx - 2\sum_{i=1}^{n} x_i = 0 \tag{4.11}$$

Therefore,

$$x = \frac{1}{n}\sum_{i=1}^{n} x_i \tag{4.12}$$

which is of course the mean.

Gauss tried to place the least squares principle on a solid foundation, and deduced that the measurements must be distributed according to the normal frequency distribution. The proof of this is beyond the scope of this book. Both Gauss and Laplace also tried to justify least squares without referring to the arithmetic mean, and, much later, Fisher placed the least squares principle within the context of his method of maximum likelihood.

Even if the observations are not normally distributed and independent, the least squares principle still provides a simple method of assigning values to unknown quantities when the number of observations is greater than the number of unknowns. There is no other method that has been found superior to least squares.

4.3.2 Weights

The weighting of observations simply means accounting for the fact that the observations may be of unequal reliability.

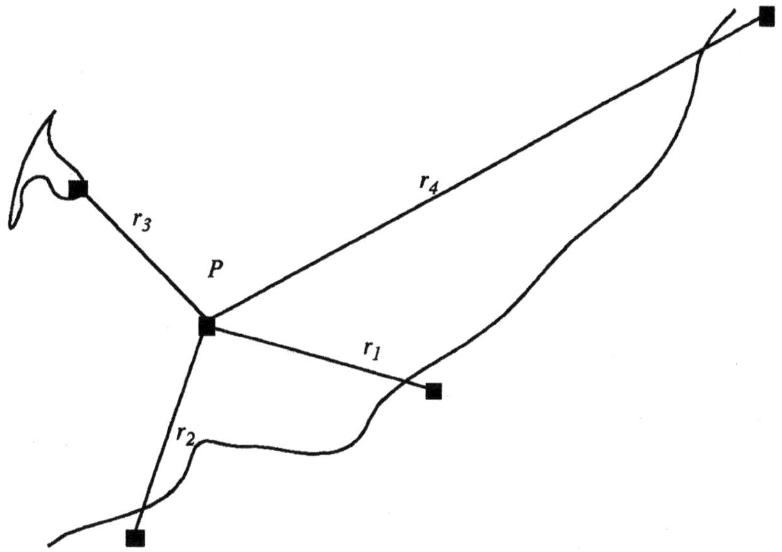

Figure 4.3 Four-way fix.

In Figure 4.3 the vessel is fixed at P by the four ranges r_1, r_2, r_3 and r_4. Suppose that the system being used is Argo, and that there is an error in the velocity of propagation due to poor calibration. The effect on r_4 will be twice as great as the effect on the other three ranges, because r_4 is approximately twice as long as all the other ranges. If r_4 is allowed to interact with the other three ranges in equal proportions, the position will be moved unnecessarily far from the truth. If we drop r_4 altogether, we risk losing redundancy. So we need to devise a method of leaving r_4 in the computation but at the same time having less influence on the position.

The method used is to assign weights to the observations, the weight being the inverse of the variance of the observation. Thus we might assign low standard errors to ranges r_1, r_2 and r_3 (because they are shorter and likely to be more reliable) and a higher standard error to r_4. Typical standard error values for Argo might be 5m for the shorter ranges and 10m for the longer ones. The weights then become 1/25 for the shorter ranges and 1/100 for the longer one. Note that the weight decreases as the square of the standard error increases.

4.3.3 Variation of coordinates

Variation of coordinates is a particular least squares strategy adopted almost universally in offshore positioning software. It is not difficult to explain or derive, and we can best do this by an example which makes full use of matrix algebra.

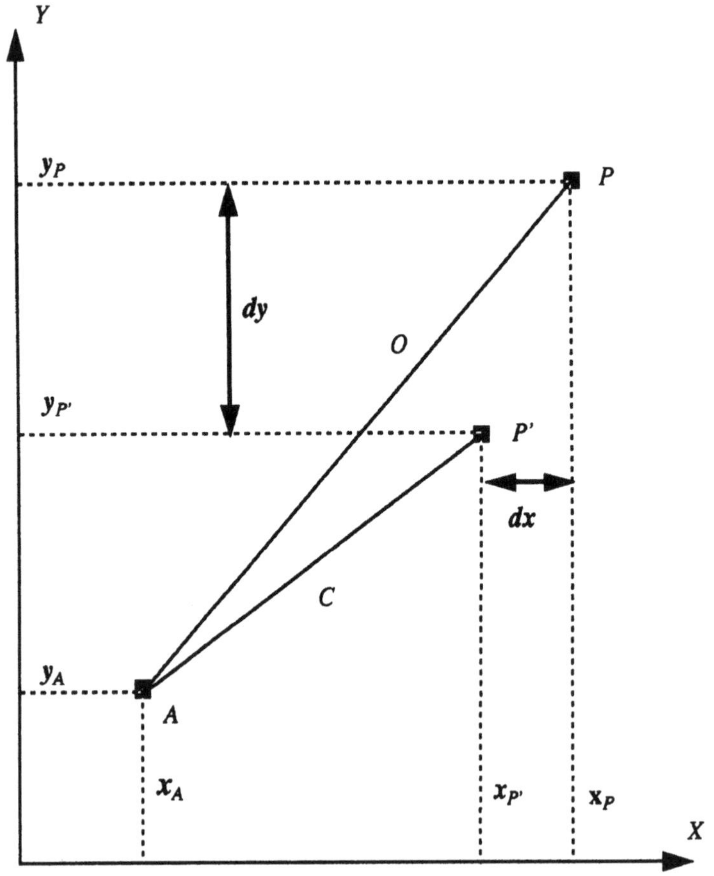

Figure 4.4 Derivation of observation equation.

In Figure 4.4 a range is observed between a shore station at A and the vessel at P. At this stage P is unknown, but we have established an approximate position (a 'seed') at P'. So the true position of P is given by

$$x_P = x_{P'} + dx \tag{4.13}$$

and

$$y_P = y_{P'} + dy \tag{4.14}$$

In Figure 4.4 the computed range c is known. What we need to know is how c changes when we move from P' to P:

$$c^2 = (x_{P'} - x_A)^2 + (y_{P'} - y_A)^2 \tag{4.15}$$

Differentiating equation (4.15) gives

$$2cdc = 2(x_{P'} - x_A)\,dx_{P'} + 2(y_{P'} - y_A)\,dy_{P'} - 2(x_{P'} - x_A)\,dx_A - 2(y_{P'} - y_A)\,dy_A \tag{4.16}$$

But the position of A is fixed so we can disregard the last two terms of equation (4.16), giving

$$dc = \frac{(x_{P'} - x_A)}{c}dx + \frac{(y_{P'} - y_A)}{c}dy \tag{4.17}$$

We can now say that

$$c + dc = o + v \tag{4.18}$$

where v is the true residual of the observed distance o. So we can write

$$\frac{(x_{P'} - x_A)}{c}dx + \frac{(y_{P'} - y_A)}{c}dy = (o - c) + v \tag{4.19}$$

Now equation (4.19) is a linearized **observation equation.** We have shown the derivation of a typical observation equation for a range. That range may be a radio navigation range, an in-spread acoustic range or even a GPS pseudo-range; the derivation is the same. We can also develop observation equations for azimuths, angles, hyperbolic lane counts and so on. Each derivation follows the same argument as above; only the shape of the equation is different.

If we have four ranges, and denote the first and second terms of equation (4.19) by *kdx* and *ldy*, we end up with a set of four linearised observation equations:

$$k_1dx + l_1dy = (o_1 - c_1) + v_1 \tag{4.20}$$

$$k_2dx + l_2dy = (o_2 - c_2) + v_2 \tag{4.21}$$

$$k_3dx + l_3dy = (o_3 - c_3) + v_3 \tag{4.22}$$

$$k_4dx + l_4dy = (o_4 - c_4) + v_4 \tag{4.23}$$

Equations (4.20) — (4.23) form a set of **simultaneous equations** which we must now solve for *dx,dy*, v_1, v_2, $v3$ and v_4. The solution must also satisfy the condition that the sum of the squares of the residuals is a minimum. Note that we are going to solve for dx and dy, not x and y themselves. We add the results of the computation to the x and y we started with, and the new coordinates become the starting point for the next iteration. This adding of increments to the starting coordinates gives the method its name. We start by writing the observation equations in matrix form:

$$\hat{A}\hat{X} = \hat{B} + \hat{V} \tag{4.24}$$

where

$$\hat{A} = \begin{bmatrix} k_1 & l_1 \\ k_2 & l_2 \\ k_3 & l_3 \\ k_4 & l_4 \end{bmatrix} \quad \hat{X} = \begin{bmatrix} dx \\ dy \end{bmatrix} \quad \hat{B} = \begin{bmatrix} (o_1 - c_1) \\ (o_2 - c_2) \\ (o_3 - c_3) \\ (o_4 - c_4) \end{bmatrix} \quad \text{and } \hat{V} = \begin{bmatrix} v_1 \\ v_2 \\ v_3 \\ v_4 \end{bmatrix} \tag{4.25}$$

The only matrix that remains to be defined is the weight matrix. Recalling that the weight assigned to each observation is equal to the inverse of the variance, and assuming that the observations are uncorrelated, we can write

$$W = \begin{bmatrix} \frac{1}{\sigma_1^2} & 0 & 0 & 0 \\ 0 & \frac{1}{\sigma_2^2} & 0 & 0 \\ 0 & 0 & \frac{1}{\sigma_3^2} & 0 \\ 0 & 0 & 0 & \frac{1}{\sigma_4^2} \end{bmatrix} \tag{4.26}$$

The weighted observation equations are now

$$\sqrt{(W)}\hat{A}\hat{X} = \sqrt{(W)}\hat{B} + \sqrt{(W)}\hat{V} \tag{4.27}$$

The square root of the weight matrix in equation (4.27) does not need to be computed, because equation (4.27) is now normalized to give

$$\hat{A}^T W \hat{A} \hat{X} = \hat{A}^T W \hat{B} \tag{4.28}$$

Now equation (4.28) is a **normal equation**, and minimizes $W\hat{V}^2$.

We solve this equation in matrix form by finding the inverse of $\hat{A}^T W \hat{A}$ and multiplying both sides by it:

$$(\hat{A}^T W \hat{A})^{-1} (\hat{A}^T W \hat{A}) \hat{X} = (\hat{A}^T W \hat{A})^{-1} \hat{A}^T W \hat{B} \tag{4.29}$$

The inversion at the normal equation stage is the part of the computation that is most computer-intensive. The time taken to invert increases as the square of the number of parameters. This does not matter much when we consider only the position of the vessel because there are only two parameters, but if we have to find the positions of a large number of nodes at the front end of a triple streamer dual source configuration there may be over 50 parameters involved.

Thus equation (4.29) has given us the solution for *dx* and *dy*. To find the residuals we simply substitute back into equation (4.24).

Now we have the most likely position of the vessel, and we have the residuals for all of the observations. What of the reliability of the position? It turns out (very conveniently for us) that the variance — covariance matrix of the parameters dx and dy is simply the inverse of

$$(\hat{A}^T W \hat{A}) \tag{4.30}$$

which we have already computed as part of the solution. In fact the inverse should be multiplied by the **unit variance**, but if the weights for the observations have been correctly estimated the unit variance will be 1.

Now for an example computation. Let us take four stations and their positions as in Table 4.2:

Table 4.2 Station coordinates

STATION	x(m)	y(m)
1	49434.23	96300.85
2	55385.90	95322.63
3	55816.54	90517.73
4	48147.72	89819.25

The stations' locations relative to each other, and the position of the vessel at P, are shown in Figure 4.5.

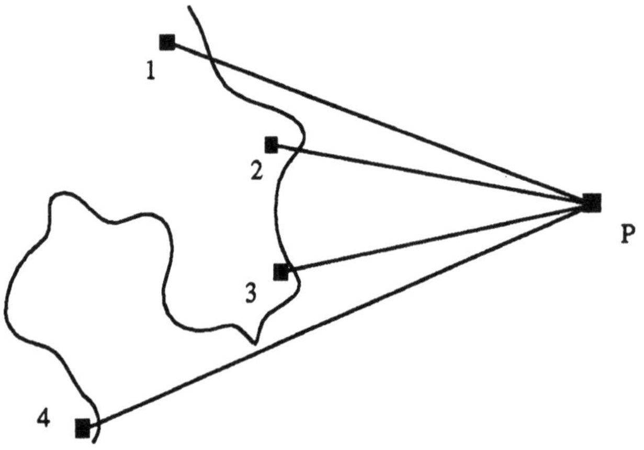

Figure 4.5 Locations of stations.

The seed position for P is estimated to be: $x = 58564.0$m $y = 96754.0$m.

The four observed ranges and their standard errors, and the computed ranges to P are shown in Table 4.3:

Table 4.3 Ranges and standard errors

Station	Range (m)	SE (m)	C (m)
1	9143.5	2	9141.0
2	3491.6	2	3485.6
3	6813.2	2	6814.7
4	12516.3	5	12513.6

By subtracting column 4 from column 2 in Table 4.3 we get

$$\hat{B} = \begin{bmatrix} 2.5 \\ 6.0 \\ -1.5 \\ 2.7 \end{bmatrix} \qquad (4.31)$$

We use equation (4.19) and equation (4.24) to construct

$$\hat{A} = \begin{bmatrix} 1.00 & 0.05 \\ 0.91 & 0.41 \\ 0.40 & 0.92 \\ 0.83 & 0.55 \end{bmatrix} \qquad (4.32)$$

The weight matrix is

$$W = \begin{bmatrix} 0.25 & 0 & 0 & 0 \\ 0 & 0.25 & 0 & 0 \\ 0 & 0 & 0.25 & 0 \\ 0 & 0 & 0 & 0.04 \end{bmatrix} \qquad (4.33)$$

The normal equation yields

$$\begin{bmatrix} 0.525 & 0.216 \\ 0.216 & 0.266 \end{bmatrix} \begin{bmatrix} dx \\ dy \end{bmatrix} = \begin{bmatrix} 1.930 \\ 0.361 \end{bmatrix} \tag{4.34}$$

The inverse of the first matrix in equation (4.34) is

$$\begin{bmatrix} 2.860 & -2.323 \\ -2.323 & 5.646 \end{bmatrix} \tag{4.35}$$

So our solution is

$$\begin{bmatrix} dx \\ dy \end{bmatrix} = \begin{bmatrix} 2.860 & -2.323 \\ -2.323 & 5.646 \end{bmatrix} \begin{bmatrix} 1.930 \\ 0.361 \end{bmatrix} = \begin{bmatrix} 4.68 \\ -2.45 \end{bmatrix} \tag{4.36}$$

The dx and dy are now added to the original coordinates of P to give the adjusted position of $x = 58568.7\text{m}$ $y = 96751.6\text{m}$

In practice the whole computation will then be repeated using the new coordinates of P as the seed. If this is done it will be seen that the new dx and dy are very close to zero and the computation is said to have **converged.**

We now substitute the dx and dy values from above back into the equation (4.24) to get the residuals vector:

$$\hat{V} = \begin{bmatrix} 4.6 \\ 3.2 \\ -0.4 \\ 2.5 \end{bmatrix} \tag{4.37}$$

The equation for the unit variance is:

$$\sigma_0^2 = \frac{(\hat{V}^T W \hat{V})}{(n - m)} \tag{4.38}$$

where n is the number of observations and m is the number of parameters, in this case 2.

It turns out that $\sigma_0^2 = 4.07$, which means that for this fix our chosen standard errors were unduly pessimistic. However, it is best not to change the standard errors (assuming we have chosen reasonably sensible values) until at least one or two lines have been shot and we are better able to estimate the performance of the observations. Assuming that the average unit variance turns out to be close to unity, we can use equation (4.38) to estimate the reliability of the position. The standard deviation of the x-coordinate is 1.7m and that of the y coordinate is 2.4m. So we have a 68% chance that the position is within 1,7m (x) and 2.4 (y) of what we calculated, or a 95% chance of being within 3.4m (x) and 4.8m (y). We can then go on to draw the error ellipse for the position.

This is the basic method used not only to calculate the vessel position but also all the streamer and source positions. All we do is increase the number of parameters in the unknowns vector and write down suitable observation equations for the new parameters.

4.4 Streamer and source shaping

4.4.1 Streamers

Streamer modelling is a subject which often provokes argument amongst surveyors, and the last word on it will certainly not be said here. But there are a few guidelines we can follow.

The first point is that streamers, like any other body, want to follow the path of least resistance through the water. For this reason alone every part of the streamer tends to track along its own length, with the result that the after parts of the cable follow the same path as the front.

If the towing vessel makes a sudden, sharp deviation of course a wave is very often set up which travels down the entire length of the streamer. The amplitude of the wave may even increase under certain conditions. This is analogous to picking up one end of a straight length of rope and giving it a sharp flick.

Various systems have modelled the shape as a parabola, a hyperbola and a cubic spline. It is now most common to use a polynomial fit for the streamer, with the polynomial being of order 3 to 5. It seems unlikely that the streamer would take up curves of a high enough frequency to warrant using an order greater than 5, while using an order less than 3 tends to produce high compass residuals. One way in which the polynomial can be built is as follows.

In Figure 4.6 P represents a compass on the starboard streamer of a dual-streamer configuration. Two axes X and Y are drawn through the vessel reference point as shown. The direction of Y may be the course made good, the gyro heading, or the line heading. X and Y form a local grid with the NRP as origin. If we specify a fifth order polynomial, we can write down the position of P as:

$$x_P = a + by_P + cy_P^2 + dy_P^3 + ey_P^4 + fy_P^5 \quad (4.39)$$

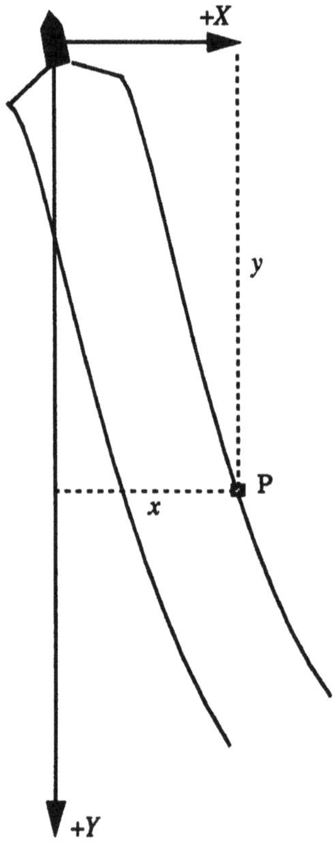

Figure 4.6 Streamer shaping.

There are six coefficients (a, b, c, d, e, f) in equation (4.39) which need to be found, Therefore we need at least another five similar equations to solve for the six unknowns. Usually there is such a proliferation of compasses, laser, acoustic and tailbuoy sensors that we are never short of redundant observations. If we want to include stretch factors we will need to model them and increase the number of parameters further.

In equation (4.39) we have described a **model** for the streamer shaping. In practice the model would be a little more complex, but it would still follow the same format. We then develop observation and normal equations in exactly the same way as described in 4.3.3 to solve for the polynomial coefficients. Note that if we adopt this approach we end up with a polynomial that we can use for any point on the streamer, provided that we know its Y-offset. So having obtained the shape, we can find where every group, acoustic node and compass is on the streamer. Extending the argument and solving for the port streamer coefficients, we can also determine the streamer separation at any point. We do not solve for the compass and acoustic node positions directly, since their positions are not required.

4.4.2 Sources

We can use a similar method for source positioning:

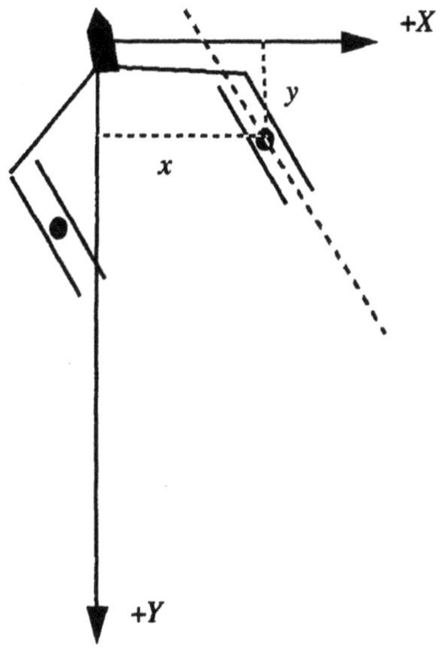

Figure 4.7 Source shaping.

Figure 4.7 illustrates a dual-source spread with two gun strings in each source. The centre sources are represented by the black circles, and the same local X,Y coordinate system is used as for the streamer shaping.

Because the gun strings are so short, we can model them as straight lines so that:

$$x = a + by \tag{4.40}$$

In Figure 4.7 the dotted line drawn through the centre of the starboard source is the line whose equation is equation (4.40).

Note that in both equation (4.39) and equation (4.40) we are dependent on knowing the y

offset for our observations. We are not directly in possession of this information, but we do know exactly where each acoustic node and laser target is on any gun string or streamer. We can then differentiate equation (4.39) and equation (4.40) to determine the slope at each node and therefore compute the y-offset.

4.5 Summary

The method of least squares does not give the true solution, but then neither does any other known solution. Least squares does give the most probable solution and it has a very sound statistical basis.

The method is very easy to program and does not require a very large machine, although to produce a real-time system capable of positioning the vessel, streamers and sources does require a fast floating point processor.

Once good values for the standard errors have been estimated the method can be completely automated, including data rejection (as we shall see in the Chapter 5).

There is never any ambiguity in the result; the method always gives a unique solution.

It is quite easy to program a solution for any type of observation, and different systems can easily be combined. For example, the vessel NRP can be positioned using a combination of Syledis and Argo ranges, together with a DGPS latitude and longitude, with the great advantage that all LOP's can be individually weighted in the solution.

The one real drawback is that the accuracy is greatly affected by the choice of mathematical model used, and the precision is greatly affected by the weight matrix. This means that the model must be chosen to represent the physical reality as closely as possible, and the standard errors need to be carefully estimated.

5 Kalman filters

5.1 Introduction

5.1.1 Integrated systems

Many of the processing systems currently available treat each part of a marine seismic spread as a collection of discrete units — individual compasses on individual streamers, individual acoustic transponders on individual gun strings and so on. Even acoustic networks are treated individually — a network at the front of the spread, one at the tail of the streamers and perhaps another in the centre. This 'individualism' even goes so far as to treat each time event on its own.

What such systems have failed to grasp is that ALL of the elements in a seismic spread are interconnected — they are all attached to the vessel(s) in some way, and therefore they cannot react independently to external forces such as wind, tide and currents. All parts of the spread will, over the length of a line, move in the same direction and at the same speed as the vessel reference point since they are all connected to it. Similarly, the navigation data collected at a single event cannot be treated independently - that event's data is connected (by time) to all the data from all of the events preceding it, and to those succeeding it.

The types of errors which will occur if a non-integrated solution is used are listed below. If navigation data are despiked, adjusted and reduced to individual point positions the following can, and in many cases will, happen:

- Errors in parameters such as offsets may remain undetected. For instance, an acoustic range which has been despiked around a 'good looking' mean may contain a reflection error — in fact it may well be the 'spikes' which are the good values. This has been seen to occur countless times in real data. If such a range were introduced into a system which took account of the integrated nature of the problem and included compass, vessel and laser positions the problem would almost certainly be detected.
- Individual range despiking is operator intensive and is not statistically correct.
- It becomes almost impossible to ascertain the final quality of the data. There is no way in which a meaningful error ellipse can be attached to a receiver group position unless that group's position has been derived by a rigorous, integrated least squares method of adjustment. The propagation of variances and covariances between the different adjustments is complicated and will hardly ever (if at all) be implemented.
- A line which has been discarded simply because the data 'looks noisy' may well turn out to be salvageable. No sound judgement can be made about a data set unless that data is subjected to proper statistical testing.
- Many systems which employ point adjustment routines are dependent on continuity of that data. It is often critical to have a particular range or ranges present in order to be able to compute the network — if there are more unknowns than observations, the network cannot be adjusted. It therefore follows that if those ranges do not appear in the data set,

they must be approximated in some way. How can this be done? Should the last good range be used as an approximation? Should we avoid computing the net at all and use the positions computed from the last shot? Should a "computed range" be used - one that fits in well with the rest of the data? None of these options are correct, and none of them will produce the right answer for the network.

An integrated system overcomes the above problems in the following ways:

- Offset errors will quickly become apparent because all of the observations become inter-related (unless, of course, there are very many offset errors).
- Ranges are never despiked and there is no pre-processing. Instead each observation is subjected to a statistical W-test prior to entry into the computation.If the observation fails the test it is rejected; if it passes it is accepted into the computation.
- The final quality of the data can easily be assessed because the integrated system already has an associated variance/covariance matrix containing covariances for the whole system.
- Judgements on what is good data and what isn't are made automatically by the W-test. In some cases more than 50% of the data might be rejected or missing from the raw data set, and the line may still be acceptable.
- Continuity of data is no longer a problem. One of the great attributes of a Kalman Filter is that it can predict a variable's value using all previous data available to it. In other words, the value of variable xn at event n can be predicted using the previous (n-1) values of x. It is entirely possible for the filter to run for 3 or 4 shots with no new data at all, without causing a major deterioration in accuracy.

5.1.2 Kalman filters

The author once heard a Kalman Filter described as a card which could be inserted into the back of a PC and which then despiked incoming data! Even worse, this statement was from a surveyor who had been in the offshore business for about ten years! Nothing could be further from the truth. It is a software filter and is simply an extension of least squares. The Kalman filter is a method of filtering, smoothing and predicting data which can be illustrated as follows.

In Figure 5.1 , the circles labelled $P_1, P_2, ..., P_i$ represent a vessel's NRP as derived by point least squares solutions at times $t_1, t_2, ..., t_i$. These solutions are arrived at using only the observations current at each shot and are in no way related to each other.

What we require to do is find the points marked by black squares which represent the **optimal or most probable position** of the NRP taking into account not only the observations current at a particular shot but also all the observations up to and including that shot. It is very important to realize that under no circumstances are we trying to fit an approximate curve to a set of known real positions. We don't know where the vessel positions are, and we never will know their true positions, but the Kalman filter is a rigorous least squares optimal filter which will lead to a maximum likelihood solution. The fact is that vessels do not travel in a series of disjointed hops, they transit along a smooth curve which in the case of a seismic vessel shooting a prime line will approximate to a straight line.

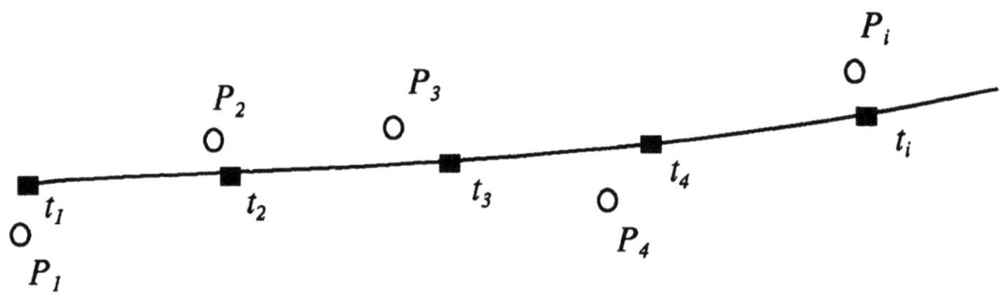

Figure 5.1 Vessel's track.

To carry out an ordinary least squares computation using all the available data up to and including the current shot would require solving a massive set of simultaneous equations and inverting matrices of mammoth orders. However, were such a solution to be used it would give exactly the same answer as a Kalman filter.

In seismic surveying we are not only interested in the position of the vessel; in fact, the vessel position is not really important at all. What we really need to know are the positions of the sources and the streamer receiver groups. The filter can encompass both source and streamer positioning, at all times maintaining the basic premise that the spread is a single system, all parts of which are physically joined together, and which must therefore use an integrated algorithm.

An optimal estimator is an algorithm that processes all the available data to yield the 'state' of a system, whilst at the same time satisfying some predetermined optimal criterion. In our case, the 'state' is a vector of parameters called the **state vector** which contains the vessel position, velocity, drift, streamer and source shape coefficients. In addition, if the vessel is towing acoustic transponders from sleds and those transponders do not form part of a source or streamer, an additional pair of positional parameters can be included for each sled node position. The information processed by the filter includes all observations such as Argo ranges, DGPS positions, laser and acoustic ranges, compass bearings, and so on. It also includes knowledge of the ship's dynamics. This does not imply detailed equations modelling wind and wave action on the ship together with its roll and pitch characteristics, but is instead a simple polynomial treatment of the dynamics; the detail is not required because the time interval between shots is usually only a few seconds.

5.1.3 Kalman filter equations

Again using Figure 5.1 , consider what is happening at times t_1 and t_2.

Suppose that $\hat{X}_1$ is the state vector at time t_1, and that a vector of observations $\hat{\partial}_1$ is measured.

Then we can write a functional model

$$F_1(\hat{X}_1, \hat{\partial}_1) = 0 \tag{5.1}$$

Similarly, at time t_2 we can write

$$F_2(\hat{X}_2, \hat{\partial}_2) = 0 \tag{5.2}$$

These equations can be linearised to

$$\hat{A}_1\hat{X}_1 + \hat{C}_1\hat{v}_1 - \hat{b}_1 = 0 \tag{5.3}$$

$$\hat{A}_2\hat{X}_2 + \hat{C}_2\hat{v}_2 - \hat{b}_2 = 0 \tag{5.4}$$

The above equations will then be recognised as the standard least squares solution equations, and each equation is known as a primary or **static model**. The static model also has a weight matrix $\hat{W}_1, \hat{W}_2, etc$ which is derived from the observational covariances in the usual way.

In addition to the static model, we also need to define another model which describes how the state vector changes with time - a **dynamic model**. We can write a second set of equations relating state vector 2 to state vector 1:

$$F_{12}\left(\hat{X}_1, \hat{X}_2, t_1, t_2\right) = 0 \tag{5.5}$$

This equation is then linearized to

$$\hat{X}_2 = \hat{M}\hat{X}_1 + \hat{Y} \tag{5.6}$$

where $\hat{M}$ is the **transition matrix,**

and $\hat{Y}$ is the **noise matrix**, representing the unknown errors in the dynamic model.

Associated with the dynamic model is a covariance matrix $\hat{C}_Y$ which reflects the precision of the dynamic model.

The filtering problem is to find least squares estimates for $\hat{X}_1, \hat{X}_2, etc$, so we need to minimize the equation

$$\hat{v}_1^T W_1 \hat{v}_1 + \hat{v}_2^T W_2 \hat{v}_2 + \hat{Y}^T W_M \hat{Y} \tag{5.7}$$

subject to the static and dynamic constraints imposed by equation (5.3), equation (5.4) and equation (5.6).

We can now follow the process through from start to finish, by assuming that at time ti we have an optimal estimate of the state vector $\hat{X}_i$ with a corresponding covariance matrix $\hat{C}_{\hat{X}_i}$.

At time t_{i+1} we receive new data in the form $\hat{\partial}_{i+1}$ which are related to the state vector by the equation

$$F_{i+1}(\hat{X}_{i+1}) = \hat{\partial}_{i+1} \tag{5.8}$$

There is also a dynamic model of the form

$$F(\hat{X}, t) = 0 \tag{5.9}$$

There are two main steps in the filtering process:

- Predict the state vector and its associated covariance matrix using the dynamic model.

- Alter the prediction using the observation model and the statistics of both the static and dynamic models.

Linearizing equation (5.9) gives us equation (5.10), i.e.

$$\hat{X}_{i+1} = \hat{M}_{i,i+1}\hat{X}_i + \hat{Y} \quad (5.10)$$

The prediction is carried out using

$$\hat{X}_{i+1} = \hat{M}_{i,i+1}\hat{X}_i \quad (5.11)$$

because the noise matrix $\hat{Y}$ is unknown. The composition of the transition matrix will be explained later.

Prediction of the covariance matrix involves knowledge of both error propagation through the prediction process and the statistics of the noise matrix, resulting in a simple equation

$$\hat{C}_{\hat{X}_{i+1,i}} = \hat{M}_{i,i+1}\hat{C}_{\hat{X}_i}\hat{M}^T_{i,i+1} + \hat{C}_{\hat{Y}} \quad (5.12)$$

This is the first stage of the Kalman filter completed, since we now have the **predicted state vector** and **predicted covariance matrix**. We can now compute the second stage by first putting equation (5.3) and equation (5.4) into a more generalized form:

$$\hat{A}_{i+1}\hat{X}_{i+1} = \hat{b}_{i+1} + \hat{v}_{i+1} \quad (5.13)$$

At this stage it should be remembered that equation (5.13) often cannot be solved by the usual method, because there may well be more parameters in the state vector than there are new observations. The new observation data for time t_{i+1} enters the Kalman Filter along with a matrix $\hat{G}$, the **gain matrix**. Essentially the gain matrix determines the extent to which the predicted state vector must be changed to accommodate the new observations. The gain matrix is computed from

$$\hat{G} = \hat{C}_{\hat{X}_{i+1,i}}\hat{A}^T_{i+1}\left(\hat{A}_{i+1}\hat{C}_{\hat{X}_{i+1}}\hat{A}^T_{i+1} + \hat{C}_{\hat{\partial}}\right)^{-1}. \quad (5.14)$$

Equation (5.14) uses $\hat{C}_{\hat{\partial}}$, the covariance matrix of the observations themselves, and is the stage at which inversion occurs. It will be seen that the size of the matrix to be inverted is equal to the number of new observations; the inversion of a matrix which may be of order 150 or higher requires a large amount of floating point power and a very fast clock speed.

Once the gain matrix has been computed, the optimal estimate of the new state vector can be computed from

$$\hat{X}_{i+1} = \hat{X}_{i+1,i} + \hat{G}(\hat{b}_{i+1} - \hat{A}_{i+1}\hat{X}_{i+1,i}) \tag{5.15}$$

and the new covariance matrix from

$$\hat{C}_{\hat{X}_{i+1}} = (\hat{I} - \hat{G}\hat{A}_{i+1})\,\hat{C}_{\hat{X}_{i+1,i}} \tag{5.16}$$

5.1.4 Summary

In this introduction we have examined the reasons for using an integrated solution, why a Kalman Filter is used and how the mechanics of the Kalman Filter work. In the next section we shall look at the application-specific aspects of the process.

5.2 A typical filter — overview

5.2.1 State vector

In seismic surveying, the state vector in equation (5.17) could be used.

Explanation of the various terms follows:

- The vessel X and Y coordinates are the increments in the X and Y direction, since the last filtered shot. We are working on a local 3D Cartesian grid, with the X direction aligned true east and the Y direction true north. Within the limits of the prospect, and assuming that the origin is chosen reasonably close to the origin, movement in the vertical direction will be negligible.

- The velocities in the X and Y directions are those for the NRP. These are necessary to help predict where the NRP will be on the following shot.

$$\begin{bmatrix} Vessel-X \\ Vessel-Y \\ Velocity-X \\ Velocity-Y \\ Drift \\ Source-coefficients \\ Streamer-coefficients \\ Sled-X \\ Sled-Y \end{bmatrix} \tag{5.17}$$

- The drift is the angle between the course made good and the corrected gyro heading (i.e. corrected for gyro error).
- The sled-X and Sled-Y elements are the increments in the X and Y positions of each sled (tailbuoy) in the spread since the last shot. There will be one pair of these increments for each sled.

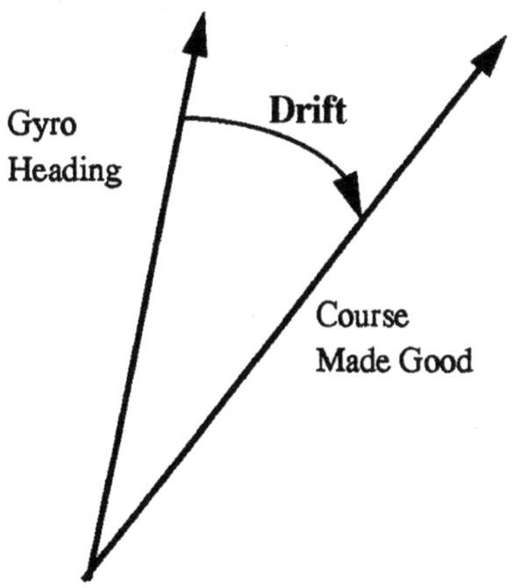

Figure 5.2 Modelling drift.

- The sources are modelled as straight lines; for two sources there will be four coefficients.
- For a dual-streamer spread using fifth order streamer polynomials, there will be twelve streamer coefficients.
- In a spread where the vessel is towing acoustic pods from vessel-mounted sleds, there will be additionally two parameters for each sled node, representing increments to the X and Y coordinates of those nodes.

As an example, for a triple-streamer, dual-source spread utilizing four sleds, there will be 2 + 2 + 1 + 18 + 4 + 6 = 33 parameters in the state vector.

The sleds or tailbuoys can be modelled separately or as part of their parent body, i.e. as part of the streamer or source to which they are attached. The latter is the author's preference since then the system is more integrated. This approach also reduces the number of parameters in the state vector.

5.2.2 The cycle

At the start of the line, all absolute positioning data for the vessel is deskewed to the reference point and a point solution is computed by iterating through a variation of coordinates routine until the X and Y parameters in the state vector move by less than 0.01 m. For subsequent cycles the filter parameters for the previous shot are taken as the start point and this step is ignored. An initial course and speed vector is estimated from the current gyro heading and the expected time between shots. Initial estimates for the source and streamer parameters are taken as zero with the exception of the least significant parameter which is designated as the A- or X-offset for that body — see Figure 5.3 :

The position at the first shot is shown in Figure 5.3 .

An initial estimate for the covariance matrix is also required —- see details later.

We then compile observation equations for all positioning systems. All C - O corrections, delays, propagation coefficients, etc. are applied at this point.

The state and covariance matrices are then predicted as described in section 5.1.3 above.

We now combine the predicted state vector with the design (observation) and constants matrices to form a vector of predicted residuals. These residuals are tested against a null hypothesis. The hypothesis tests whether or not the ratio of the predicted residual to its predicted standard deviation falls inside a specified range. If it does, the value of that element will be incorporated into the filter. If not, it will be rejected. It is entirely possible for every single observation to fail this test (although extremely unlikely) — the filter will still perform and the program should not grind to a halt. This statistical test (the W-test) provides a rigorous means of validating an observation before it goes into the filter — which is why software of this type can process on-line and in a single pass.

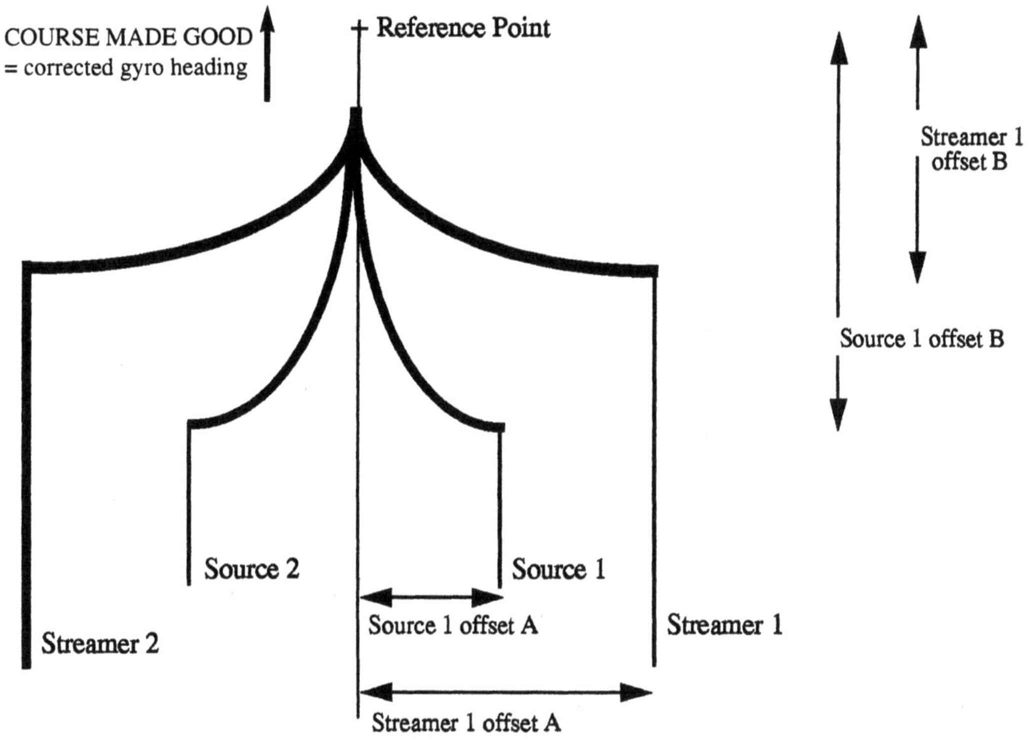

Figure 5.3 Position at first shot.

Observations which have passed the W-test are now used to form the gain matrix and the new state vector. Having computed the new state matrix, the residuals and covariance matrices are found.

The source and streamer coefficients in the state vector are used to compute the active source and all the group positions — they can be written directly to a P190 file together with the reference position.

It is very important to realize that (except for the sled positions) the actual positions of the various sensors are never computed — the observations are all used to determine either the reference position or the source and streamer coefficients.

It can be seen that the entire spread is treated as a single body; after all, both the streamers and the sources are physically connected to the vessel, and over a period of time all points on the arrays must move at the same speed and in the same direction as the vessel reference point.

5.3 Observation equations

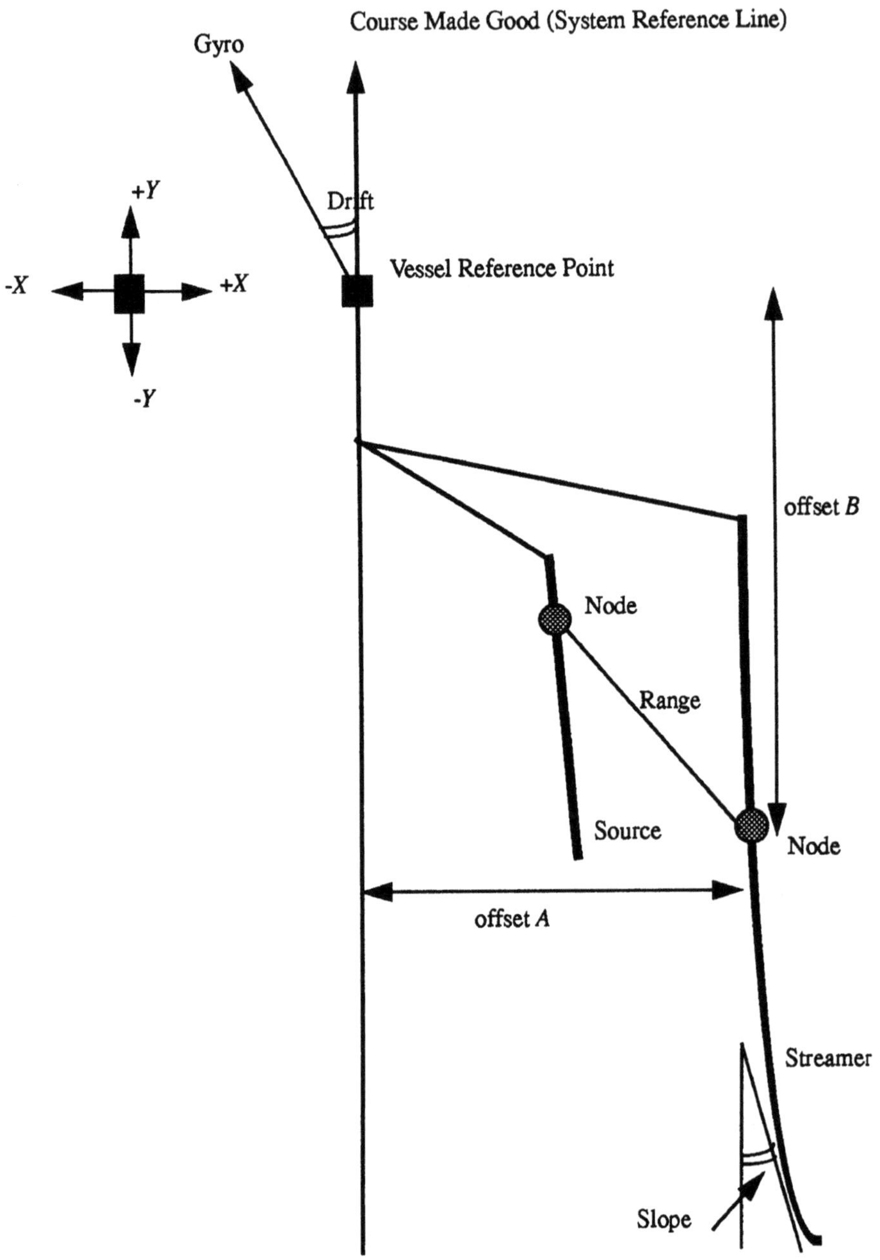

Figure 5.4 Equation terminology.

When reading the notes below on observation equations, terminology refers to Figure 5.4.

5.3.1 Distances and azimuths

For ranges from radio navigation stations to a vessel or tailbuoy antenna, the global 3D Cartesian system is used (e.g. that associated with the AGD84 datum). For a distance from point 1 to point 2,

$$D = \sqrt{((X_1 - X_2)^2 + (Y_1 - Y_2)^2 + (Z_1 - Z_2)^2)} \tag{5.18}$$

This gives the chord distance, which must then be reduced for curvature correction to give the arc distance:

$$D_T = D + \frac{D^3}{24R^2} \tag{5.19}$$

where D_T = arc distance

and R = radius of the earth

For distances within the spread (where ranges will not usually be more than (say) 4000m), the curvature correction is ignored.

For true azimuths,

$$A_{12} = \text{atan}\left(\frac{a}{((Z_2 - Z_1)\cos(\phi_1) - b\sin(\phi_1))}\right) \tag{5.20}$$

where

$$a = -\sin(\lambda_1)(X_2 - X_1) + \cos(\lambda_1)(Y_2 - Y_1) \tag{5.21}$$

and

$$b = \cos(\lambda_1)(X_2 - X_1) + \sin(\lambda_1)(Y_2 - Y_1) \tag{5.22}$$

For azimuths within the spread, using local 3D coordinates, the formula can be simplified to:

$$A_{12} = \text{atan}\left(\frac{(X_2 - X_1)}{(Y_2 - Y_1)}\right) \tag{5.23}$$

5.3.2 Ranges to vessel

All ranges to the vessel should be deskewed to the reference point, using the following form:

$$offsetAngle = \text{atan}\left(\frac{offsetA}{offsetB}\right) \tag{5.24}$$

$$dist = hypot(offset(A, offsetB)) \tag{5.25}$$

The design matrix terms are then

$$-\left(\frac{(X_2 - X_1)}{computeddist}\right) \tag{5.26}$$

and

$$-\left(\frac{(Y_2 - Y_1)}{computeddist}\right) \tag{5.27}$$

where *computeddist* is the computed distance between the station and the vessel receiver.

The constants vector term is

$$computeddist - deskewed \tag{5.28}$$

where *deskewed* is the observed range corrected for C — O, velocity factor, propagation velocity and frequency, corrected for skew using the *offset Angle* and *dist*. The development of each individual formula is beyond the scope of this book; suffice it to say that there is a different formula for each type of range and pseudo-range.

The reason for deskewing back to the reference point is so that the vessel reference position can accommodate any number of receivers onboard the vessel. The contribution to vessel position made by each range of each system is determined by the standard error attributed to that range. This enables (for instance) calibration for partial Argo lanes to be made against a Miniranger beacon without change to the parameters except for attributing a low standard error to the Miniranger and a high standard error to the Argo lane being calibrated.

5.3.3 Ranges to sled

Observation equations for ranges to the sleds from shore stations are developed in exactly the same way as in section 5.3.2 above, except that there is no deskewing — all ranges are taken direct to the receiver. It is assumed that there is only one receiver per sled.

5.3.4 GPS and DGPS

GPS and DGPS positions may be reported either as latitude and longitude or as range and bearing from the reference point. In the case of latitude and longitude the position is first converted to the survey datum if necessary. Then the true range between the predicted reference point and the GPS position is computed and the range and bearing used in the normal way. The ranges and bearings must first be computed in terms of the system reference line, which is the course made good, before entry into the observation equations.

5.3.5 Relative ranges

Relative ranges can emanate from the vessel, a sled, a source or a streamer and can terminate on any of those four bodies.

For a range to or from the vessel, the transponder's offset A and offset B are first adjusted by the drift angle as follows:

$$x = -offsetB \times \sin(drift) + offsetA \times \cos(drift) \tag{5.29}$$

$$y = offsetB \times \cos(drift) + offsetA \times \sin(drift) \tag{5.30}$$

For a range to or from a sled node, the transponder's offset A and offset B are taken directly from the vector holding the sled positions (this vector is updated on each shot by the increments computed as part of the state vector).

For a range to or from a source or streamer, the transponder's offset A and offset B are first adjusted by using the slope of the body at the transponder. For instance, using a fifth-order polynomial, the X offset is given in terms of the Y offset from the reference point as:

$$X = a + bY + cY^2 + dY^3 + eY^4 + fY^5 \tag{5.31}$$

and

$$\frac{dX}{dY} = b + 2cY + 3dY^2 + 4eY^3 + 5fY^4 \tag{5.32}$$

The nominal transponder positions are then found by substituting slope for drift in equation (5.29) and equation (5.30).

Once all the start and end points for the transponders have been adjusted, the nominal ranges are found (nominal in this case being the expected range from the current source and streamer shaping as computed from the current state vector) and the observation equations are set up in the normal way.

5.3.6 Compasses

For a compass, the nominal Y-offset of the compass along the streamer first needs to be found in terms of the course made good, which is the system's reference line. Once again the slope is computed and then equation (5.29)and equation (5.30) used.

The compass is then corrected for magnetic declination, fixed correction and directional correction and the course made good is subtracted to orientate the compass into the system reference. The corrected compass reading is then related to the streamer polynomial function by using the streamer slope and the compass tangent.

5.3.7 Secondary vessels

Secondary (slave) vessels can be treated as a sled; we need to provide two more parameters in the state vector for the X- and Y-offsets from the primary reference point to the reference point on the slave vessel, then add in source and streamer coefficients for the secondary vessel's sources and streamers. This technique works well for the slave vessel being alongside the primary vessel or ahead/astern of it. It is unnecessary to compute drift for the secondary vessel; for any ranges and/or azimuths between the vessels, such as acoustic network ranges, all we need to do is take account of the continuously varying X- and Y-offsets between the two reference points; streamer and source shaping takes place in exactly the same way as on the primary vessel.

5.3.8 General

Only the observation equation elements for ranges to the vessel have been properly described here. The remaining types of observation are given in general terms, but the following general points should be noted with regard to the *A*- and *B*-offsets for individual nodes and compasses:

- The state vector gives only the polynomial coefficients for the streamers and sources, together with increments for the sled positions and the reference point, the velocity and the drift.
- Any node on a vessel must first be corrected by the drift.
- Any node on a sled uses the filtered position for that sled.
- Any node on a source or streamer is corrected using the differentiated slope of that body.

5.4 Weighting

5.4.1 Observations

All observations can be individually weighted. It would be easy to 'hardwire' weighting for different systems into the program, but the system would then lose all its flexibility. However, it is very important to note that **correct weighting of observations is absolutely crucial.**

A least squares computation will generally give an answer; that answer will be the most probable value for an unknown parameter or vector of parameters. But the validity of the answer depends on two things:

- The proximity of the model being used to the truth.
- The correct weighting of the observations.

We use the standard errors assigned to each observation in the following way:

$$W = \frac{1}{se^2} \tag{5.33}$$

where se^2 is the square of the standard error. The higher the standard error, the smaller the weight and the less the influence of that observation on the result. Suggested standard errors are given below:

Syledis	5 m
Argo	5 m
DGPS	5 m
Trisponder	3 m
Hyperfix	10m
Geoloc	10m
Sonardyne acoustics	1 m
Digicourse acoustics	1 m
Mars Laser	1 m
Syntron compasses	1 or 2 deg
Digicourse compasses	1 or 2 deg

If an observation fails the W-test, it is rejected from the filter.

5.4.2 Initial state covariance

At the start of line the state covariance matrix must be assigned values as follows:

For the X and Y position increments, the variances and covariances are taken straight from the inverted normal equation.

For the X and Y velocities, $0.1\,\mathrm{m/s^2}$.

For the drift, 0.5 deg/sec.

For the source and streamer polynomials,

$$(1 \times 10^{-exponent})^2 \tag{5.34}$$

where **exponent** = **order** * 3

and **order** is the order of the polynomial for each coefficient.

For the sled position increments, $0.1\,\mathrm{m}^2$.

5.4.3 Dynamic covariance (driving noise matrix)

For the drift, dynamic variance = 0.05 * dt^2

For the sources and streamers, dynamic variance =

$$(index \times 10^{-exponent})^2 \times dt^2 \tag{5.35}$$

where **index** = 0.01 and **exponent** = **order** * 3. There is assumed to be no correlation.

For the position and velocity elements,

$$\begin{bmatrix} \frac{a}{4}(dt^4) & \frac{a}{2}(dt^3) & 0 & 0 \\ \frac{a}{2}(dt^3) & \frac{a}{4}(dt^4) & 0 & 0 \\ 0 & 0 & a(dt^2) & \frac{a}{2}(dt^3) \\ 0 & 0 & \frac{a}{2}(dt^3) & a(dt^2) \end{bmatrix} \tag{5.36}$$

5.5 Transition function

5.5.1 General

As described earlier in section 5.1.3, the transition matrix is used as in equation (5.10) and equation (5.11). However, the filter is designed around intended straight-line sailing. The proximity of the vessel's track to a straight line is valid even when shooting for coverage, because we are only looking at a few shots at a time. (Even if circular shooting were employed, the radii of the spiral track would be so large as to make this assumption still valid).

We can expand the position and velocity using simple polynomials, so that

$$X_{t+dt} = X_t + \dot{X}_t dt + \frac{1}{2} a_X dt^2 \tag{5.37}$$

$$Y_{t+dt} = Y_t + \dot{Y}_t dt + \frac{1}{2} a_Y dt^2 \tag{5.38}$$

$$\dot{X}_{t+dt} = \dot{X}_t + a_X dt \tag{5.39}$$

$$\dot{Y}_{t+dt} = \dot{Y}_t + a_Y dt \tag{5.40}$$

The terms containing a represent the driving noise and are responsible for the corresponding entries in the dynamic covariance matrix above. The transition function can then be rewritten as:

$$\begin{bmatrix} X_{t+dt} \\ Y_{t+dt} \\ \dot{X}_{t+dt} \\ \dot{Y}_{t+dt} \end{bmatrix} = \begin{bmatrix} 1 & 0 & 0 & 0 \\ 0 & 1 & 0 & 0 \\ 0 & 0 & 1 & 0 \\ 0 & 0 & 0 & 1 \end{bmatrix} \begin{bmatrix} X_t \\ Y_t \\ \dot{X}_t \\ \dot{Y}_t \end{bmatrix} + \begin{bmatrix} \dot{X}_t & \frac{a_X}{2} \\ \dot{Y}_t & \frac{a_Y}{2} \\ a_X & 0 \\ a_Y & 0 \end{bmatrix} \begin{bmatrix} dt \\ dt^2 \end{bmatrix} \qquad (5.41)$$

thereby reducing the transition matrix to the identity matrix.

5.6 Data rejection

5.6.1 The *W*-test

Once we have arrived at the end of Stage 1 of the filter, i.e. once the **predicted state vector** and the **predicted covariance matrix** have been found, we are in a position to test the observations for possible rejections.

The **predicted residuals vector** is found simply from:

$$\hat{V}_{predicted} = \hat{A}\hat{X}_{predicted} - \hat{b} \qquad (5.42)$$

and the **predicted covariance of the residuals** is found from:

$$\hat{C}_{res} = W + \hat{A}\hat{C}_{predicted}\bar{A}^T \qquad (5.43)$$

A vector is then constructed which contains for its elements each predicted residual divided by the square root of its predicted covariance. Each element is then tested to see if it lies within a specific range provided by a look up table. A few selected confidence levels and their look up values are as follows:

- 90% -1.65 — +1.65
- 95% -1.96 — +1.96
- 99% -2.58 — +2.58
- 100% -3.00 — +3.00

As an example, suppose that the confidence level is set to 99% and that an acoustic range has a predicted residual of 2.3m and a predicted covariance of 1.24m.

The required ratio is $\dfrac{2.3}{\sqrt{1.24}} = 2.06$, so the observation will be accepted.

Now let the predicted residual be 3.6m and the predicted covariance be the same, at 1.24m.

Now the required ratio is $\dfrac{3.6}{\sqrt{1.24}} = 2.90$

so the observation fails as it is outside the maximum threshold of 2.58. Now that observation will be rejected from the resulting computation.

5.7 Residuals and statistics

5.7.1 Computation of residuals

Once the state vector has been found, we can find the residuals vector with ease:

$$\hat{v} = \hat{A}\hat{X} - \hat{b} \tag{5.44}$$

5.7.2 Statistics

Having obtained the residuals vector, we can then enter data into the statistics arrays. There could be a separate array for each element, i.e. every compass, absolute range, acoustic range, separation, etc., has its own array. The data loaded into each array at each shot might be the shot number, the raw value, the residual and the weight.

Once the data has been loaded into the statistics array, we can compute the mean raw value, mean residual, standard deviation, delta standard deviation, maximum and minimum raw values and the current rate of change. All of these statistics should be made available as numeric or time plot displays.

5.7.3 Separation

Separations can be computed between any designated nodes in the spread; using the polynomial coefficients the node positions themselves can be back-computed.

5.8 Group computations

5.8.1 Groups

The receiver groups can be computed as follows:

For each streamer we have a set of polynomial coefficients describing the streamer shape, and we know the Y-offset of the centre near group and the group interval.

We compute the streamer slope at the centre near group by differentiating the polynomial as in equation (5.31) and equation (5.32). Then the nominal offset is adjusted using

$$y = offsetB \times \cos(slope) + offsetA \times \sin(slope) \tag{5.45}$$

Now we can use the polynomial to find the X- offset corresponding to the computed Y-offset. At this stage the X and Y coordinates are still in the local system — i.e.the reference point is the origin and the Y-axis is in the direction of the course made good.

So now we rotate the (X, Y) pair back into the local Cartesian system, then rotate and translate it to the global system. From there we can convert back to eastings and northings, and the procedure is repeated all the way down the streamer.

5.8.2 Sources

The source positions are found using the same technique as above.

5.9 Summary

Kalman filtering is a very powerful tool for positioning offshore. It can accept different types of data, solve for a large number of parameters and produce a reliable and precise result. In particular, the following points are worthy of note:

- Kalman filters require no pre-processing.
- Data rejection is both automatic and based on sound statistical principles.
- The system is integrated — one of the benefits of this can be seen in the following example. The traditional network of acoustic transponders at the front, middle and back of the streamers has been replaced with a 'ladder' of transponders extending right down the streamers. Only an integrated system would be capable of solving the streamer shapes using such an acoustic net, because there are far fewer observations than unknowns if the nodes themselves were to be computed. But since we are only solving for the streamer polynomial coefficients, there are actually enough acoustic ranges to compute the shapes without compasses. When used together with compasses, the network becomes very strong because the observations are evenly spaced down the streamers and not bunched

together with few observations between the bunches.

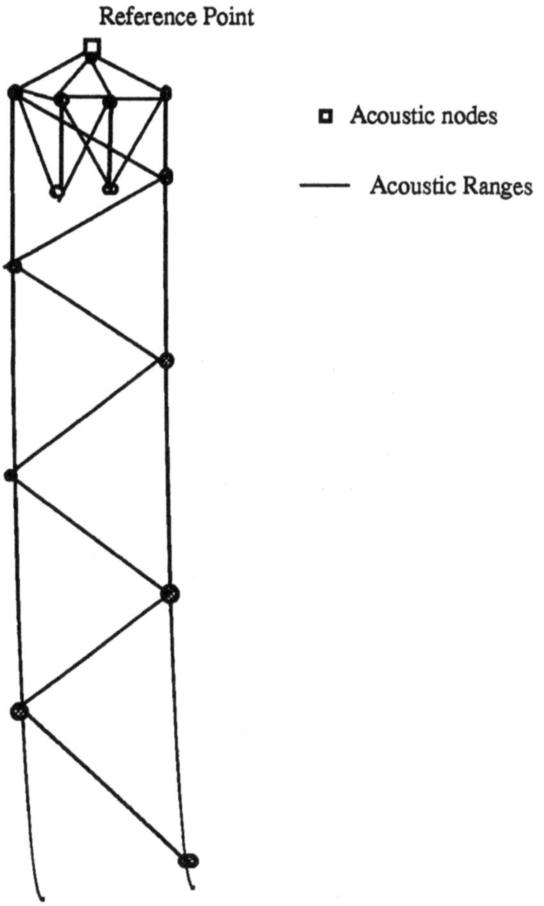

Figure 5.5 Simplistic acoustic array.

- The network in Figure 5.5 is very simplistic, and would be augmented by compasses, live tailbuoys and by a front-end laser network. Such a net has been tested in the field with excellent results.

6 Radio positioning systems

6.1 Introduction

There has been such a shift towards GPS and DGPS in the past five years that many radio positioning systems now sit on shelves gathering dust. However, there may yet be a place for such systems if GPS becomes unreliable due to increased selective availability or if the system becomes too expensive through licensing or otherwise. In some parts of the world radio systems are still used as the prime positioning method, but even these are quickly being superseded by DGPS. The following will give the reader some information on the major systems still in use. Calibration of these systems is covered in a later chapter.

6.1.1 Low-frequency systems (sections 6.2 — 6.5)

In this section radio positioning systems in the lower end of the frequency spectrum up to and including the 2MHz systems are examined. The accuracy and computational aspects of each of the reviewed systems are looked at in detail

While discussing the various systems, it is as well to remember that position-fixing at sea is a dynamic operation. The vessel is always moving, and, unless it is flat calm with no swell, is always subject to pitch, roll and yaw. The antenna is placed at the highest point of the vessel to take advantage of low interference from other ship's systems such as radar and to gain the longest line of sight range. The antenna is also subject to pitch, roll and yaw magnified greatly by its height above the sea surface, and so to require a precision of less than 2 — 3m is rather pointless.

All radio positioning systems require transmitting stations ashore or on fixed structures at sea, and in turn these stations must be coordinated to better than the system resolution otherwise accuracy will be degraded. The general situation of having several shore stations transmitting synchronously or simultaneously, with the vessel receiving data at its own antenna is the same for all systems; only the mode of transmission and computation are different.

Classification of systems by frequency is convenient; the lower-frequency systems are associated with longer ranges, less precise positioning and more elaborate and expensive shore stations. The higher-frequency systems are shorter range, more precise and more mobile.

Although some of the systems have no method of solving lane ambiguity, they all have the ability to count lanes.

For all systems described, the quoted accuracy is that claimed by the manufacturers.

6.1.2 High-frequency systems (sections 6.6 — 6.7)

This section reviews two high frequency radio positioning systems - Micro-fix and Syledis. They are generally shorter range, more mobile and more precise than the lower frequency systems. The systems fall into two distinct categories - microwave and UHF.

These systems are representative of their group. A brief history is given for each system, followed by a list of specifications. The description for each system details the mode of operation, including deployment details, circular polarization, pseudo-random codes and propagation.

6.2 Pulse/8-II

6.2.1 History

The Pulse/8 system was introduced in 1975 and is a hyperbolic and/or ranging system based on Loran C. It has been used extensively for both construction and seismic surveys, and was the preferred system used by clients in the North Sea until the advent of DGPS and the deployment of Syledis chains on offshore platforms. The latest version, Pulse/8-II, is an accurate system capable of providing positioning for most offshore oil and gas operations.

6.2.2 Specifications

Manufacturer:	Racal Marine Systems Limited
Support services:	UK, Holland, Norway, Eire, Malaysia, Indonesia
Frequency:	100kHz
Range:	800km
Resolution:	10ns
Accuracy:	5 — 10m
Mode:	Hyperbolic and/or ranging
No. of users:	Unlimited
Operating Areas:	North Sea, South China Sea, Java Sea

6.2.3 Description

Pulse/8-II obtains lines of position (LOPs) by measuring the difference in arrival times of synchronized pulse transmissions from master and slave stations, and the difference in phase of the carrier waves within each pulse. There are eight pulses within each pulse group, separated by 1000 μs, and each pulse is 280μs long. Each chain is recognized by its unique pulse repetition rate. Since each pulse is 280μs long, and the frequency is 100kHz, it follows that there are 28 cycles in each pulse. The receiver incorporates new technology allowing it to track the signal on the seventh and tenth cycles of the pulse. This ensures that tracking is maintained near the pulse's peak amplitude and also that the system's precise positioning capability is maximized, since dual tracking is employed. Users select the required pulse repetition rate on the receiver, and the receiver then acquires, indexes and tracks signals entirely automatically to provide position-fixing information 24 hours a day at up to 800km range. The unwanted skywave from one pulse affecting subsequent pulses is avoided because each pulse is identified by a phase coding, making the initial cycles positive or negative. So a skywave reflection from the other pulses will cause destructive interference and the offending signal will be rejected.

In the ranging mode, a caesium local frequency oscillator is synchronized with the start of each group repetition interval, enabling a direct time interval measurement between the transmission of pulse groups at the chain stations and the arrival time at the vessel receiver.

6.2.4 Hyperbolic mode

Each station has a 92m mast (i.e. 1/32 wavelength) which emits over 1kW rf power.

Used in the hyperbolic mode we require at least three stations to provide two LOPs. Of course, to provide the necessary redundancy at least four stations are always required in practice.

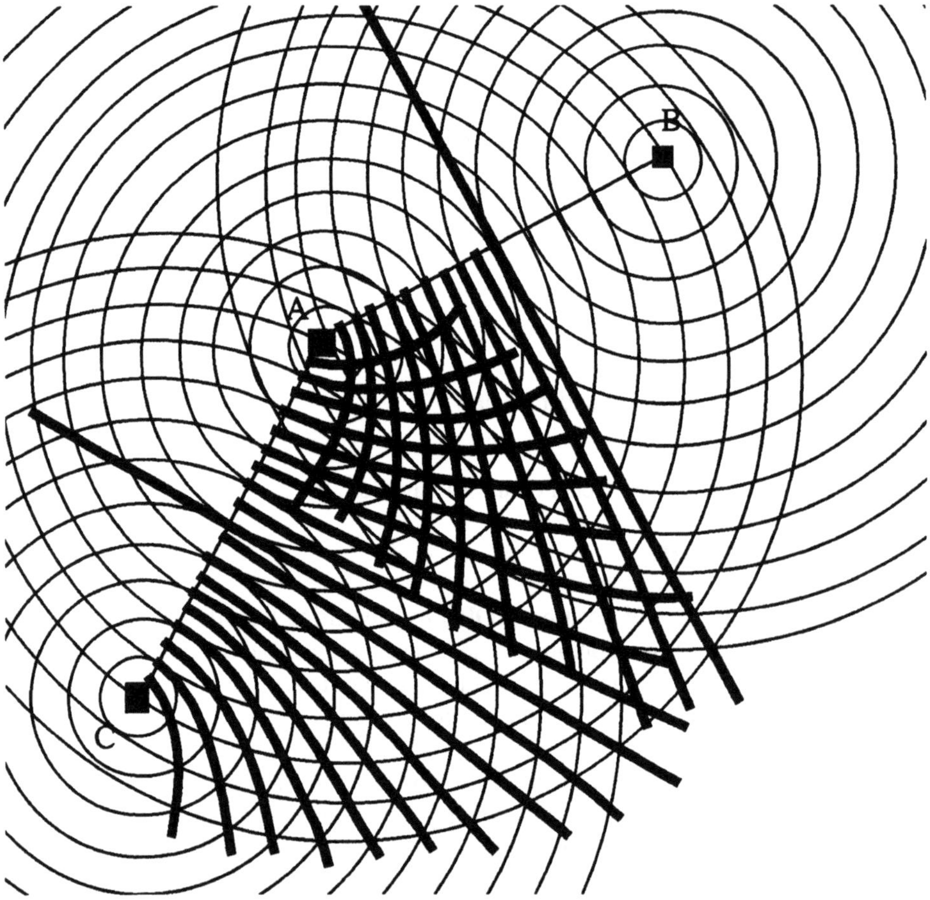

Figure 6.1 Hyperbolic mode.

In Figure 6.1 there are three Pulse/8-II stations. Station A is the master and B and C are two slaves. The circles emanating from each station represent circles of zero phase difference.

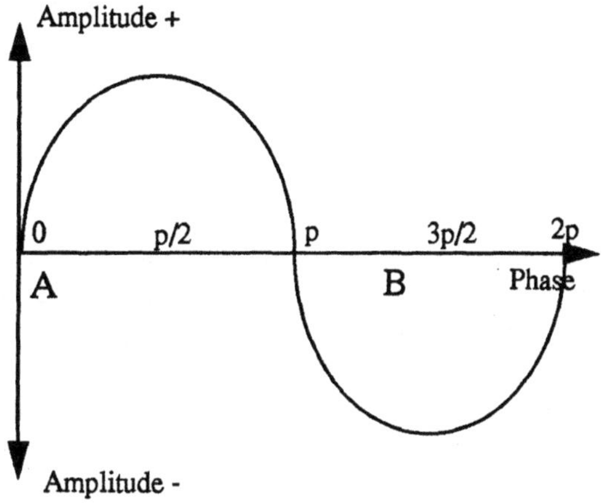

Figure 6.2 Transmission of one pulse.

In Figure 6.2 the transmission of one cycle within one pulse is depicted.

Whether the cycles goes positive first or negative first is immaterial — the phase difference is zero at the start of the pulse and when the phase is pi. The difference is that between the phase at the start of transmission A (which is always 0) and the phase at the receiver B. So in Figure 6.1 the circles are zero phase difference circles from each station, and are all half a wavelength apart. In Pulse/8-II's case the velocity of propagation is approximately 299655 km/s, the frequency is 100kHz and following the formula:

$$\lambda = \frac{V}{f} \tag{6.1}$$

It can be seen that the wavelength is 2996.6m so the zero phase difference circles are 1498.3m apart.

In the hyperbolic mode the vessel is **passive**, and an LOP is defined as being a line of **equal phase difference**. These lines can be constructed by joining the intersections of the zero phase difference circles from each master/slave pair, as in Figure 6.2 . Note that this will give **hyperbolae** rather than circles for position lines, and also that the distance between adjacent hyperbolae on the baseline is one half wavelength, whereas the further away from the baseline travelled, the further apart the hyperbolae become. This is called **baseline expansion** and is a very real drawback with hyperbolic systems. The resolution of the system doesn't change, but as it is defined in terms of time or phase difference as the lanes get wider so it deteriorates.

So with hyperbolic systems two stations are always required to define a single LOP.

Note also that the phase difference measurement is only capable of resolving within a half-wavelength, so for **resolving the ambiguity** some other method of counting the number of whole lanes is needed . For the old Decca and Hifix systems there was no such counting system and one simply had to know where the vessel was to within a lane by using another system. Also, such systems were susceptible to lane drops due to skywave and poor weather. However, the Pulse/8-II system is able to obtain the difference in arrival times of the pulsed transmissions, and this difference is a measure of the total number of lanes.

The system also makes use of an automatic pattern control (APC) system . The APC system uses remote monitor stations linked to the main control centres via dedicated communications channels. Incoming data is analysed automatically by computers to ensure that pattern time differences stay within ±10ns. Pattern time difference corrections are, if necessary, applied to the appropriate transmitter by modulating radiated pulses from one of the Pulse/8-II transmitters.

6.2.5 Ranging mode

In the ranging mode the receiver effectively becomes a slave station and receives transmissions direct from all of the shore stations. The local oscillator allows synchronization with the start of each group repetition interval, so that direct time measurements can be made. The coarse time measurements again give the whole number of lanes, while fine phase measurement gives the partial lane increment. In this case the LOPs are circles, because they emanate from one station and not two. This implies (very importantly) that the lanes are always the same distance apart (at half a wavelength). So the ranging method of use is obviously always going to be more precise except when on a baseline, when the two methods will be the same.

6.2.6 Summary

Pulse/8 is a well-tried, well-tested system that has been in use for many years. Its accuracy before the new enhancements in Pulse/8-II was never much better than 20m, but the manufacturers now claim accuracy to better than 10m, which puts it in the same league as Argo and DGPS. Its main limitation is that the chains are permanent, expensive, non-portable and emit high power radiation.

For calibration procedures see Chapter 10.

6.3 Argo

6.3.1 History

Manufactured by Cubic Western Data, USA. In use for very many years, and used extensively around Australia for seismic surveys.

6.3.2 Specifications

Manufacturer: Cubic Western Data

Support services:	Worldwide
Frequency:	1.6 — 2 MHz
Range:	250 — 350km
Resolution:	0.01 lane
Accuracy:	0.05 lane
Mode:	Ranging (can also be used in hyperbolic)
No. of users:	7 (using four ranges)
Operating Areas:	Worldwide

6.3.3 Description

Argo (which stands for Automatic Ranging Grid Overlay) is nearly always used in its ranging mode, and works using a time-multiplexed system.

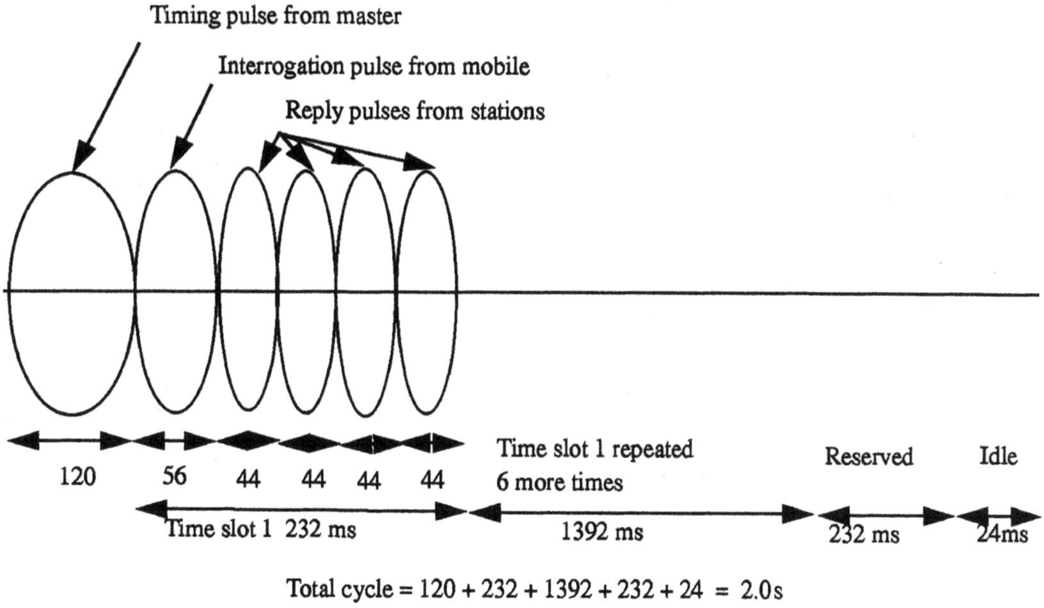

Figure 6.3 Argo operation.

Figure 6.3 illustrates how a complete cycle operates within a 2 second timeframe. In fact, this is the rate at which the Argo receiver sends data to the onboard processing systems — once every 2 seconds.

At the start of the time frame a timing pulse is transmitted by the master station — this synchronizes the other shore stations. The first mobile then transmits an interrogation pulse and each of (up to) four stations reply in turn. This part of the cycle is repeated by any other mobiles up to a maximum of seven altogether.

The measurements are made by measuring phase on the peaks of the pulse envelopes of the outgoing mobile signal and the incoming station signal. The difference in phase gives the partial lane count which is resolved to 0.01 lane.

Typical velocity and frequency values are as follows:

$$V = 299650\text{km/s}$$

$$f = 1614.1\text{kHz}$$

from which we deduce that:

$$\lambda = 185.6452512\text{m and the lane width is } 92.822626\text{m}.$$

Note that Argo operated in this mode is a **circular system** and the lane width remains constant.

The shore stations consist of 30m masts equipped with ground planes, and the antenna on the vessel is a 7 — 11m whip.

6.3.4 Velocity

When using Argo around the coasts of Australia a number of problems must be confronted. Note that the lane width has been computed to six decimal places in the the above section. Consider the case where Argo is being used at ranges exceeding 200km (or 2000 lanes) from the base stations. An error of 0.01m in the calculation of lane width would cause a range error of 20m at that distance. Since the velocity and frequency of the system is known it may be argued that the lane width can be computed accurately without any problem. The trouble is that the velocity is **not** known accurately, and will vary with temperature and relative humidity, both of which vary considerably in areas such as the North West Shelf of Western Australia. The sort of problem that can occur is illustrated below:

Figure 6.4 illustrates an ideal situation, where the position of the vessel at P is determined by four ranges from stations surrounding the prospect, three on the mainland and one on an island to the north-west. If a systematic error exists in those ranges, such as an incorrectly determined velocity of propagation, then all the ranges will be too short or too long. However, although the residuals will increase, the actual position will not be changed appreciably.

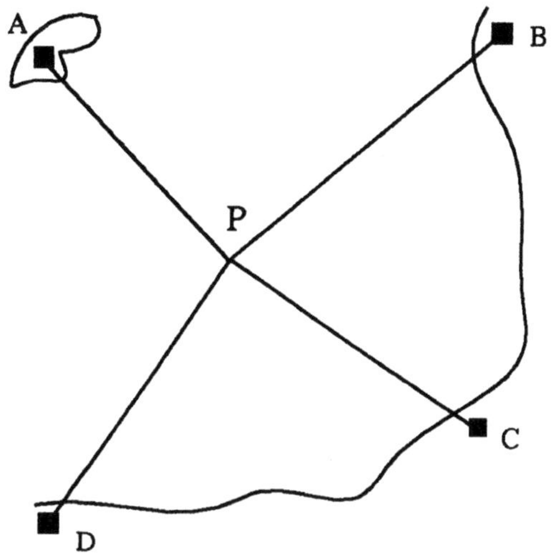

Figure 6.4 Argo chain ideal configuration.

Now consider Figure 6.5 .

The stations in this case lie along a predominantly straight coastline. This is precisely the configuration that can be expected when working around the coast of Australia. If a systematic error exists in the four ranges, that error will be translated directly into the vessel's position, causing it to slide up and down the line PP'. The ironic point is that although the position varies considerably, the residuals do not change anywhere near as markedly, making the false position much harder to detect. Hence the great importance placed on ensuring that there are no systematic errors in the system. This is particularly true when using Argo because it is usually being used for working a long way offshore (otherwise a higher-frequency system would be used) and the systematic error increases linearly with range.

There are two ways we can go about preventing the problem. The first is to measure the refractive index of the air through which the waves are travelling, because one way of defining the refractive index is:

$$n = \frac{c}{u_m} \tag{6.2}$$

where n = refractive index

c = velocity in vacuo (constant)

u_m = velocity in air.

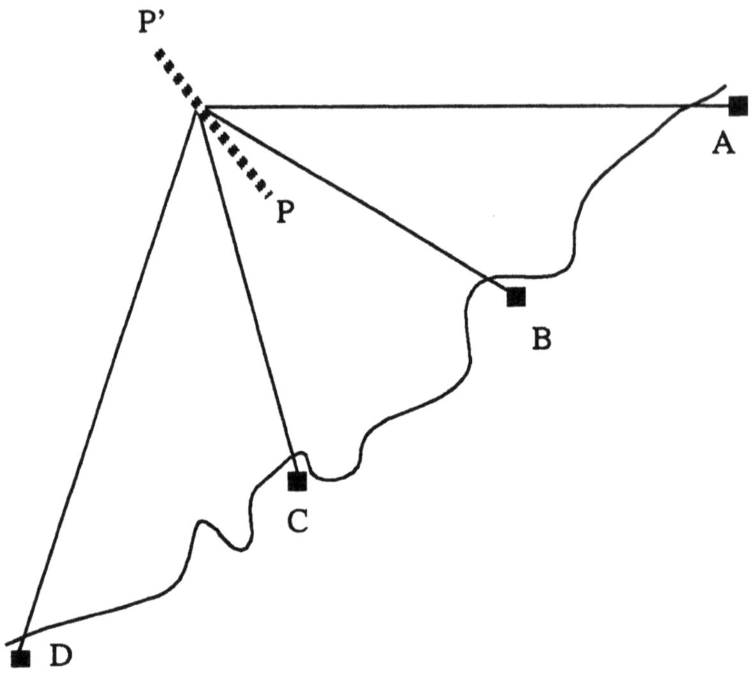

Figure 6.5 Argo chain — straight line configuration.

If the refractive index is measured, then the real velocity can be calculated. Because of the ranges involved, the refractive index has to be measured both at the vessel and at the base stations. The refractive index can be calculated using a shortened form of the Froome and Essen equation:

$$(n-1) \times 10^6 = \frac{103.49}{T}(P-E) + \frac{86.26}{T}\left(1+\frac{5748}{T}\right)E \qquad (6.3)$$

where n = refractive index

P = atmospheric pressure in mmHg

E = water vapour pressure in mmHg

T = temperature in K

The formula gives the refractive index in ppm, and we can then compute the required velocity.

In practice this method is unreliable. The relative humidity and temperature change rapidly over any 24 hour period, and every few hours a new velocity would have to be measured, computed and injected into the processing system. This practice in itself is dangerous because the more often the parameters are changed the more likely that blunders are made.

A secondary navigation system such as Miniranger or Micro-Fix should be used to calibrate the Argo on a (say) weekly basis. Such calibration, together with the method used to calibrate the partial lane counts on an Argo system, will be given later. The partial lane count is the fraction of a lane to be added to an observed lane count from a station to make that lane count equate to the correct range, and is a computed - observed (C - O) value due to instrumental delays at both the station and the vessel.

6.3.5 Summary

Argo is a system in wide use in Australian waters. If meticulous detail is paid to installation and calibration, it can offer accuracies in the 5 — 10m range. It is generally trouble-free, robust and offers excellent repeatability. The variation in velocity experienced particularly in northern waters where there are rapid seasonal and diurnal temperature and relative humidity variations means that the system performance will **definitely be degraded** if regular calibration checks are not carried out.

A drawback is that the system is quite costly to mobilize and the stations require to be manned.

6.4 Hyperfix

6.4.1 History

Hyperfix is a third generation 2MHz system, the first two generations being Hifix and Hifix/6. It is a high-accuracy phase comparison system and has been in use worldwide for about 10 years.

6.4.2 Specifications

Manufacturer:	Racal Marine Systems Ltd., UK
Support services:	Worldwide
Frequency:	1.6 — 2.5 MHz
Range:	250 — 700km
Resolution:	0.001 lane (rounded to 0.01 lane)
Accuracy:	0.01 lane (under the optimum conditions)
Mode:	Hyperbolic and ranging
No. of users:	Unlimited in hyperbolic mode
Operating areas:	Worldwide

6.4.3 Description

A Hyperfix chain consists of from three to 18 transmitting shore stations. One of these stations transmits a trigger signal of 40 ms which controls the timing of the chain. At each station the phase of that station's transmitted signal can be referenced to the signal from any other station — this is called **phase-locking**.

The transmitting stations are equipped with a monopole sectional antenna varying in height from 10 — 30m to suit the operating conditions. The stations may be left completely unmanned, but there should be a chain manager ashore to service batteries, generators, etc. The vessel antenna is a short broadband active receiving antenna which needs no pre-tuning to the chain frequency.

Measurement is by phase comparison; at 2MHz the baseline lane-width is about 75m, so the system has a displayed resolution on the baseline of about 0.75m. As with any hyperbolic system, lane expansion increases with distance off the baseline.

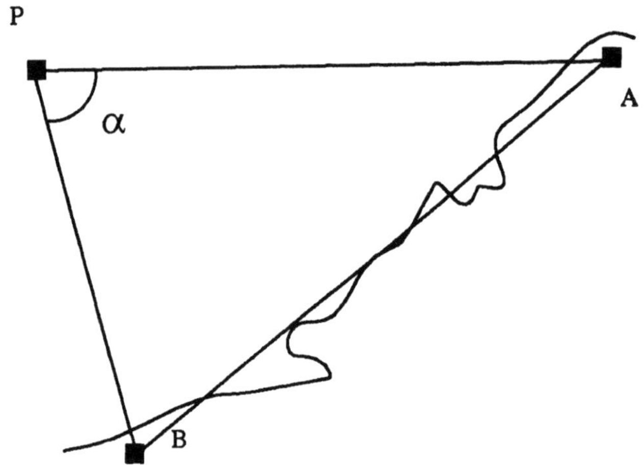

Figure 6.6 Lane-width computation.

Referring to Figure 6.6

$$L_P = L_B \operatorname{cosec}\left(\frac{\alpha}{2}\right) \tag{6.4}$$

where α = angle subtended at P by the baseline stations A and B

L_P = lane-width at P

L_O = lane-width on the baseline AB.

For example, let $L_O = 75.0$m and $\alpha = 50$ deg

$$L_P = 177.5\text{m}$$

Again, if $L_O = 75.0$m and $\alpha = 30$ deg

$$L_P = 289.8\text{m}$$

So as the angle α becomes smaller (and the distance increases from the baseline), so the lane-width increases. Therefore if the accuracy is quoted as 0.01 lane by the manufacturer, this equates to 0.75m on the baseline, 1.8m in the first example and 2.9m in the second example. Experience has shown that it is unrealistic to expect an accuracy of better than 5m near the baseline; in the examples above the accuracy would be nearer 10m and 20m respectively. Remember that these quoted accuracies are for a single LOP. If it is possible to get good geometry for the chain it may be able possible to improve the accuracy by using 4 or 5 well-conditioned LOPs. The accuracy may well be even worse if the stations have to be sited on a straight coastline.

The signal in Hyperfix is not pulsed, as in Pulse/8-II.

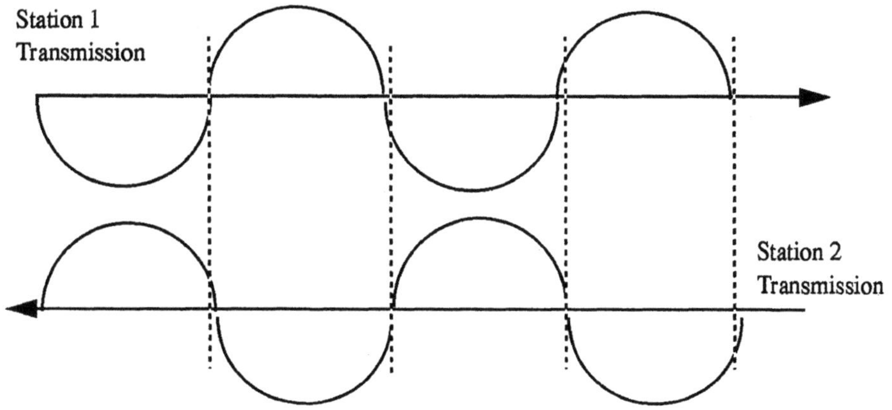

Figure 6.7 Hyperfix transmission.

Figure 6.7 shows the transmissions from two stations; station 2 is phase-locked to station 1 and it can be seen that the dotted lines represent points of zero phase difference. When these points are connected graphically hyperbolic LOPs are formed.

The phase resolution of Hyperfix has been discussed; what about the absolute lane count? Hyperfix operates on two frequencies between 1.6MHz and 3.4MHz. Racal recommend an ideal separation of 2-10%. The two signals can be mixed and the differenced frequency used to solve the lane ambiguity. For example, let f_1 be 2.000MHz and f_2 be 2.040MHz, a separation of 2%. The differenced frequency is 40kHz, equivalent to a lane-width of 3750m on the baseline. So the phase of the differenced frequency can be measured to raise the ambiguity from 75m to 3750m on the baseline. This ambiguity resolution can be automatic or manually driven.

Frequencies at the 2MHz level are in the range that can be significantly affected by weather. In 1982 — 1983 when Hyperfix was in its infancy, it was common to experience lane drops on patterns during a squall and then have to steam back inshore to an area within the coverage of another system such as Trisponder to enable the lane count to be recomputed. One would then sail back to the prospect in time for the next squall, and so on. Since then, many improvements have been made to the system so that it suffers far less from interference; in any case, it is now possible to recompute lanes from stand-alone or differential GPS without requiring a secondary radio navigation system as backup.

6.4.4 Summary

Hyperfix is one of the better 2MHz systems, and, like Argo, when care is taken with the set-up of stations and calibration, it is capable of good, reliable performance. There is a new development in Hyperfix — that of integrated differential GPS. This allows for the combination of differential satellite positioning with conventional shore-based techniques. The advantage of this type of system is that the (generally) more stable Hyperfix can be used to smooth the DGPS positions.

6.5 Geoloc

6.5.1 History

Geoloc is a medium-to-long range system produced by the French company Sercel. It belongs to the 2MHz family of systems but uses an entirely different mode of operation. It is probably best used in conjunction with another system such as DGPS. It has been used successfully on 2D surveys around Australia.

6.5.2 Specifications

Manufacturer:	Sercel S.A., France
Support services:	Worldwide
Frequency:	600 kHz band based on 2MHz
Range:	Over 1000km

Accuracy:	$2m \pm 1.5 \times 10^{-5}$ Range
Mode:	Ranging and Hyperbolic
No.of users:	Unlimited in hyperbolic mode
Operating Areas:	Worldwide

6.5.3 Description

Geoloc is a land-based 2MHz system using spread spectrum techniques. Conventional systems use a high radiated energy level concentrated in a narrow bandwidth, with the receivers using a matched bandwidth and high sideband selectivity.

With Geoloc, the normal base band signal goes through an additional coded modulation, the rate of which is far higher than the baseband flow. This results in the energy being spread over a wide spectral band. The receiver uses a decoder to correlate the incoming signals with a locally generated coded signal identical to that of the transmitter, and accurately synchronized to it using atomic clocks.

The main advantage of this system is the low power output — about 25 mW during daytime and 250 mW at night. At this level the signals are well below the ambient noise level; as such they really require no frequency licensing and they are free from most forms of jamming. They are also less susceptible to skywave at night because of their long pseudo-random noise (PRN) modulation.

Geoloc uses both PRN group delay measurements (similar to Syledis) and frequency carrier phase measurements from the incoming signals. The bandwidth is 666kHz based on a 2MHz spectrum centre. The receiver is passive and the system can therefore be used by several users.

To compensate for phase delay jitter caused by multipath and differing land/sea paths, a correction model is used in the system. This consists of an accurate determination of refractive index computed on board; it also compensates for local time group and phase delay distortions arising from differing land/sea paths. The quoted accuracy of $2m \pm 1.5 \times 10^{-5}R$ equates to about 10m at 500km range.

6.5.4 Determination of position, course and speed

For seismic purposes it is recommended that the system be operated in the 'H' mode – this is considered the most accurate by Sercel.

Both transmitters and receivers are equipped with time references - these being ultra-stable caesium clocks. A continuous updating of position is accomplished by observing the very slow-drifting pseudo-ranges from the shore transmitters. A periodic update is made by injecting position from another system such as DGPS. Obviously, the secondary system needs to be as accurate as Geoloc otherwise the correction method will not work. The timing of the updates is operator-controllable; therefore if the DGPS becomes unstable for an hour or so, the Geoloc will quite happily continue to give reliable positions. The reliability starts to deteriorate after about 2 — 3 hours of no updating.

Having measured the pseudo-ranges to the shore stations, the vessel's position can be computed in the normal way, except that several other variables must be included in the observation equations.

Firstly, all receivers and transmitters must be synchronized to the system's time zero. This is taken to be the time of yet another caesium clock, maintained ashore at an easily reached location. When each station is set up, a portable timekeeper is used to transfer the time from the master clock to each of the station clocks. This is easily achieved by connecting the two clocks and synchronizing them using inbuilt software. The vessel's receiver clock is synchronized in the same way. All the stations and the vessel are also equipped with the less accurate portable timekeepers which act as backups.

Having synchronized the system to one datum, the clock drifts need to be measured. This is done by using the secondary system - usually DGPS. Every (say) 30 minutes, the DGPS position is scanned for quality, and, if acceptable, is injected into the Geoloc receiver. The Geoloc uses the DGPS position to compute what the pseudo-ranges should be. The observed pseudo-ranges are used with the current lane-width, clock error and drift rate to compute Geoloc's version of those ranges. Comparison of the two sets of pseudo-ranges results in an update of the clock offsets and their drift rate. The offsets are generally less than 10 microseconds and the drift rate a few tenths of a microsecond per day.

6.5.5 Station accuracy

In a system such as this the station coordinates' accuracy is crucial. If the vessel works close to the shore stations, clearly any inaccuracy on the station coordinates will affect the system accuracy. But at typical user ranges of 700 — 800km, as the vessel moves through the prospect the change in azimuth between the station and the vessel is small, and an error in the station coordinates will not be noticed. This is because any error will be taken up by the clock drift during the course of a few DGPS updates.

It therefore becomes clear that the accuracy of the DGPS base station controls the accuracy of the survey. For this reason, if Geoloc is used for further surveys, it is very highly recommended that the **same base station coordinates** are used for repeatability.

6.5.6 Summary

When used in conjunction with DGPS, Geoloc is a very stable, reliable system. However, its accuracy will only ever be as good as the DGPS system which is used for updating the clock offsets and drifts.

6.6 Micro-Fix

6.6.1 History

Micro-Fix is another positioning system developed by Racal. The system is the direct descendant of Trisponder and was introduced in 1984. Micro-Fix has been enhanced since that time and is now able to overcome many of the problems inherent in Trisponder.

6.6.2 Specifications

Manufacturer:	Racal Marine Systems Limited
Support services:	Worldwide
Frequency:	5.48 GHz or 5.52GHz
Range:	80km (line of sight)
Resolution:	0.01 m
Accuracy:	1 — 2m
Mode:	Ranging (and hyperbolic)
No. of users:	16
Operating areas:	Worldwide

6.6.3 Description

Micro-Fix is a short-range position-fixing system, operating in the 5GHz band. It uses a low power solid state transmitter and provides a working range of 80 km providing there is direct line of sight between transmitter and receiver. A basic system consists of two principal units: a microwave transponder (T/R) and a control measurement unit (CMU). Two or more T/Rs are sited at known shore locations, and a master station is located on the vessel — the master consists of a CMU connected to a T/R. The master interrogates all remotes in turn and measures the time interval between the departure of the outgoing pulse and arrival of the received reply. There is automatic adjustment to compensate for turn-around delays (i.e. the system is 'self-calibrating').

In Figure 6.8 a range is defined as the distance between the master and remote. The total time taken for each complete measurement is

$$t_1 + t_2 + \mathrm{TAD} \tag{6.5}$$

where TAD is the turn around delay due to passage through the cabling and electronic circuitry at the remote station. In Micro-Fix the TAD is calculated dynamically and continuously and is automatically subtracted from the total time. The TAD is typically about 17.8 microseconds, equivalent to a one-way range of about 2660 m.

Since the operating frequency is so high, there are no phase lag effects because the transmitted rays are many wavelengths above the surface. So, since there is no phase lag and the system compensates automatically for instrumental delays people have been tempted to install the system and operate it without any form of calibration. To do this of course is quite insane, as it immediately destroys the surveyor's first rule of checking everything by independent means. It would make a great deal of sense to deploy the chain and then run a check-calibration with the stations in situ, from somewhere close to the prospect if possible. This will then check three points:

- The integrity of the self-adjusting system.
- Whether or not the assumed velocity of propagation is correct.
- Whether or not the station coordinates are correct.

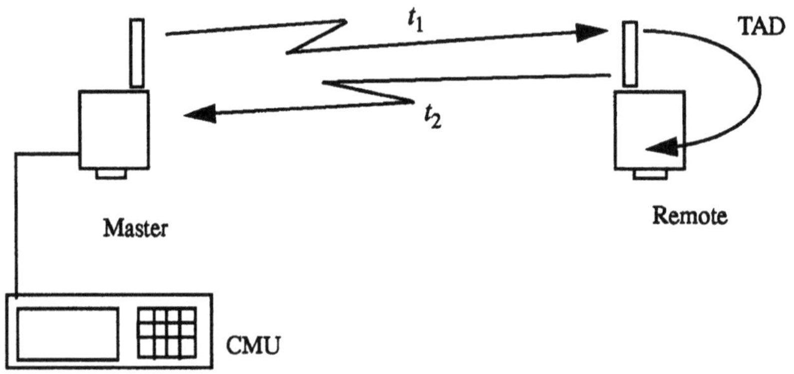

Figure 6.8 Micro-Fix.

In Figure 6.9 the chain is formed by stations A, B and C. There is a (convenient) platform sited close to the prospect, and the master station could be temporarily assembled on the platform to validate the chain.

The actual adjustment for the TAD is achieved by the remote T/R sending back two reply pulses; the difference in time between the pulses is a measure of the TAD, so the master can calculate the true range.

6.6.4 Circular polarization

There is a very real problem with microwave systems — that of reflection.

Figure 6.10 shows the return signal from a remote station back to a master. If the sea surface is calm, there will be a reflected ray as well as a direct ray. If the reflected path is of similar length to the direct path, and if the reflected signal arrives out of phase with the direct signal, there will be destructive interference and consequently no range.

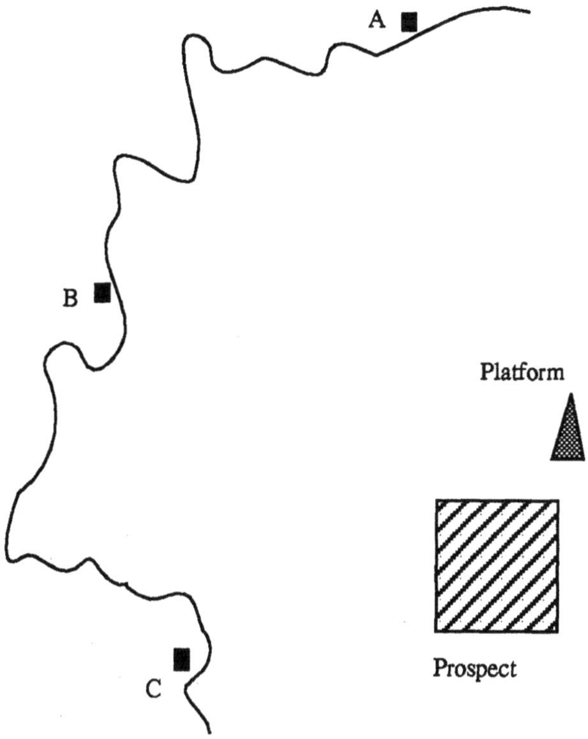

Figure 6.9 Micro-Fix chain.

This phenomenon is known as **range holes**. The positions of the range holes can be calculated as:

$$S_n = \frac{2h_1h_2}{n\lambda} \tag{6.6}$$

where h_1 and h_2 are the remote and master antenna heights respectively. n is an integer, and so it will be seen that:

- Range holes are dependent on antenna heights.
- There can be many range holes for each master/remote pair.

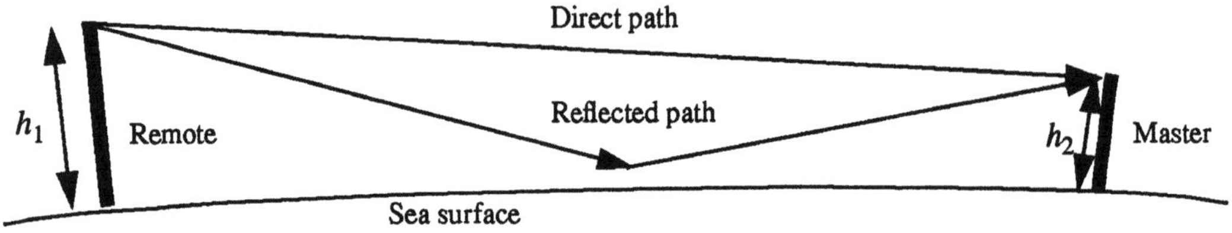

Figure 6.10 Range hole.

For example, suppose the two antennae are at heights 100m and 25m. The wavelength is 0.06m, and starting with (n = 20) and decrementing n, range holes are found at 25.03km, 26.34km, 27.81km, and so on. Trisponder attempted to overcome the problem by using twin antennae at the vessel, with the antennae spaced a few metres apart vertically. The purpose was that the reflected path to only one of the antennae would cause a nulling phase shift, so the signal from the other antenna could be used, but this was never entirely successful.

Micro-Fix uses a unique antenna design to overcome range holes. A combination of the waveguide, radome and helix causes a twisting of the electric field, and the resulting field can be described as a screw thread. The receiving antenna will only accept a signal with the correct helical direction. Signals from a reflective surface undergo a phase reversal and are heavily attenuated at the antenna. Of course, a double reflection would cause the helix to be in the correct direction, but the theory here is that the resultant signal would be so heavily attenuated that it would be rejected by the CMU's tracking gate.

6.6.5 Summary

Micro-Fix is an excellent system for short-range work up to 80km. It is reliable, accurate and each mobile can accommodate up to eight ranges. Because of the very short wavelength of just a few centimetres, the system can be affected by weather. The system is easily mobilised, easily installed and is relatively inexpensive. The remote stations are powered by 16 — 30v dc and only use 12 watts when operating. In stand-by mode power consumption is only three watts.

6.7 Syledis

6.7.1 History

Syledis, in its various forms, has been in use for many years. It is manufactured in France by Sercel, S.A., and is in use worldwide as a precise positioning system.

6.7.2 Specifications

Manufacturer:	Sercel S.A., France
Support services:	Worldwide
Frequency:	408 — 434MHz and 420 — 450MHz
Range:	80 — 120km
Resolution:	0.1m
Accuracy:	3 — 5m
Mode:	Various
No. of users:	Depends on configuration
Operating Areas:	Worldwide

6.7.3 Description

Syledis is a time-sharing system, and as such can operate in a number of different configurations.

6.7.4 Active range range mode

In the active mode (see Figure 6.11) a mobile interrogates each beacon, which then replies to the mobile.

LOPs are circles, so the geometry is best.

The maximum number of users is limited by the maximum number of time slots available in the format (30 slots maximum, 60 slots in SB5 beacon).

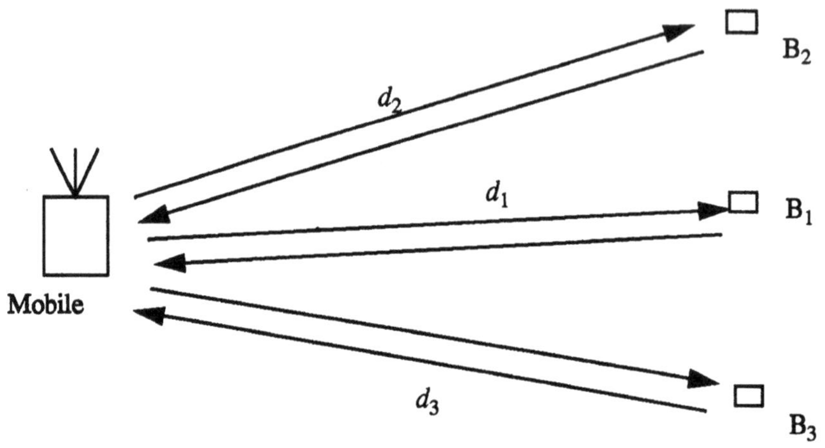

Figure 6.11 Syledis active mode.

6.7.5 Passive pseudo-range mode

In this mode the mobile is a receive-only unit. The beacons are all synchronized with each other; one of the beacons is designated master and transmits to all other beacons and mobiles. The slave beacons retransmit the signals which are received by the mobiles.

In Figure 6.12 B_1 has been designated the master beacon. It transmits to beacons B_2 and B_3 and to the mobile. Beacons B_2 and B_3 then also transmit to the mobile.

The resulting LOPs are hyperbolae because they represent distance differences rather than distances (or time differences to be exact). Therefore the geometry is not as good as with the range range system, but the number of users is virtually unlimited.

6.7.6 Other modes

Various combinations of the above two modes can also be used. One of these is the inverted mode, in which the position of the mobile is computed by a station ashore. Another mode is the compound network, in which active and passive mobiles are combined.

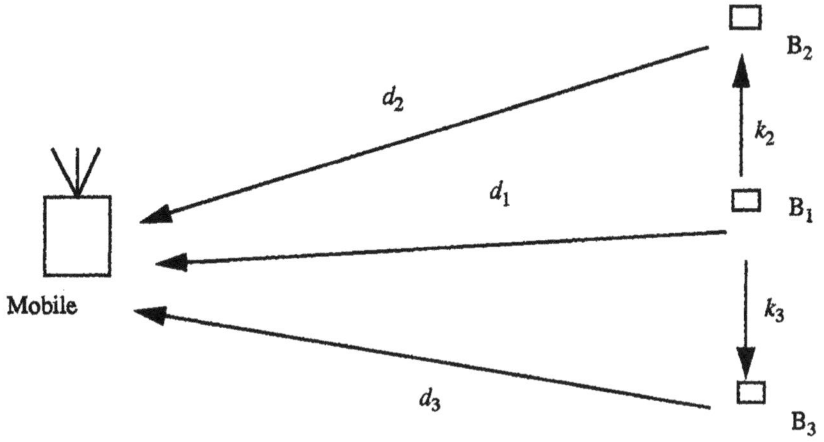

Figure 6.12 Syledis passive mode.

6.7.7 SB5

The SB5 is a recently introduced beacon which has the following features:

- A single code generator delay-lock circuitry (multiplexed) used to handle all the time slots in a Syledis format, for transmission as well as reception.
- An auxiliary control and display unit (BCC5) used to fully configure the unit, even in terms of frequency settings.
- Time measurement circuits which compute the time offsets between the codes delay-locked in each format slot, thus returning pseudo-range values for each slot.
- 60 available slots (the older versions had only 30 slots).
- Two digital data ports.

SB5's flexibility allows up to 15 mobiles in range range mode, compared with only four in the older B-type systems.

6.7.8 The Syledis pseudo-random code

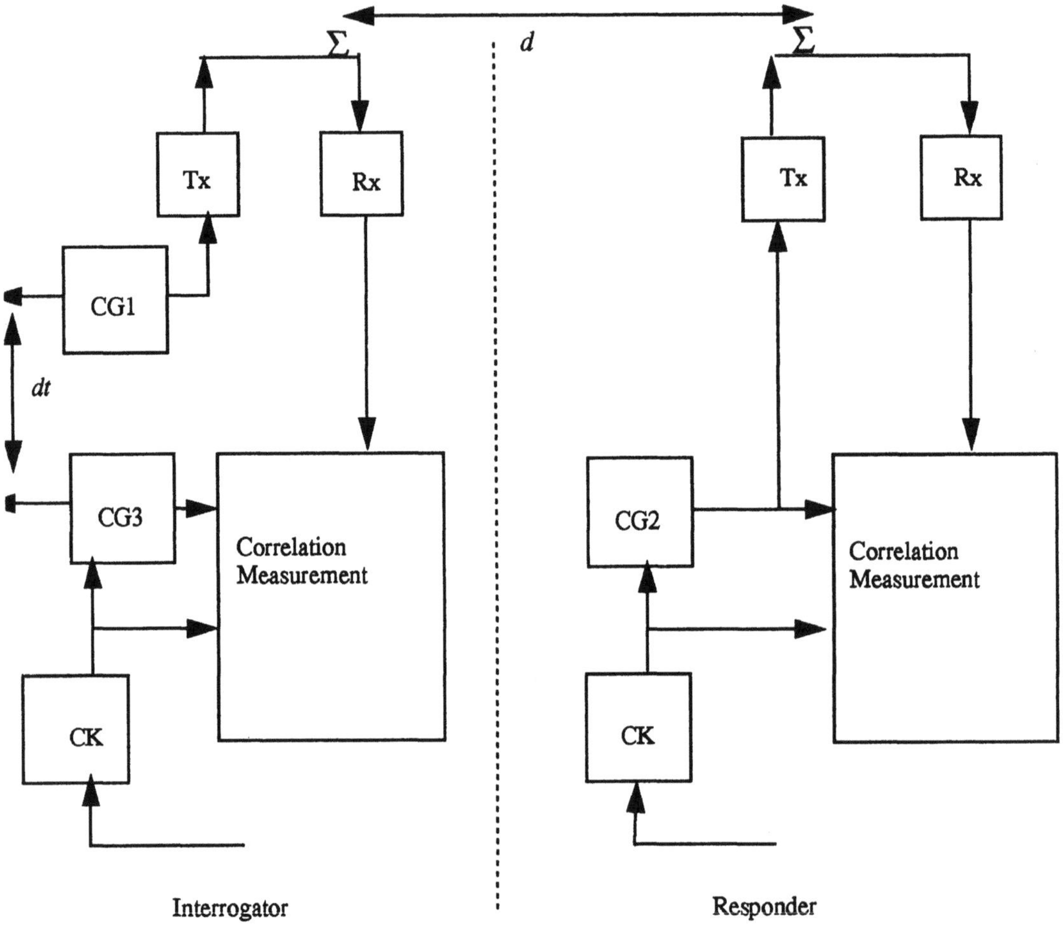

Figure 6.13 Pseudo-random code.

Figure 6.13 shows the basic interrogator/responder set-up. A free-running pseudo-random code generator CG1 modifies the interrogator's signal and sends it to the responder. Here it is amplified and fed to measurement circuitry that correlates the received code from CG1 with a local replica generated by CG2. The signals from the correlator are used to eliminate any offset between the incoming code and the local code by shifting the frequency of CG2's clock. The time offset (dt_1) between the code generated by CG1 and that generated by CG2 is therefore given by:

$$dt_1 = \frac{d}{v} + k_1 \qquad (6.7)$$

where d is the distance from interrogator to responder

v is the propagation velocity

k_1 is the instrumental delay experienced at both the interrogator and the responder.

The code generated at CG2 modulates the responder's transmission signal, and is received and amplified by the interrogator. The same correlation and time offset removal technique is then performed, this time against a local code generated by CG3. The time offset dt_2 is then given by:

$$dt_2 = \frac{d}{v} + k_2 \qquad (6.8)$$

The total time offset dt is then:

$$dt = dt_1 + dt_2 = 2\frac{d}{v} + k \qquad (6.9)$$

where k is the sum of k_1 and k_2.

Therefore if we measure dt, the true range can be computed. However there are some points to take note of:

Point 1

- The pseudo-range output by the SB5 is a two-way pseudo-range, and it is uncorrected for delays.
- The pseudo-range output has an ambiguity of 20km, leaving an ambiguity of 10km in the one-way range.

Point 2

- The 10km ambiguity, is brought about because the pulse code length is 66.6 microseconds, equating to a two-way range of 19956m at 420MHz. Because the time measurement is made **inside the pulse code** there is a 10km limit on ambiguity. This is not usually a problem when computing the vessel position, but it does mean that some extra thought is required when computing an active tailbuoy position.

Consider the following example:

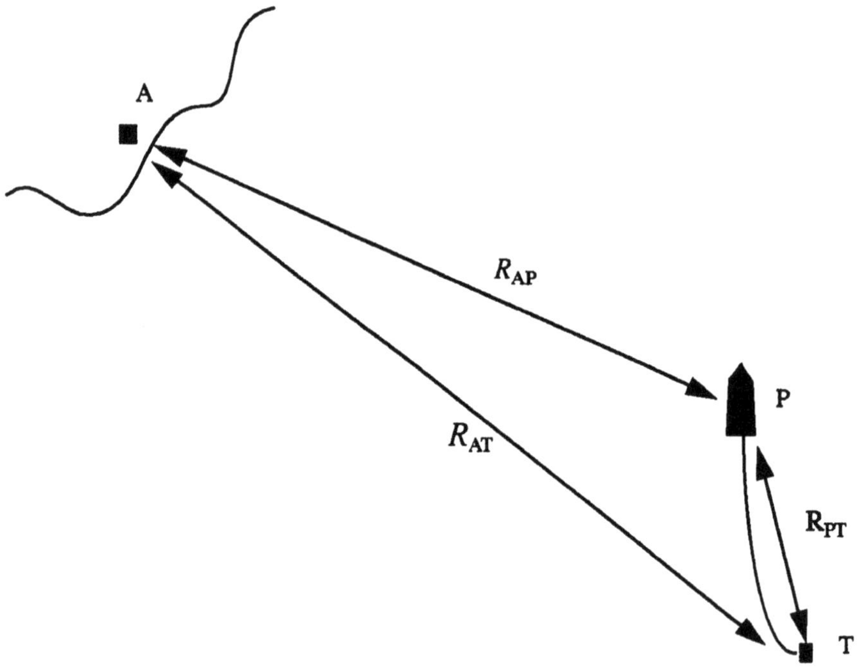

Figure 6.14 Ambiguity.

In Figure 6.14 there is a vessel towing a single streamer heading in a northerly direction. There is a mobile on the vessel and beacons on the tailbuoy and ashore at A. The problem is to determine the range from the tailbuoy to A, R_{AT}. Suppose there are the (partial) slot configurations shown in Table 6.1.

Table 6.1 Slot configuration

Slot	Pseudo-range
4	R_{AP}
11	R_{PT}
12	$R_{AP}+R_{AT}+R_{PT}$

Note that R_{AT} is not directly measured, but that instead there is the "ring-a-round" range $R_{AP}+R_{AT}+R_{PT}$. The ranges must be extracted from all three slots the correct instrumental delays applied first of all. R_{AP} and R_{PT} are both two-way ranges, but the ring-a-round range goes only once around the circuit from ship to shore to tailbuoy to shore. Having applied the instrumental delays the 2-way ranges are then halved and the ambiguities for all three ranges computed. This implies that the location of the tailbuoy must be known roughly to compute the ring-a-round ambiguity. Once the ambiguities are found the true range from tailbuoy to A can finally be found. This procedure must then be repeated for all shore stations and for all tailbuoys.

6.7.9 Propagation

The Syledis frequency of around 400 MHz yields a wavelength of about 0.7 m; this is short enough to ensure that there is no skywave interference, but long enough to enable small variations in the velocity of propagation. The velocity changes with varying temperature and pressure along the transmission path. For this reason it is necessary to calibrate the system regularly.

6.7.10 Summary

Syledis has been used since the 1970s for precise positioning. The system has been refined to give very stable, reliable positioning at the 5 m accuracy level up to ranges of about 100 km. Under certain circumstances, such as when "ducting" occurs, it is possible to get good positioning at ranges far in excess of 100 km, but this is by no means always the case. The shore stations are easily set up and can use directional Yagi antennae (similar to TV antennae) to concentrate power output over chosen arcs, or omni-directional whip antennae. There is no need for elaborate ground nets with this system.

7 GPS

7.1 Introduction

The NAVSTAR global positioning system is a satellite-based positioning system implemented and operated by the US Department of Defense (DOD). NAVSTAR stands for Navigation System using Time and Ranging, and it was designed to provide continuous, all-weather navigation and positioning capabilities throughout the world. Work on the system began in 1973 as a result of the merger of the US Navy's Timaton Program and the US Air Force's 621B Project. These two programs had been established in the mid-1960s to develop a passive navigation system using measured ranges.

There are three 'segments' in the system — the satellites (space vehicles or SVs), the control system and the users.

The system is fully deployed with 21 operational Block II satellites and three in-orbit spares. The original Block I satellites were used as a test constellation, although two of them (SVs 1 and 2) have now become Block II satellites.

The satellites were launched using Delta II rockets and were put into a near-circular orbit at a nominal altitude of 20183 km.

The satellites are arranged in 6 orbital planes inclined at 55° to the equator, and the orbital period is 12 sidereal hours, equal to one half of the earth's rotation. (A sidereal day is the time taken for the earth to rotate on its axis once with respect to the First Point of Aries. It is 23h 56m 04.09054s of mean solar time. Solar time is reckoned from the time taken for the earth to travel once around the sun).

The satellites are tracked by 4 monitor stations in Guam, Hawaii, Alaska and California. Each monitor station accumulates range data from the satellites, together with local meteorological data and transmits this data to the master control station in Alaska, where the ephemeris and clock errors are generated.

7.2 Observational concepts

7.2.1 Transmissions

Each GPS satellite transmits on two frequencies, L_1 and L_2. The L_1 carrier frequency is 1575.42MHz and has a wavelength of about 19cm. The L_2 carrier frequency is 1227.60MHz with a wavelength of approximately 24cm. The L_1 frequency is modulated by a precision (P) code and a Coarse Acquisition (C/A) code whereas the L_2 frequency has the P code as its only modulation. The codes are referred to as pseudo random noise (PRN) codes and serve as a very precise time reference that enable receivers to compute the time of transmission of the satellite signal. The timing information needed for precise positioning is generated by precise atomic clocks carried on board each satellite. The satellites send navigation messages containing ephemerides, clock behaviours, health and status of satellites, almanacs and other general information; the navigation messages are transmitted on both frequencies.

The C/A code is a PRN sequence with a period of 1 ms and a frequency of 1.023 MHz. This frequency corresponds to a range of about 300m, which can be improved by interpolation. The ambiguity of about 300km corresponding to the 1 ms period is avoided automatically by the receiver sampling several satellites at once.

The P code is a PRN sequence with a frequency of 10.23 MHz (the GPS base frequency). It has a period of 38 weeks and is therefore completely unambiguous. Each satellite uses a one-week segment of the total code, and is re-set each week. One bit in the P code frequency corresponds to about 30m, which can be improved again by interpolation.

7.2.2 Positioning from space

For many hundreds of years man has sought to obtain his position on the earth's surface using the sun, moon and stars. The main technique before the advent of satellites was optical astronomy. In simple terms we required to find the position of the local zenith among a group of stars. The problem with this technique is that the observations are made with reference to the local vertical, which varies with the earth's gravity field (and the shape of the geoid). The problem is complicated at sea because the observing platform is moving, and real-time positioning is all but impossible. It becomes completely impossible on cloudy nights!

The basic concept of satellite positioning is shown in Figure 7.1 :

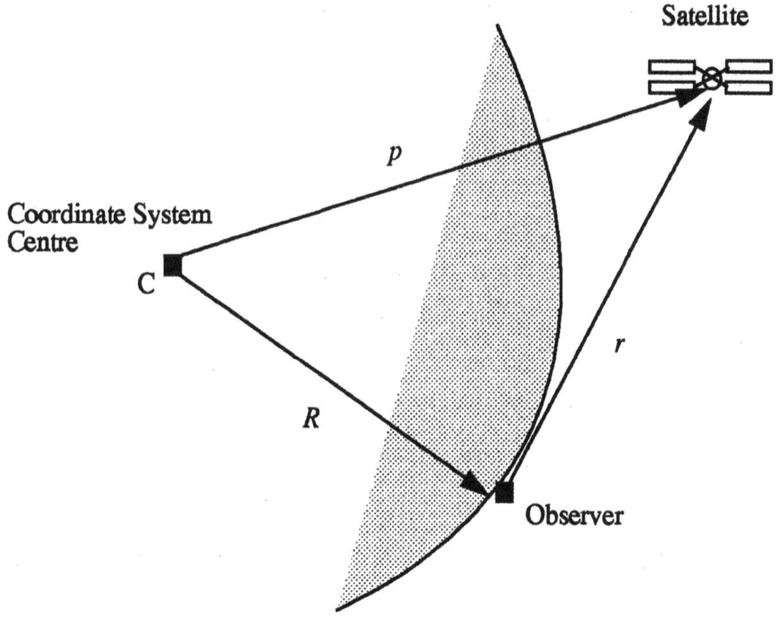

Figure 7.1 Satellite positioning.

In this diagram C is the centre of our (three-dimensional) reference system, usually at or close to the centre of mass of the earth. We require to find the vector *R* (in terms of *X,Y* and Z coordinates) with respect to C. We are given the vector *p*, which is the satellite's position with respect to C, in the satellite's navigation message, and so we have to measure the range vector *r* between the satellite and the receiver.

Having said that the satellite's position is known, its position as sent in the navigation message is actually a predicted position. The prediction of satellite positions is quite complicated, and relies on a specialized knowledge of celestial mechanics and dynamics. The proper name of the function of the variation of *p* with time is 'ephemeris', and these ephemerides are computed and predicted by the US DOD. It is possible for the users of the system to improve the transmitted ephemerides, and also to produce their own ephemerides. To do this the users need to establish their own network of tracking ground stations. It is worth mentioning that if users set up their own ground station network, the way is open to use any satellites that transmit signals and that can be tracked — not just the GPS satellites. All that is required is for the ground stations to compute and predict the satellites' ephemerides and then transmit the data to the users. It is not a requirement for the satellites to transmit the data if some other means can be found. Of course, the users must be able to track the satellites themselves and must have some means of measuring ranges to them.

In an ideal world the length and direction of the range vector r would be measured. Surprisingly, this is not impossible; one way would be to determine the direction using a theodolite and electronically measure the range. The method does not work in real time because it is difficult to continuously measure the right ascension and declination of the satellite, the two angles necessary to determine its direction. In fact, the angular measurement is useful, as NASA maintains catalogues of the thousands of objects in orbit around the earth by photographing them from known ground stations and recording their positions against the star background.

Direct-ranging to satellites has met with more success. If the direction to the satellite is not known, ranges to three non-coplanar satellites simultaneously are needed to determine the position of a moving receiver (if the receiver is stationary, the ranges could be measured at different times).

In GPS there are three main measurement modes employed — pseudo-ranging, carrier phase measurement and Doppler observations; pseudo-ranging is the main mode.

7.2.3 Pseudo-ranging

In order to make a pseudo-range measurement to a particular satellite, the receiver must generate the same PRN code as that satellite. It is then the time difference between these identical codes (one received, the other locally generated) that provides the basis for the pseudo-range measurement.

A simplified sequence of events for the GPS measurement is as follows:

- The receiver generates its own replica PRN code which is exactly the same as the received code except for the time base. The receiver does not contain a clock of anything like the accuracy of the precise atomic clocks carried by the satellites, and it cannot be synchronized into GPS time exactly. This time bias is introduced into the computation as a fourth unknown (the other three being the X,Y and Z coordinates of the receiver's position), and is the reason why four satellites and not three are needed to solve for position.

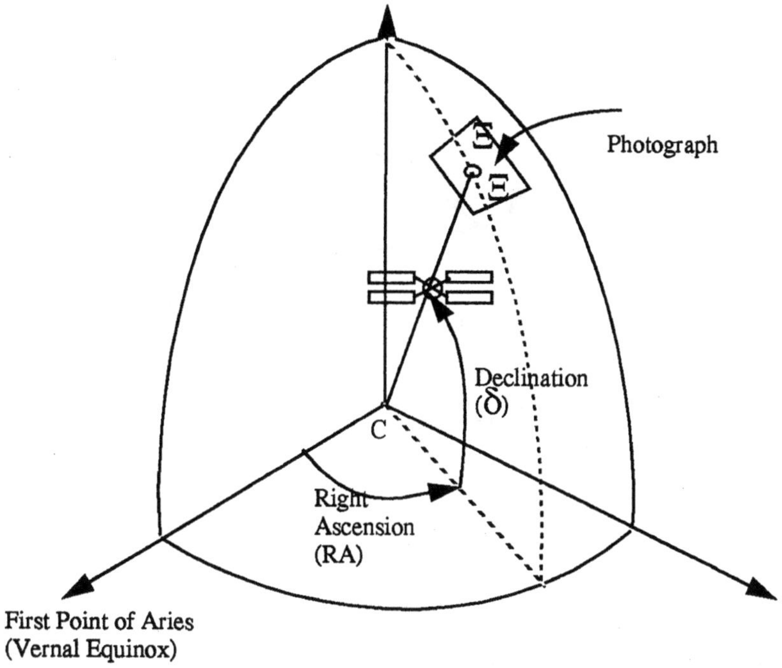

Figure 7.2 Satellite photography.

- Alignment or cross-correlation between the received and replica codes leads to a measure of the transit time or pseudo-range, which at this stage still includes the receiver clock bias.

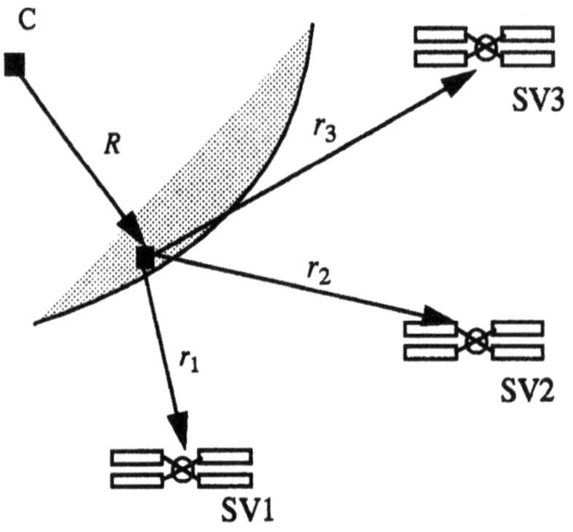

Figure 7.3 Direct ranging.

Pseudo-ranging is a very powerful positioning technique, and is the main mode of GPS measurement in real time. There are one or two points to consider, the first of which is how many satellites are required for a position? There are three positional coordinates to find (*X*,*Y* and *Z*) and one unknown clock bias. This makes a total of four unknowns, therefore at least four satellites are required for a position. For redundancy at least five satellites are needed, since using only four is exactly analogous to using a two-range Syledis or Argo fix. It is possible to constrain the height of the antenna if the height can be accurately measured, and this will reduce the number of satellites required for redundancy to four, since there are now only three unknowns to solve for. If the height is to be constrained, it must be known on the GPS datum (WGS84), and the best way of measuring the antenna height is to allow the height to be unconstrained and observe it over a period of about 24 hours. This 24-hour period must be one of good constellation and observing conditions; it must also be when the vessel is on the prospect. The reason for the latter is because the geoid (approximating to mean sea level) is not a uniform shape and the spheroidal height on the prospect may be very different from that at the mobilization port or during passage.

Once the height has been measured it can be used as an observation in the processing algorithm; having said this, the use of height aiding can be dangerous for the following reasons:

- Position biases can be introduced by using the wrong height. Furthermore, these biases change with satellite geometry.

- Height aiding must be used with an appropriate empirical standard deviation — say 5m. This is because the vessel is subject to heave and therefore its height is constantly varying. In areas where the tidal range is great there would also be a considerable effect. In general terms, every metre of error in the height will cause 1 — 2m of error in the horizontal position of the vessel.

One important point is that the height aiding is introduced in the processing software, and is not used in the derivation of the pseudo-ranges. This implies that if the wrong height is used (and the correct height is known!), then provided the raw data is logged the positions can be re-processed.

7.2.4 Doppler

The Doppler effect is named after Christian Doppler, who discovered it in the nineteenth century. The classic example of the effect is when an observer stands by the roadside and listens to the siren of an approaching ambulance or police car. As the vehicle approaches, the pitch of the siren is high and remains so until the vehicle passes the observer, at which point it drops noticeably and remains low as the vehicle recedes.

The explanation is that the frequency is higher during the approach, since the source and the observer are moving closer together, and the frequency is lower as they move apart. This is demonstrated in Figure 7.4 .

The Doppler equation is

$$f_r = f_s\left(1 - \frac{1}{c} \times \frac{dS}{dt}\right) \tag{7.1}$$

where f_s = frequency at the source

c = velocity of light

f_r = received frequency and dS/dt is the range rate.

Returning to our example, the closer the observer is to the road the faster the change in pitch of the siren, as shown in Figure 7.4. Thus if the speed of the vehicle is known and the frequency is monitored, the observer's position can be established in a two-dimensional coordinate system where one axis is the road. This is the principle of Doppler positioning and is directly transformed to satellites in space.

The position of a receiver can be established by measuring the Doppler shift of the received signals. These data are combined with accurate coordinates of the satellite (which of course are available in the navigation message) and used to determine the receiver's position. Once again, a single satellite's Doppler data will provide only two coordinates, so multiple satellites are needed to provide 3D positioning.

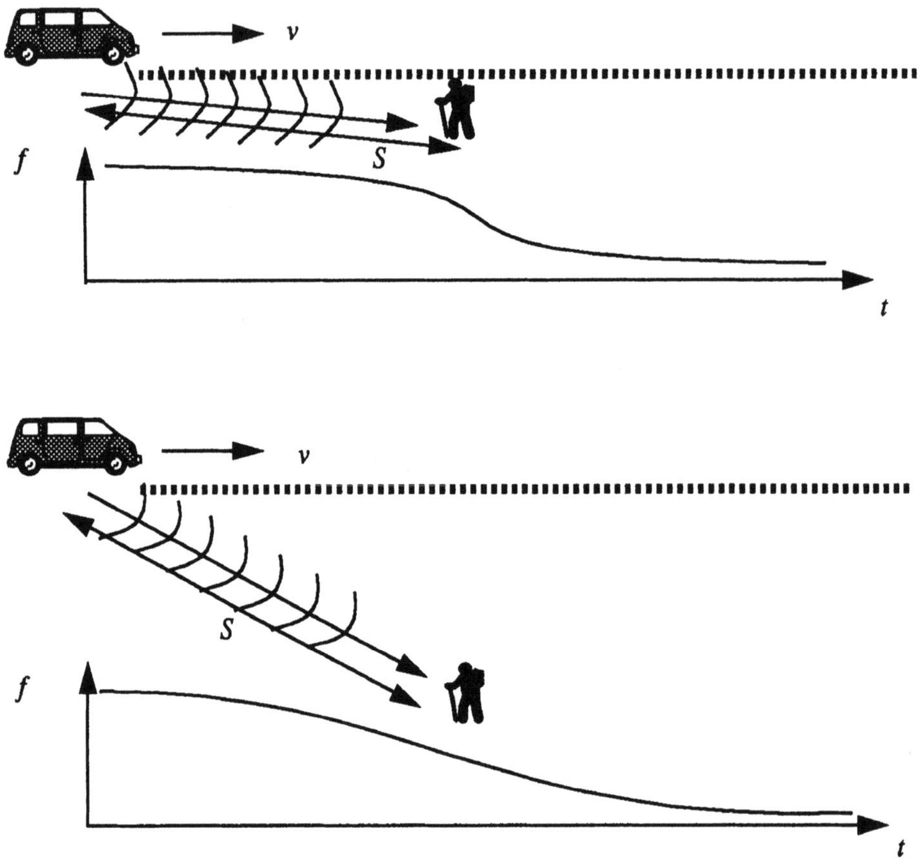

Figure 7.4 Doppler effect.

The old Transit system depended upon Doppler measurements for its positioning, as only one satellite was ever in view at a time. The accuracy of a Transit fix was in part dependent on the altitude attained by the satellite during its transit across the sky. If a reasonable altitude was not achieved, the Doppler shift would be very small and there would be a resulting loss in accuracy.

Unfortunately, GPS is not such a good candidate for Doppler measurement as was Transit, because the satellites pass at a considerably higher altitude (20200km as against 1100km), so there is a much reduced Doppler shift. However, some instruments use Doppler as a means of smoothing the pseudo-range data to good effect.

7.2.5 Interferometric phase measurements

Although not well understood in general, it is the phase measurement mode which provides the highest accuracy in stand-alone GPS positioning. The interferometric method used most is the carrier phase measurement. This is the mode used for the final measurement in rig positioning, and for setting out new geodetic stations ashore. It is not used per se for dynamic positioning, but is used in some instruments in conjunction with pseudo-ranges to smooth dynamic positions.

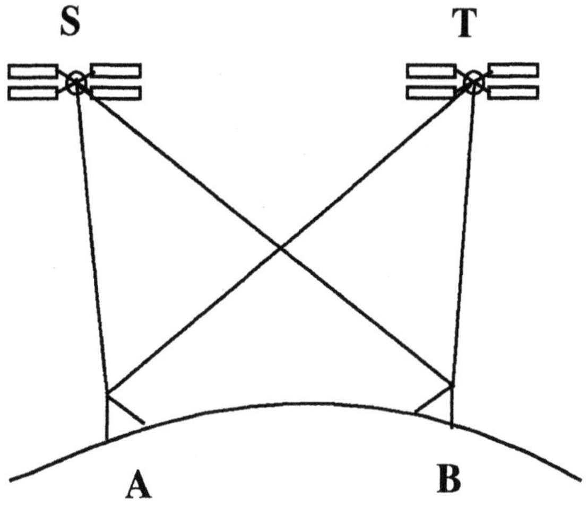

Figure 7.5 Interferometry.

In this method signals are received from multiple satellites at two or more receivers, and the C/A code is regenerated in the receivers. The regenerated code is used to clear the navigation modulation and leave the reconstructed carrier frequency L_1. In Figure 7.5, if the receiver and clock biases are neglected, the range from satellite S to receiver A is given by

$$S_A = (M_A \times \lambda_{L1}) + \frac{\lambda_{L1}}{2\pi}\phi_A \tag{7.2}$$

where

λ_{L1} = the wave length (approximately 0.19m) of the reconstructed carrier L_1, corresponding to 1575.42 MHz.

M_A = an unknown integer number of wavelengths

ϕ_A = the observable phase of the beat frequency.

Introducing the clock bias into equation (7.2)

$$S_A = (M_A \times \lambda_{L1}) + \frac{\lambda_{L1}}{2\pi}\phi_A + (\Delta t_S - \Delta t_A)\,c \tag{7.3}$$

$$S_B = (M_B \times \lambda_{L1}) + \frac{\lambda_{L1}}{2\pi}\phi_B + (\Delta t_S - \Delta t_B)\,c \tag{7.4}$$

and, subtracting equation (7.3) from equation (7.4),

$$(S_B - S_A) = (M_B - M_A)\,\lambda_{L1} + \frac{\lambda_{L1}}{2\pi}\phi_{AB} - (\Delta t_B - \Delta t_A)\,c \tag{7.5}$$

where

$S_B - S_A$ = range difference

$M_B - M_A$ = unknown number of wavelengths (2π ambiguity)

ϕ_{AB} = (interferometric) phase measurement.

$\Delta t_B - \Delta t_A$ = difference between the clock biases at B and A.

However, if measurements are made simultaneously from stations A and B, then the resulting 'double range difference' contains no clock biases at all. This is known as 'double differencing'.

The main problem to be overcome is the unknown number of wavelengths, or the 2π ambiguity. But this is resolved by making M_B - M_A an unknown in the least squares solution. Successive observation cycles will all contain the same ambiguity provided that there are no cycle slips.

Sophisticated processing software of the type produced by Trimble and Ashtech (among others) will detect cycle slips and produce results accurate to within a few centimetres over a baseline of several kilometres in length using only an hour's simultaneous observations.

If baselines arranged in a geometric shape such as a triangle are simultaneously observed, values can be obtained for the baselines and then these values can be used in a least squares adjustment, as shown in Figure 7.6.

7.3 Selective availability

When GPS was first conceived, it was decided that the P code would be encrypted and only available to selected users (such as the military). The C/A code would be free of encryption for civil and other unauthorized users. The C/A code users would, however, be required to pay a licence fee of some sort, possibly on an annual basis.

However, things have turned out very differently. In the first place, the achievable accuracy with the C/A code vastly exceeded expectations, and secondly the use of carrier phase measurements and, more importantly for the offshore industry, differential use had not been thought of.

Therefore the US Government was prompted to review the situation and decided to introduce selective availability (SA). The two main areas in which SA works are the ephemeris and the clocks. The ephemeris broadcast on the C/A code is deliberately degraded, so that the satellites' predicted positions are less accurate. The clock frequencies are deliberately jittered so that the clock bias between the satellite and the receiver is continually changing. Both of these problems can be largely overcome by using differential GPS — this is looked at later in the chapter. The effect on stand-alone GPS is that the single point positioning using pseudo-ranges is degraded to an accuracy of 100m or more when SA is switched on. Only the Block II satellites are equipped to use SA — the Block I vehicles do not have the necessary hardware and software on board.

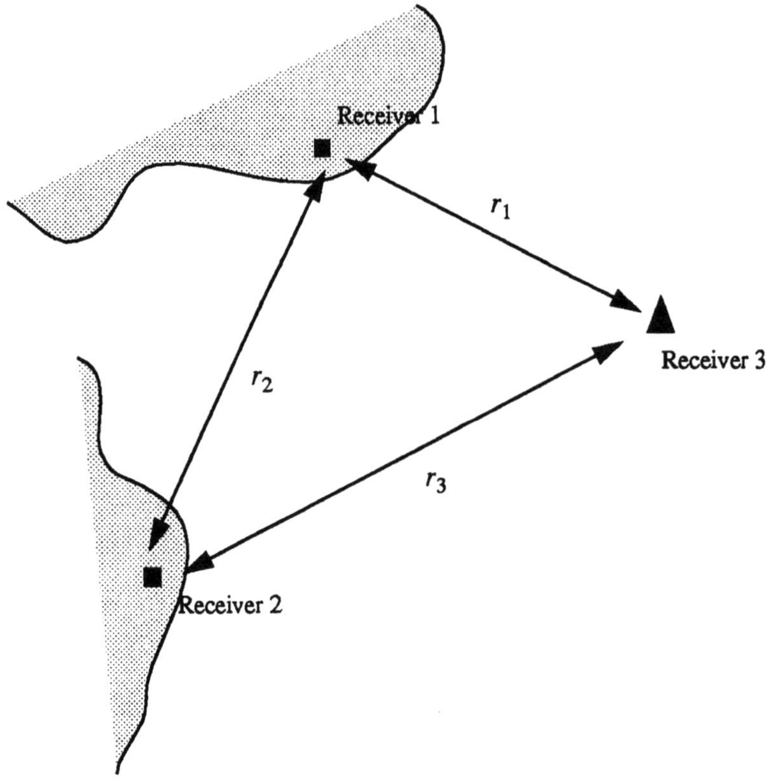

Figure 7.6 Phase measurement.

7.4 Some terms explained

Ambiguity — The unknown integer number of cycles of the reconstructed carrier phase contained in an unbroken set of measurements from a single satellite pass at a single receiver.

C/A code — the coarse/acquisition (or clear/acquisition) code modulated on to the GPS L1 signal. This code is a sequence of 1023 pseudo-random binary biphase modulations on the GPS carrier at a chipping rate of 1.023 MHz, thus having a code repetition period of 1 millisecond.

Clock offset — constant time difference between two clocks.

Delay lock — the technique by which the received code (generated by the satellite clock) is compared with the internal code (generated by the receiver clock) and the latter is shifted in time until the two codes match. Delay lock loops can be implemented in several ways; tau dither and early-minus-late gating.

Differential (relative) positioning — the determination of relative coordinates of two or more receivers which are simultaneously tracking the same satellites. Dynamic differential positioning is a real-time calibration technique achieved by sending corrections to the mobile user from one or more monitor stations, and this forms the subject of an entire chapter in this book.

Differential processing — GPS measurements can be differenced between receivers, satellites and epochs. Note that this term is NOT the same as differential positioning, which is something completely different. The present convention for differential processing of GPS phase measurements is to take differences between receivers (single difference), then between satellites (double difference), and then between measurement epochs (triple difference).

- A **single difference** measurement between receivers is the instantaneous difference in phase of the signal from the same satellite experienced by two receivers simultaneously.
- A **double difference** measurement is obtained by taking the difference between the single difference for one satellite and that of a chosen reference satellite.
- A **triple difference** measurement is the difference between a double difference at one epoch of time and the same double difference at the preceding epoch.

DOP — stands for dilution of precision. At the end of the least squares computation for position we end up with a covariance matrix. The trace of the covariance matrix is another name for its leading diagonal, and the matrix is always symmetrical about the leading diagonal. Different kinds of DOPs are simply the square roots of the traces of different sub-matrices of the overall covariance matrix of the parameters. These can be listed as follows:

- **GDOP — geometric** (three position coordinates plus the clock offset in the solution)
- **PDOP — position** (three coordinates)
- **HDOP — horizontal** (two horizontal coordinates)
- **VDOP — vertical** (height only)
- **TDOP — time** (clock only)
- **RDOP — relative** (normalized to 60 seconds).

Doppler shift — the apparent change in frequency of a received signal due to the rate of change between the transmitter and receiver.

Elevation mask — the angle below which satellites should not be tracked - usually 15° but sometimes relaxed to 10°. The angle is set to minimize reflection and multipath, and to lessen the effect of refraction through the atmosphere.

Ellipsoidal height — the height of the antenna above the chosen ellipsoid (spheroid). This is NOT the same as the height above mean sea level, which is approximately the geoidal height.

The difference between the two can be tens of metres.

Ephemeris — a list of locations of a celestial object as a function of time. Available as a real-time broadcast ephemeris or a post-processed precise ephemeris.

Epoch — the measurement interval or data frequency - usually this is 15 seconds in static GPS.

Fundamental frequency — the fundamental GPS frequency ofs 10.23MHz from which all other frequencies are derived. The L_1 frequency is 154 times the fundamental at 1575.42MHz, and the L_2 frequency is 120 times the fundamental frequency at 1227.60MHz.

GPS ICD-200 — the GPS Interface Control Document, a government document that contains the full technical description of the interface between the satellites and the user. GPS receivers must comply with this specification if they are to receive and process GPS signals properly.

JPO — the Joint Program Office for GPS located at the USAF Space Division in El Segundo, California. The JPO consists of the USAF Program Manager and Deputy Program Managers representing the Army, Navy, Marine Corps, Coast Guard, Defense Mapping Agency and NATO.

Microstrip antenna — GPS antennae made from a precisely cut two-dimensional flat piece of metal foil glued to a substrate.

Multipath — interference similar to 'ghosting' on a TV screen which occurs when the satellite signal arrives at the antenna via different paths, causing a phase change.

PDOP — position dilution of precision, a unitless figure expressing the confidence level of the user's position. A figure of about 3 or less is considered good for positioning, while a figure of over 7 would be considered poor. PDOP is directly related to satellite geometry, and small figures are obtained when the satellites in view are widely separated.

Point positioning — a geographic position produced from one receiver in a standalone mode. At best, position accuracy obtained from a standalone receiver is 15-25m, depending on satellite geometry. With SA switched on it is considerably worse.

PRN — Pseudo-random noise, a sequence of digital 1s and 0s which appears to be randomly distributed like noise, but which in fact can be exactly reproduced.

Pseudo-range — a measure of the apparent propagation time from the satellite to the receiver antenna, expressed as a distance. Pseudo-range is obtained by multiplying the apparent propagation time by the speed of light. It differs from the actual range by the amount that the satellite and user clocks are offset, by propagation delays and by other errors.

RINEX — Receiver INdependent EXchange format. A set of standard definitions and formats to provide the free exchange of GPS data and facilitate the use of data from any GPS receiver with any software package. The format includes definitions for three fundamental GPS observables: time, phase and range. A complete description of the format is in the Commission VIII

International Coordination of Space Techniques for Geodesy and Geodynamics *GPS Bulletin*, May — June 1989.

RTCM — Radio Technical Commission for Maritime Services. Commission set up to define a differential datalink to relay GPS correction messages from a monitor station to a field user. RTCM SC-104 recommendations define the correction message format and 16 different message types.

Selective Availability (SA) — a DOD program to control the accuracy of pseudo-range measurements, whereby the user receives a false pseudo-range which is in error by a controlled amount. Differential GPS techniques can reduce these effects for local applications.

SPS — the Standard Positioning Service is the level of static and/or dynamic positioning capability that will be provided by GPS, based on the single-frequency C/A code. The accuracy of this service is to be set at 'a level consistent with national security'.

7.5 Operational considerations

As previously mentioned, stand-alone GPS is not suitable for offshore dynamic positioning because of its inaccuracy. Even without SA the accuracy would hardly be good enough for 2D surveys. But it is used in earnest as a highly accurate surveying tool for locating, coordinating and checking shore stations. Even if Argo or Syledis (for example) are chosen in preference to DGPS for offshore positioning, the chain's shore stations would probably be checked using GPS.

The other main use for stand-alone GPS is in static offshore positioning — i.e. for semi-submersible and jack-up rigs, monopods and jackets.

When such observations are carried out, they should be subject to the following constraints:

- At least two receivers should occupy known stations ashore, plus one receiver on the offshore structure.
- All receiver antennae should be sited such that they are free from possible multipath; this will usually mean an elevation of at least one metre above the nearest flat surface, and no surrounding structures elevated higher than 15° when viewed from the antenna.
- All satellites in view should be tracked by all receivers; this is because the stations may be far enough apart that all the satellites viewed by one station may not be viewed by another at the same time.
- The antenna height must be measured accurately.
- The elevation mask should be 15°, and the epoch specified as 15 seconds.
- Observations for temperature, pressure and relative humidity should be taken every 30 minutes. The processing software will contain a meteorological model which attempts to compensate for tropospheric errors and which uses observed meteorological data.

- The observing period should be at least two hours, timed to coincide with a period of good satellite availability and low position dilution of precision (PDOP).
- The transformation parameters from the satellite datum (WGS84) to the survey datum must be known and approved.
- The usual check of read after write should be performed on completion of the observations.

Once the observations have been logged, they should be processed using single, double and triple differenced solutions. Once the baselines have been derived, they can be assigned a priori variances and used in a standard least squares adjustment.

7.6 DGPS — introduction

Differential GPS (DGPS) may be defined as the dynamic positioning of a mobile station in real time by corrected GPS pseudo-ranges. The corrected pseudo-ranges are generally Doppler- or phase-smoothed.

The system requires at least two GPS receivers to operate, together with some form of radio link between them. These segments can be defined as:

The reference station — a receiver whose antenna is set up over a precisely known coordinated station ashore.

The mobile station — a receiver set up on the vessel (or on a tailbuoy or gun float).

The link — a radio link set up between the two receivers which is able to upload pseudo-range corrections from the reference station to the mobile.

In addition, there may be a second reference station which can act either as a backup or as a source of additional correction information, and/or a monitor station set up on another precisely coordinated station.

Since 1992 DGPS has been used as the primary positioning system for 3D seismic surveys; usually a secondary DGPS system has also been specified. This shows the reliance placed on the system such that operators no longer require the backup of a terrestrial positioning method.

7.7 Basic operation

Development of DGPS for offshore dynamic positioning started in the early 1980s. Initially a reference station was set up over a known mark, and its position computed using the usual pseudo-range method with no smoothing. The difference in position between the receiver's solution and the true location was then computed, and the positional difference sent to the receiver on the vessel via modems and an HF link (HF was needed to guarantee the signal transmission beyond radio line of sight).

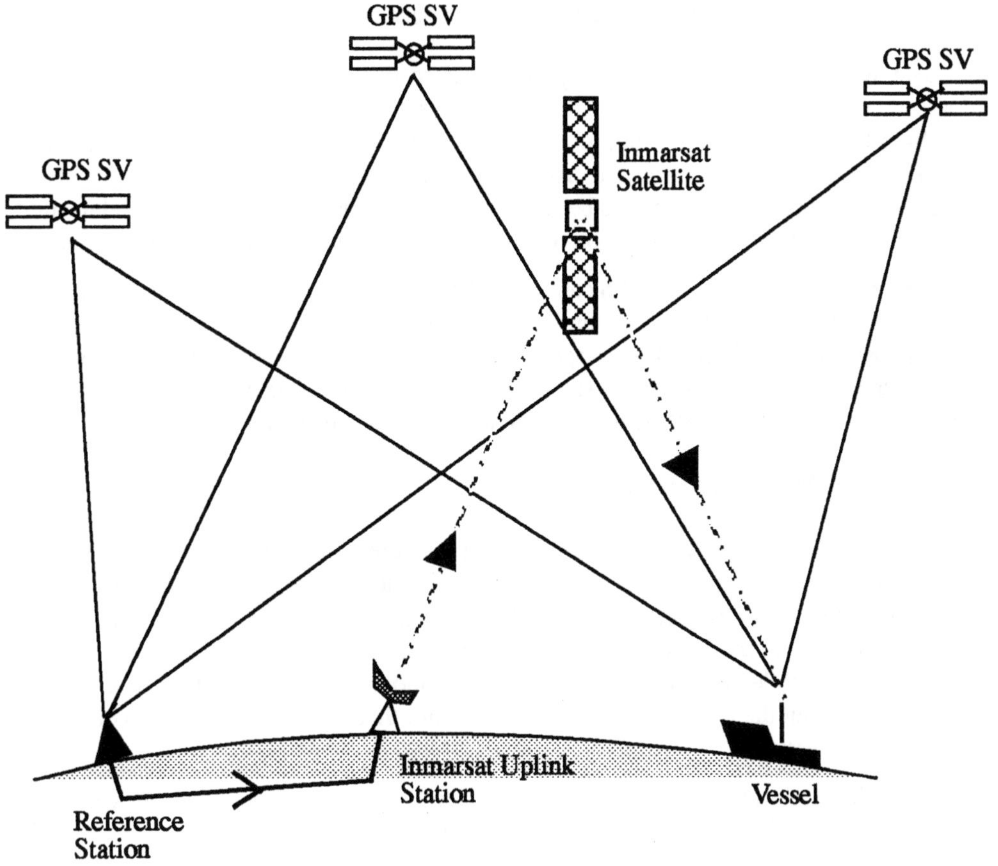

Figure 7.7 DGPS segments.

This system worked quite well but was prone to a number of problems:

- The correction update rate was only about once every 30 seconds — not a real problem in the early days because SA was not switched on, but this slow rate would not be workable at all now.
- The link was prone to skywave interference and was very weather dependent.
- There was initially no way of guaranteeing that the satellite constellation observed at the reference station (for which the corrections were valid) was the same as that observed at

the mobile. This is crucial in DGPS, because the position correction passed is completely dependent upon which satellites are being tracked.

The system was therefore developed further so that the actual pseudo-range corrections were passed on the link rather than the difference in position. This way we could be sure that only the satellites for which pseudo-range corrections existed were used in the computation at the mobile. This also opened the way for using DGPS tailbuoys. Most systems here only need a standard GPS receiver on the tailbuoys because the pseudo-range data can then be telemetred back to the vessel, where the pseudo-range corrections are applied and the computation occurs.

The real breakthrough in DGPS came with Racal's SkyFix system. This system consists of a worldwide network of reference stations linked by dual landlines (in case one breaks down) to control stations. The differential messages are uploaded at the control stations on dedicated Inmarsat channels to the various Inmarsat satellites. All that the user then requires is an Inmarsat receiver at the mobile end to receive and decode the incoming correction messages.

7.7.1 The reference station

Ideally the reference station should be set up where there is mains power, as close to the prospect as possible. The system will work at almost limitless range as far as the link is concerned (assuming an Inmarsat link), but one of the basic assumptions made with DGPS is that errors caused by the passage of the satellite signals through the ionosphere and troposphere are identical at the reference and mobile stations. This assumption is thought to hold good up to a separation of about 800km between the two stations. At ranges greater than this the assumption is less valid and the system less accurate.

The reference station antenna needs to be sited at least two metres above the nearest flat surface, and no natural or man-made features should be above the elevation mask of 15° when viewed from the antenna. These precautions are necessary to avoid multipath.

The station must be coordinated to at least second-order control in all three dimensions. This will entail establishing the station's coordinates from at least three identified primary national control and WGS84 related control points. Many contractors will offer the client a reference station that has already been set up and is in an operational state. Even in these cases the station coordinates must be checked thoroughly and the station itself inspected to ensure that the antenna has not been moved. In particular the antenna may well be mounted on a pole, and its height must also be checked. It is as well to point out here that the accuracy of the survey depends entirely on the accuracy of the reference station. In some ways the DGPS system is more prone to systematic error than a terrestrial system in that the position lines are effectively dependent on one station, whereas in the terrestrial system there will be a number of coordinated stations and it is extremely unlikely that they will all be wrong.

The reference station coordinates and all other set-up parameters should be held in ROM and backed up by battery as part of an uninterrupted power supply (UPS). Should there be a mains power failure, the UPS should be capable of keeping the station running. If there is a power failure, the system should be bootable in the correct configuration without operator intervention.

7.7.2 Multipath audit

The reference station should be put through a multipath health check as part of the contractor's responsibility. This requires approximately 48 hours' continuous logging, which is then processed to analyse all satellite pseudo-range and phase measurements. The standards provisionally accepted by United Kingdom Offshore Operators Association (UKOOA) for these tests are: 3.0m (1 sigma) on the code (pseudo-ranges) 0.05m (1 sigma) on the carrier (phase).

Position performance should also be checked using a local monitor station. This station must also be sited away from the effects of multipath on a known station, and used as a 'mobile' receiving corrections from the reference station. In this way the absolute accuracy and the reliability of the system can be checked under controlled conditions. (The alongside check outlined in Chapter 10 will go some way towards this, but won't give such good results).

In some systems adjacent reference stations can receive each other's corrections and thereby act as permanent monitor stations.

7.7.3 The mobile station

The performance of DGPS can often be quite variable, and the system can sometimes give poor results. Such performance can, however, largely be overcome through good operating procedures. It is very unfortunate that GPS in general tends to be treated with the 'black box' approach. Its operating principles are very similar to terrestrial systems, and although ultimate control of the system is out of our hands, a great deal can be done to optimize performance on the vessel.

The criteria applied to the reference station antenna should also be applied to the vessel — at least two metres above the nearest flat surface and no structure extending higher than the elevation mask (15°). Antenna location on the vessel is further complicated by the presence of a large number of transmitting antennae of various sorts — radar, main communication HF whips, other positioning systems and so on. It is recommended that the GPS antenna be above all of these if possible.

7.7.4 Redundancy

The first point to emphasize is that if DGPS is specified as the primary positioning system, it should never be used for prolonged periods with less than five satellites without height aiding or four satellites with height aiding. The trade-off against the lack of performance indicators is considered to be worth while. True residuals and associated error figures are only obtainable if there are more than four satellites.

Height aiding should only be used where its implementation is properly controlled. The input height must be the ellipsoidal height. In most receivers' software there is a 'look-up' table which is able to give geoid — spheroid separations for any position on the earth's surface. These tables are not considered to be of a high enough accuracy for use in 3D work, and the height should be input either from empirical derivations or from a standard geoid — spheroid separation model.

If height aiding is used, the height is effectively used as an observation in the position solution, and it must therefore carry some form of weighting. It is recommended that a standard error of 5 m be used for this. With height aiding enabled, the use of any less than four satellites must be prohibited since all redundancy is then lost.

The last point with regard to redundancy is that even if six or seven satellites are being used, proper redundancy can never be reached with a single reference station because all pseudo-range corrections are computed from that one station. If there is anything wrong with the reference station, then the error will be transmitted straight through to the vessel. The answer is to make use of at least two reference stations. By doing this the vessel can become a real-time monitor of reference station performance. The use of at least two reference stations provides a real level of redundancy, leading to a properly networked solution. Even the failure of one reference station should then become transparent to successful navigation.

7.7.5 Performance indicators

Having decided that redundancy should be maintained, the mobile station should produce the residuals of the solution as a minimum performance indicator (if there is no redundancy the residuals will all be zero). UKOOA recommend that the following indices be made available to the user:

- Range residuals
- Normalized error of residual
- DOP (PDOP, HDOP, VDOP)
- The SVs used in the solution

The following information is also considered useful:

- SNR/elevation
- UDRE (reference station calculated SD)
- URA (space segment broadcast range SD)
- Datalink performance

In addition the full real time correction message (RTCM) message containing range corrections, range rates, UDRE figures, SV health and reference station health and coordinates should be available to be displayed in real time.

7.7.6 The link

As mentioned previously, the current trend is towards using Inmarsat channels for upload of corrections. The sequence of events for the pseudo-range corrections is as follows:

(a) The navigation message is received and decoded in the reference station receiver (usually a 100% backup is required in all system segments — implying two reference station receivers and two antennae).

(b) The navigation computation is made and the pseudo-range corrections formed at the reference station.

(c) Corrections are loaded into RTCM format and sent via landline to a control hub. The control hub is a centralized onshore facility that allows the reception of pseudo-range data from several reference stations. It quality controls the incoming data from the reference stations before forwarding it to an Inmarsat Control Station for upload to the satellites. Racal have a dedicated Inmarsat channel specifically for this task. The landline links from the reference stations to the control hub are duplicated and where possible take different physical routes to minimize the possibility of damage.

(d) The mobile receives the data via Inmarsat, demodulates the signal and retrieves the navigation data. The pseudo-range corrections are applied directly to the mobile's own ranges. Each correction is time-tagged and is identified by its SV number, so that only pseudo-ranges and corrections with identical SVs are used in the solution. It needs to be remembered that when the mobile and reference stations are a long distance apart they might not be seeing quite the same constellation, so the fact that five satellites might be in view at the mobile may not mean that all five can be used in the solution.

7.8 Error sources

Figure 7.8 illustrates the basic processing flow, and shows the difference between raw data and phase smoothed pseudo-ranges (PSPR).

7.8.1 Incorrect height aiding

If it is realized at some point during a survey that the incorrect antenna height has been entered into the system, it should be possible to recover the error by reprocessing the data acquired to date. Height aiding is carried out in the position computation, not in the formation of the corrected pseudo-ranges, so only access to the phase-smoothed data is needed to reconstruct the vessel positions.

7.8.2 Incorrect smoothing

Phase smoothing of the data is usually operator-controllable by way of a filter. Smoothing of the data depends partly on the vessel dynamics, so if the vessel dynamics and the filter setting are incompatible, this should be noticed by the operator as poor performance indicators by the filter. If the filter setting is wrong and previous data need to be reprocessed, access to the original raw data before it is phase-smoothed will be needed. It is thought quite unlikely that work would proceed beyond one or two lines before poor performance of the filter was noticed.

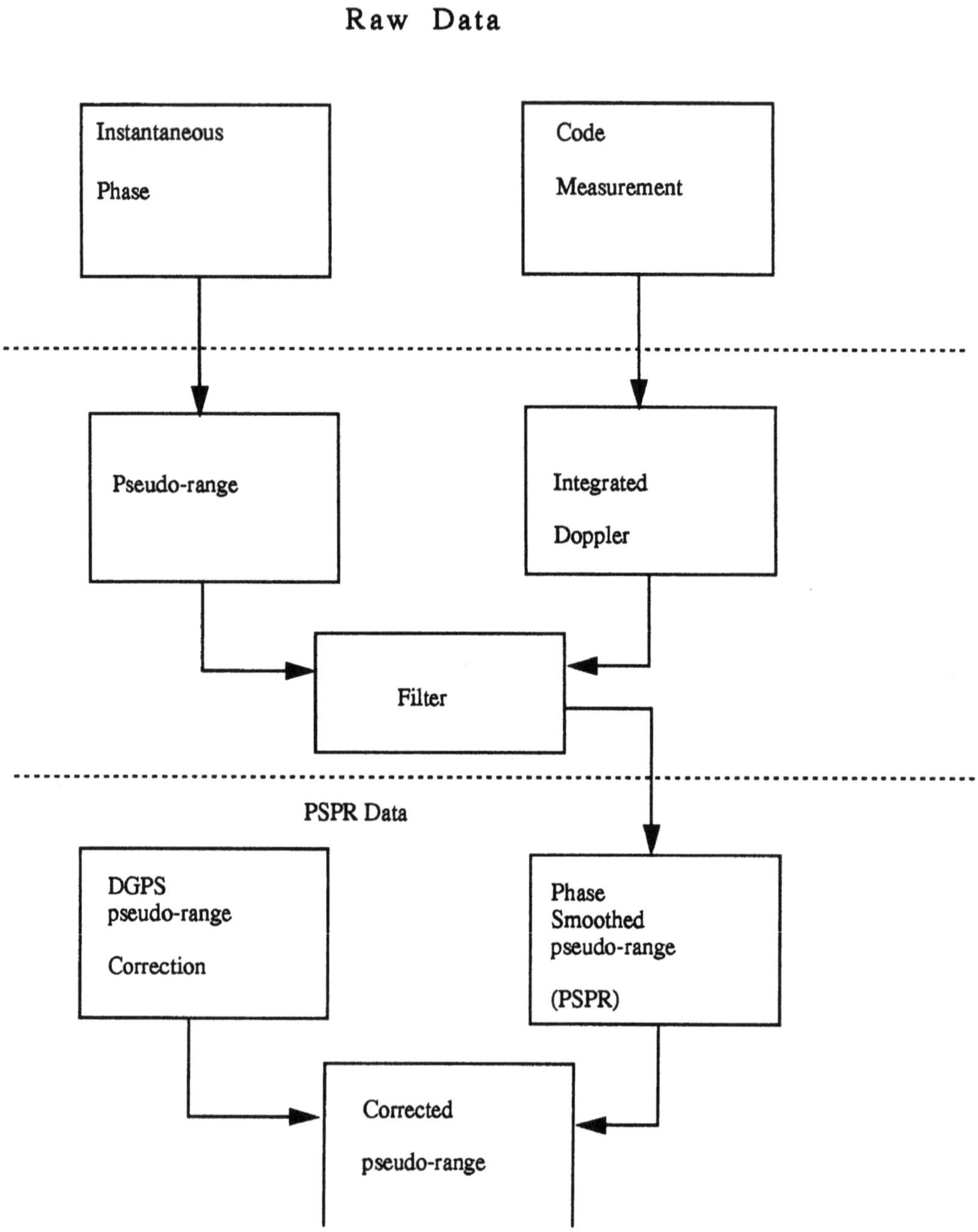

Figure 7.8 Raw v PSPR.

7.8.3 Use of unhealthy satellites

If an unhealthy satellite is used during real-time processing, then the phase-smoothed data for that satellite need to be rejected in post-processing. Of course, the receiver itself is capable of finding out which satellites are flagged unhealthy by DOD, but the trouble is that sometimes satellites become unhealthy and notification of this is not received until some days after the event.

7.8.4 Multipath errors

Multipath can occur at the reference station(s) or at the mobile. The effects are difficult to remove in post-processing, so it is preferential to remove their source in the first place. This entails siting the antennae as recommended in the preceding sections.

At the reference station the multipath effects will tend to be repeated over a 24 hour cycle and so should be easily identified during the reference station audit. Once identified, a predictive model can be developed to counter the effects in real-time.

On the vessel, unless the sea is very calm, any multipath will tend to appear as random errors, thereby increasing the noise but not being particularly detrimental to positioning. But when the sea is calm, multipath will become more constant and could have serious effects, so every effort must be made to get the antenna as high as possible and well away from multipath sources.

7.8.5 Selective availability errors

The whole idea of SA is to introduce errors into the system. It should be recalled that both the ephemeris and the clocks may be interfered with. In the case of ephemeris errors, both the reference station and mobile will see the same ephemeris at the same time unless the two are widely separated. If an error is present, it can be identified if two or more reference stations are being used. That being the case, the errors can be removed in real time if a spatially adjusted DGPS correction can be generated, otherwise they will need to be removed in post-processing.

Clock jitter will start to become effective if the pseudo-range corrections are not received at a fast enough rate. The best way to counter this is to ensure that corrections are received at an interval not exceeding 5 second,s given the present state of SA activity. To counter the effect in post-processing, the raw data from the reference and mobile stations needs to be combined at exactly the same epoch and time interval.

7.8.6 Ionospheric and tropospheric effects

If the path travelled from the satellite to a reference station is significantly different to that travelled from the satellite to the mobile, the ensuing error will be seen as an ephemeris error at the mobile and will be hard to distinguish. However, at ranges of less than 100km between mobile and reference stations the errors are not thought to be significant.

7.8.7 Link failure or corruption

If the link breaks down for some reason, the effect on navigation will be very obvious unless the navigation computer is using a very heavily damped filter. Even then, the change in position will be so great that the line will almost certainly become untenable.

7.9 Data formats

It is considered vital to log raw data for DGPS, just as we log the raw data for terrestrial positioning systems. The reason for logging raw data is so that the data can be used for subsequent re-processing or post-processing. If re-processing or post-processing is carried out, it is usually to ensure that the integrity of the survey is good. Since the whole dataset is dependent on the vessel's position there is good justification for having the raw data available. It has always been a surveyor's dictum that any survey should be able to be reconstructed from the field notes and raw data, and this still holds true.

From the above discussion on error sources it will be seen that there are two potential sources of 'raw' data — the original unsmoothed pseudo-ranges and the phase-smoothed pseudo-ranges (PSPR). Most of the blunders and systematic errors could theoretically be removed using only the PSPR data without recourse to the original pseudo-ranges.

Currently seismic contractors exchange raw data in one of the UKOOA raw data exchange formats — P2/86 or P2/91. It would seem logical that raw DGPS data should also be incorporated into the same format. There is no facility for this in P2/86 — only the derived positions can be logged here. But P2/91 is more flexible and the raw data can be incorporated into that format. To date there has been no special format to take PSPR data in P2/91, but this is expected shortly.

Figure 7.9 illustrates the basic data-handling scheme.

UKOOA recommend that both raw and PSPR data be logged as follows.

7.9.1 Data recording at the reference station

Raw data should be recorded in the manufacturer's format at the receiver cycle interval on a recording medium directly linked to the GPS receiver.

Computed data (from which data to be transmitted are selected) should be recorded at the same interval as they are computed (which may be faster than the interval at which they are transmitted.

7.9.2 Data recording at the mobile station

Raw data should be recorded in the manufacturer's format at the receiver cycle interval on a separate recording medium directly linked to the GPS receiver.

PSPR data should be presented in a P2/91 format at the event interval on the same recording medium as the other P2/91 data.

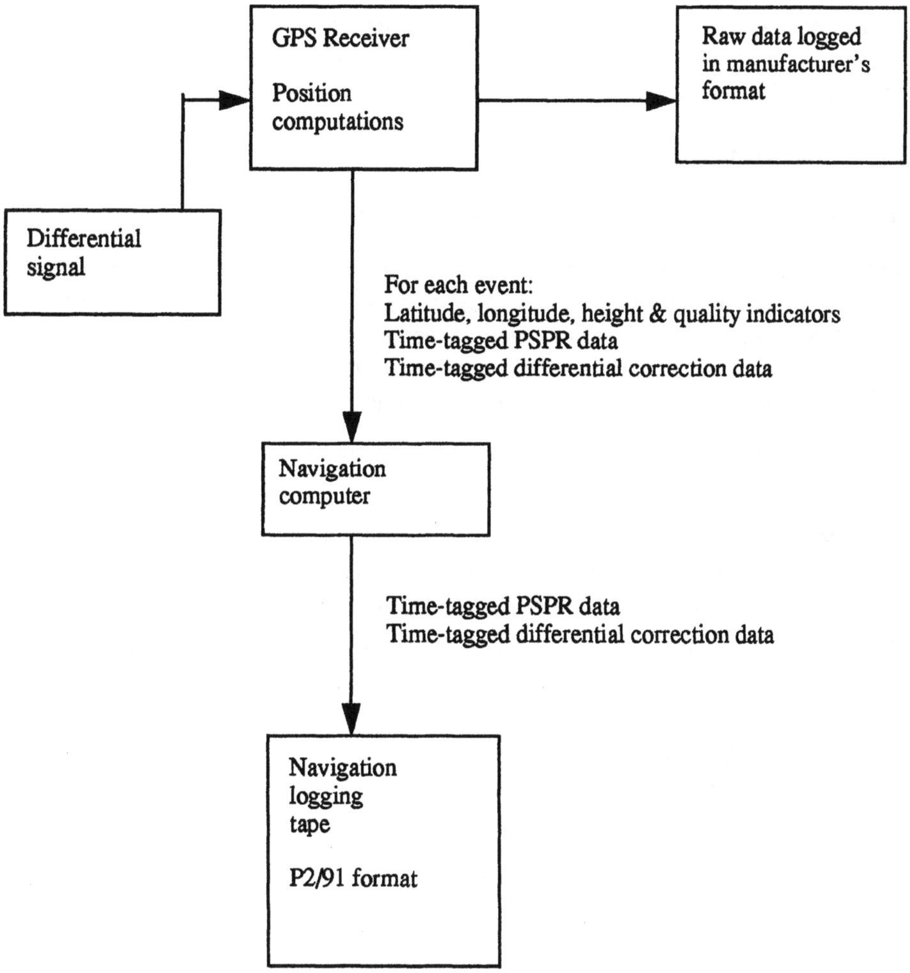

Figure 7.9 Data logging.

7.9.3 Data recording of GPS tailbuoy data

No format has yet been defined for tailbuoy data. This is an area which needs to be reviewed as a matter of some urgency.

7.9.4 Post-processing of data

Each receiver manufacturer should be requested to provide the necessary software to wholly reconstruct time-tagged, phase-smoothed pseudo-ranges (in P2/91 format) from raw data in their own format. This would effectively be a pre-processor of raw data.

7.9.5 Data volumes

The following notes are a guide to the amount of data involved in logging.

For logging raw data we can assume a 1.2 second recording interval. At the mobile station each receiver will need about 20MB of recording medium per day, assuming continuous operation. So if there are three tailbuoy receivers as well as the vessel receiver, we would need about 56MB per day. The problem here is that as yet no-one has started logging raw data from the tailbuoys because of the telemetric link problems between tailbuoys and the vessel.

At the reference station two receivers operating continuously would require about 40MB per day.

For logging PSPR data the event interval can be assumed to be about 10 seconds (i.e.shooting at 25m spacing) and each receiver on the vessel will use about 2.5MB per day. The reference station will consume about 7MB per day per receiver.

The rather large amounts of data being logged will require suitable media such as 150MB cartridge or Exabyte.

7.10 The SkyFix system

7.10.1 Introduction

The SkyFix system is produced by Racal Survey Limited, and the following notes are drawn from the Technical Description of the system with the kind permission of Racal Survey Australia Limited.

Racal Survey is the world's largest provider of precise positioning services for offshore and hydrographic survey tasks. SkyFix is a differential GPS Precise Positioning Service which uses the Inmarsat marine communications satellite system as the correction data broadcast link.

Silnce the introduction of differential GPS and particularly since the decision by DOD to proceed with selective availability, much effort and time has been expended on the use of conventional MF or HF datalinks for the transmission of correction data. These data links are only adequate where slow data rates and limited range are acceptable, and even then their availability and integrity are often poor.

The ideal datalink would have the same characteristics as GPS itself — wide area coverage, excellent availability, freedom from most sources of interference and high integrity. It would also have the ability to provide a fast message update rate to counter SA, and enable the use of multiple reference stations simultaneously. Inmarsat provides the only link to date that can reach these specifications.

7.10.2 System operation

A number of reference stations are sited to provide cover of the area of interest. In Australia there are reference stations at the following locations:

- Broome
- Perth
- Adelaide
- Sydney
- Darwin
- Cairns
- Dampier

Data from the reference stations are sent to a network control centre (NCC) where the correction data for each satellite are thoroughly checked for consistency. The NCC for the Middle/ Far East (including Australia) area is at Singapore.

A message containing data drawn from all the reference stations in the network is sent to an Inmarsat Coast Earth Station (CES), where it is uplinked into the appropriate Inmarsat satellite for broadcast. This broadcast covers a very wide area and users anywhere in the satellite footprint can use it. For example, a vessel working on the North West Shelf would usually track the Indian Ocean Satellite (IOR), but it is also in the Pacific Ocean Satellite's (POR's) footprint and could track that instead (although the altitude is lower and the signal correspondingly weaker).

Users receive the Inmarsat signal on an ordinary Inmarsat 'A' shipboard terminal unit and use the special SkyFix demodulator unit to decode the data and send it to the GPS receiver. If the vessel or rig does not have a suitable Inmarsat 'A' terminal, a SkyFix receive-only unit can be used.

At the NCC a similar receive terminal is used to simulate exactly the same process and provide an independent check that the system is functioning correctly. Later, the data recorded at the NCC may be used in post-processing for checking against data recorded in the field.

7.10.3 General features

The SkyFix satellite datalink is highly reliable. In common with other geostationary satellite communications links, it is designed to have a bit error rate of less than 1 in 10 million.

The SkyFix data rate of 1200 bps enables corrections to be sent simultaneously for a number of reference stations at a refresh rate of 3 or 4 seconds. This rate is fast enough to defeat the current level of selective availability.

The availability of the datalink is very high. There is little chance of weather effects such as rain static or thunderstorms affecting reception.

There is 24 hour monitoring and quality control by fully manned network control centres.

The useful operating range is up to 2000km from a reference station.

The datalink has an operating frequency in the L-band (1.5GHz) and the message protocol is in accordance with SC-104 recommendations. There are no diurnal effects, no seasonal effects and the availability of the link is better than 99.998% over a year.

7.10.4 Reference stations

In the case of failure of a reference station, there is usually another reference station that can be used. The data from each station are cross-checked with others at the NCC to eliminate gross errors. Selection of another reference station does not require a change of frequency or equipment, and the operator can make his own choice of stations to use.

Each reference station performs range measurements on all satellites in view. The differences between these pseudo-ranges and the expected ranges are used as the corrections.

All reference station equipment (including receivers and antennae) is duplicated with automatic switch over facilities.

The NCC can take over full control of reference station functions and operating parameters.

The following equipment is used at reference stations:

Trimble 4000 10- or 12- channel GPS receiver.

1 second measurement rate.

386/387 PC compatible, 60MB hard drive.

Operating software — Trimble DNAVR (DeltaNav Reference).

Ancillary equipment:

Modems, uninterruptible power supply.

Data produced:

pseudo-range corrections for all satellites in view.

Low elevation warning.

Range-rate corrections.

RTCM SC-104 format data: messages 1, 2, 3, 5 and16.

Type 1: pseudo-range, range-rate, issue of data time.

Type 2: Delta differential data (to cover differences in IOD).

Type 3: Reference station parameters.

Type 5: Constellation health.

Type 16: Network control message in plain language.

Data logged (on request):

Raw pseudo-range data, including carrier Doppler.

pseudo-range and range-rate corrections.

Outgoing RTCM SC-104 message to NCC.

7.10.5 Network control centre

This is where the main quality control and network checking is done. Each NCC is continuously manned, and is responsible for checking the following:

Data lines — main data lines to the uplink stations are duplicated. All terrestrial lines have automatic dial-up facility and time out on data for any reference station is automatically notified to the SkyFix monitor operator.

Validity of incoming reference data — apart from the usual checks, reference station data ares cross-compared between stations.

Outgoing data to the upload station — the data from the appropriate reference stations are transformed into a single serial data stream at 1200 bps.

Downlink signal from the Inmarsat satellite — an Inmarsat receive-only terminal feeding a set of equipment identical to that used by the SkyFix user is employed to verify signal strength and data validity. A further check is on the latency, or overall time delay, occurring between the acquisition of reference data and its availability at the user receiver. If a pre-set time, normally a few seconds, is exceeded, it is flagged as a fault condition.

Overall fix improvement — the downlinked data is used to continuously correct GPS data obtained from a local GPS monitor receiver and a record is kept of the accuracy of the corrected fix. This process is identical to that which the user employs.

Correct recording of all data — data are recorded on request from the client.

The following equipment is used at the NCC:

486/66 PC compatible with 487 coprocessor, 60MB hard drive.

1 Gbyte optical disk data recorder.

Telecommunications equipment:

Fully redundant communications facilities for each link.

Automatic PSTN line dial-up in case of dedicated line failure.

Data demultiplexer.

Off-air downlink monitor:

Racal Inmarsat receive-only terminal.

Inmarsat signal strength check.

Racal proprietary monitoring program providing real-time readout of message latency, real-time comparison of data between all reference stations, instant warning of discrepancies and graphical time series plots of corrected position.

Data processing:

Incoming reference station data is checked for veracity.

Data are time-tagged.

Inter-reference station comparisons carried out.

Individual satellites are compared between reference stations (for pseudo-ranges and range rates).

GPS signal strength is checked.

Network data:

Reference data serialized and encrypted.

An ASCII plain language Type 16 message is added if necessary.

7.10.6 Inmarsat satellite broadcast link

The upload station (CES) is not normally under Racal's direct control but is operated by the appropriate Inmarsat signatory organization. It is responsible for transmitting the data signal to the satellite and maintaining the correct level of the downlink signal. The operating authority maintains its own monitoring system for these purposes and can guarantee a datalink reliability far in excess of any other method of data transmission.

The satellite itself is provided, owned and operated by Inmarsat on behalf of its signatories. There are four main ocean regions — Atlantic E and W, Pacific and Indian Ocean areas. There are usually several satellites on each station, providing considerable back-up. Coverage is world-wide except for the polar regions.

7.10.7 Mobile station

The SkyFix signals can be received either via the vessel's own Inmarsat-A terminal or via a purpose-built Racal Survey receive-only terminal. In addition to the Inmarsat antenna and terminals the onboard equipment comprises a signal demodulator, Trimble 4000DL GPS receiver and a 386 or 387 PC and software.

Interfacing is achieved without any adverse effects on the normal use of the Inmarsat-A terminal, and the addition of Racal's DDM-2400 demodulator has no effect whatsoever on normal Inmarsat reception and the terminal can be used for transmission without affecting SkyFix reception.

Racal receive only terminal — this receive-only unit is a fully stabilised receiving dish (enclosed in the usual weather-proof dome). It has three-axis stabilization and receive performance. It needs to be interfaced to the ship's gyro to provide azimuth stabilization.

Racal portable SkyFix receive-only terminal — this terminal is small, portable and designed for use on stable platforms such as rigs. It is not dynamically stabilized in either elevation or azimuth but is designed to automatically track the satellite for best reception. It has a collapsible dish and is suitable for transport by helicopter.

Racal type DDM-2400/2404 data demodulator unit — this unit accepts the output signal from the Inmarsat terminal and extracts the data for presentation to the GPS receiver. There are five different versions of the demodulator, each designed to interface to different Inmarsat terminals.

GPS Receiver — Trimble 4000 DL — this is a single frequency C/A code instrument. It has a carrier-assisted filter to assist the pseudo-range measurement and tracks up to eight satellites simultaneously. It has a 1-second measurement rate.

Computers — the system requires a 386/387 PC running Trimble DeltaNav Navigator DNAVN.

Data recording — the system can output all GPS data including health, signal strengths, pseudo-range and range rates, DOP, elevation angles, etc. It can record the incoming RTCM SC-104 message and the corrected position.

7.10.8 Post-processing services

The recording of observed range data and corrections at both reference and user stations allows post-processing to take place after the project. Racal's Chart 300 system can provide detailed graphical information on various parameters, and position recomputation can be achieved through the Trimble Post Nav package.

7.10.9 Geodetic control

Since the successful operation of DGPS depends entirely upon the positioning of the reference stations, the determination of these positions is of the utmost importance. SkyFix is designed to have the capability of providing position to better than 5 m in absolute terms, and in order to achieve this the 3D position of the reference stations must be determined to at least one order of magnitude better than this figure.

Geodetic consistency is an important prerequisite for GPS integrity monitoring. It facilitates the straightforward application at a monitor station of measurements transferred from a reference station and also helps to minimize the risk of a positioning error occurring when changing from one reference station to another. This is particularly important if the user is moving into the province of a different control centre.

Racal's requirements for the positioning of new stations are:

- The use of dual-frequency GPS phase-differencing techniques.
- Absolute accuracy to be related to the IERS Terrestrial Reference System of 1989.
- Network solution confidence indicators to be within specifications of the dual-frequency receivers used.
- To be consistent with the established SkyFix worldwide network.

SkyFix is basically a homogeneous network within the WGS84 geodetic reference frame and the Europe/Australia net includes ten primary triangulation stations for each of which the WGS84 values are known. To adjust this network, repeated observations were made using dual-frequency GPS receivers and in some cases full P-code geodetic receivers. The baselines thus established created a link between the Far East, Western Australia and Europe. Purely for the purposes of testing strength and internal consistency this link was subjected to a least squares adjustment by variation of coordinates. In the final analysis two of the primary triangulation stations, Dongara 38 (WA) and Matera (Italy) were held fixed. For all stations included in the final analysis the mean semi-major axes of the twenty-nine 95% error ellipsoids are contained within plus and minus 0.6m, indicating that the accuracy objective has been met.

7.10.10 Quality control and network monitoring

SkyFix provides a good independent integrity check on GPS performance. Although GPS broadcasts a health word it is not always updated rapidly, nor does it signify problems that although not being of concern to the military may be important for survey users. SkyFix monitors all position-related aspects of satellite performance and notifies satellite problems to the user much faster than any other method. Statistically there is just as much chance of a reference station itself having a problem as there is of a satellite having one. SkyFix does not rely on just one reference; each SkyFix network has several, whose data is cross-checked continuously. Additionally there is psychological reassurance for those worried about using a system over which they have no control.

Checks on the DGPS message — an Inmarsat receive-only terminal is installed at the NCC which receives the broadcast SkyFix signal, demodulates it and sends it to a GPS receiver operating from its own antenna quite separately from the other station equipment, in exactly the same way as a user would do. The output of corrected position from this receiver is used as a continuous indication of system performance and is displayed to the NCC controller as a time-series plot on a VDU. It is also continuously recorded for the system archives. Another very important parameter is the time delay occurring between computation of corrections at the reference station and their arrival at the users' receiver (latency). Each outgoing message is labelled with a time tag which is compared with the time it is received back. A record is kept of these comparisons.

Checks on the data link — the SkyFix demodulator has facilities for making its own checks on the performance of the incoming data stream. It provides a continuous readout of bit error rate, which rarely exceeds 1 in 10 million (a rate of one lost bit every three hours!). This check takes into account all sources of noise on the downlink, weather effects, vessel motion, etc. and has been found in practice not to vary significantly during the day or the year. At the NCC, the comparison of outgoing and incoming messages already mentioned allows instant notification of missed messages, and a record is kept of the number of lost messages over the previous hour, a figure which rarely exceeds one or two.

Checks on the final corrected fix accuracy — the data linked SC-104 signal is used to correct a GPS receiver output and the results are continuously monitored, producing the time series plots already referred to. Standard deviations of the corrected latitude, longitude and height compared with the reference values are continuously computed and logged for all reference stations.

8 Lasers, acoustics and compasses

8.1 Introduction

So far we have looked at how to position the vessel and how to position the tailbuoys, neither of which are directly involved in CMP computations.

It has already been decided that the CMP positions must be known such that the seismic data will be placed in the correct bins. Since it is not possible to be 100% correct, it is also determined that the tolerance should be about 0.3 × bin width (1 standard deviation). This should be the same for all CMPs in the spread - the tendency at the moment is for the front CMPs to be positioned with a higher reliability than the back ones. This turns out to be mostly due to the huge amount of observations concentrated at the front of the spread. Consider Figure 8.1.

In Figure 8.1 Diagram A there is a dual-streamer configuration containing equally spaced compasses down each streamer and three acoustic networks. This is the traditional network. It has been stated repeatedly through this book that solutions should be over-determined. But in the Figure 8.1 Diagram A configuration it is very much over-determined at the front, middle and back and some what under-determined over the rest of the spread. The diagrams themselves are a little misleading because of their scale exaggeration. The distances AB and BC are often about 2000m each, whereas the areas covered by each acoustic net are of the order of 200m. If a fifth order curve is fitted to the streamers it is obvious what will happen; for two sections of curve each 2000m long there are about five observations overall — the compasses. Over the acoustic network areas there are about 25 observations, made up of acoustic and compass readings and not including the vessel and tailbuoy positions at the head and tail of the streamers. It would be very difficult to fit a curve to this configuration without heavily biasing it towards the acoustic nets and away from the main part of the streamer.

But consider Figure 8.1 Diagram B, in which the compasses are retained but the acoustic pods are staggered down the streamers to form a ladder. Now the observations are spread very evenly down each streamer, making the fit easier. The streamer separation knowledge has been strengthened by having more cross-cable observations, and yet the number of acoustic observations has actually been reduced from 28 to 16, giving the software less work to do.

These thoughts are developed further in the following pages and in Chapter 5.

8.2 Lasers

8.2.1 General

The Mars laser system has been in use on seismic spreads for many years. The transmitter has a laser diode source and associated optics which produce a 25 degree vertical fan beam depressed about 2.5 degrees from the horizontal plane. Any target within a range from 200 — 5000m is illuminated for target heights not exceeding 20m above sea level.

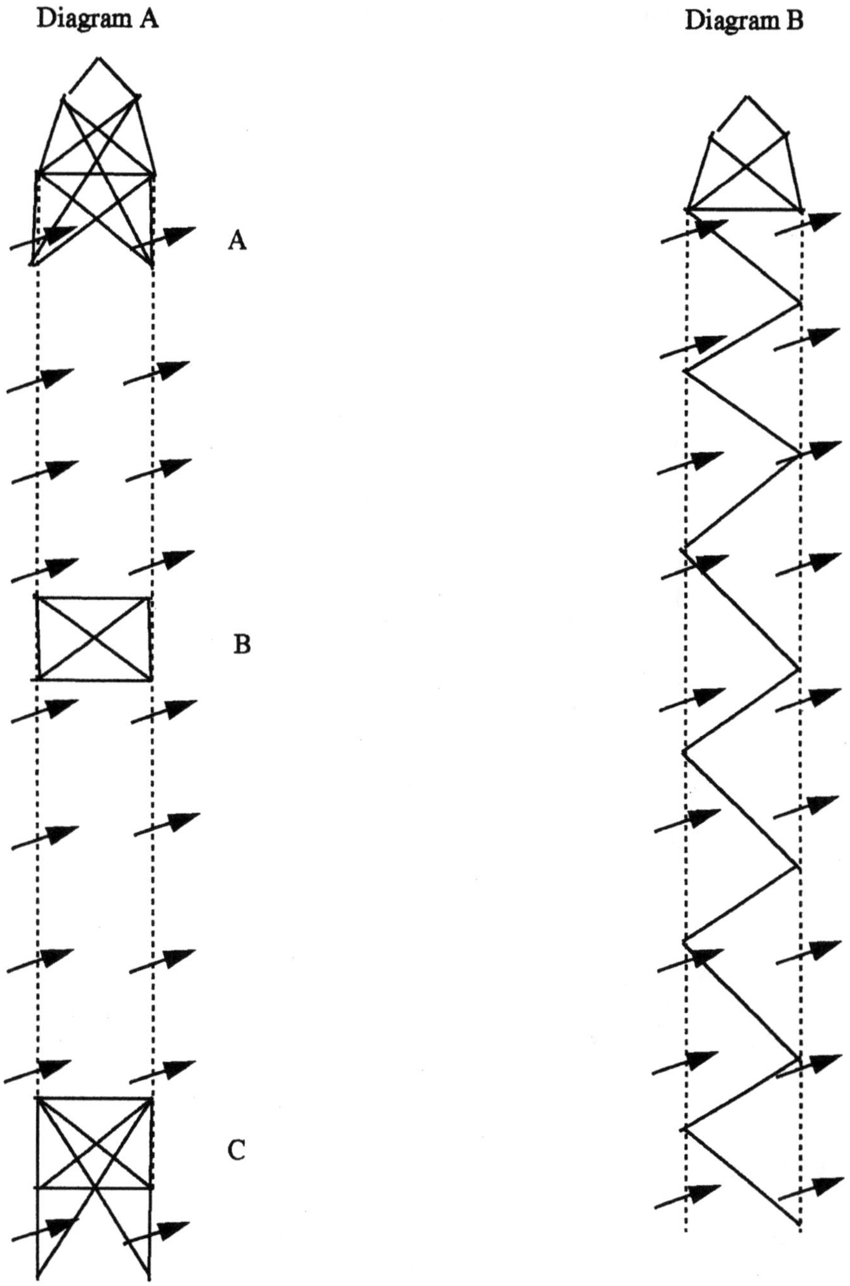

Figure 8.1 Streamer positioning.

The targets themselves are simple optical retro-reflectors, which return the laser beam along a path parallel to the transmitted path. A rotary encoder is used to measure the angular offset of the target, while its range is simply a function of time. The ranges and bearings are automatically corrected for vessel roll. A typical set-up is shown below:

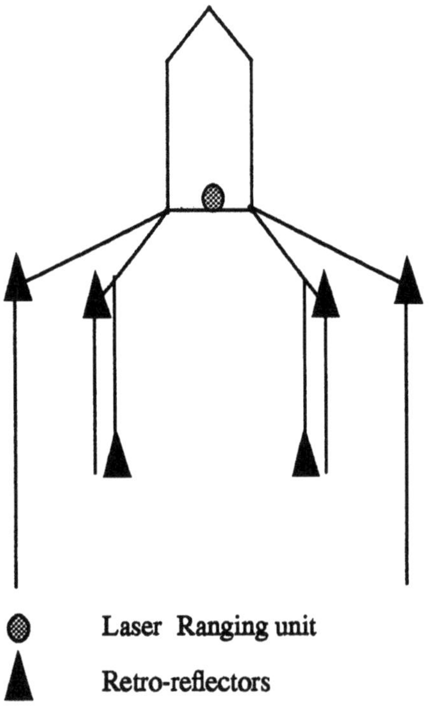

Figure 8.2 Laser ranging.

In Figure 8.2 there is a dual-streamer, dual-source configuration. Each source is made up of two gun strings. The gun strings themselves are normally supported in the water by strings of Norwegian buoys, and the targets can be directly mounted to the buoys on short aluminium poles. The streamer targets are mounted on the paravanes, which keep the streamers at the required cross-line distance from the vessel.

8.2.2 Calibration

The laser ranging unit (LRU) is mounted on the stern of the vessel. Although its output distances are true distances, the azimuths which it outputs to the navigation computer are relative. The LRU therefore needs to be installed such that it reads relative to a known true azimuth, the obvious choice being the ship's head.

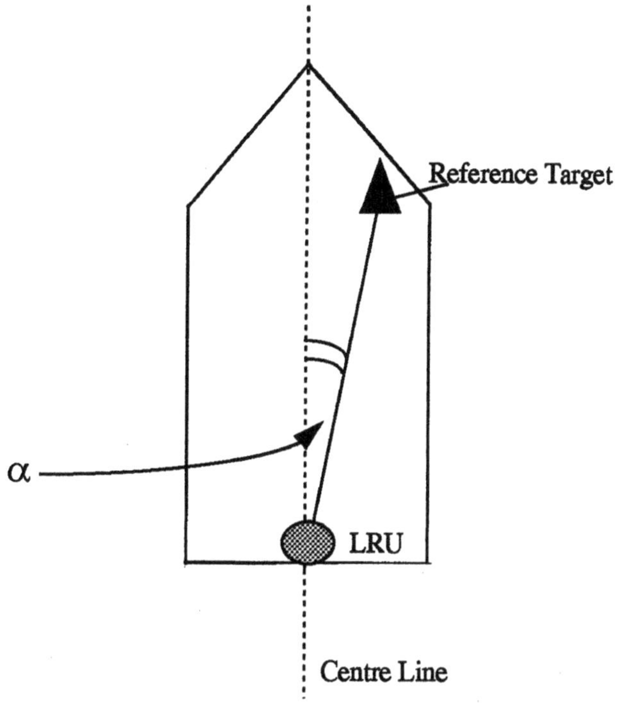

Figure 8.3 Laser calibration.

Once the LRU has been installed, a reference target should be mounted somewhere forward on the vessel's superstructure as shown in Figure 8.3. The angle α can then be measured using a theodolite, accurate to a few minutes of arc. Once at sea and prior to the start of work the LRU can then be referenced on the known target. This should be repeated before the start of each line - it is a very easy operation. Now the angle α can be added to the laser's output angles to give azimuths relative to the ship's head; adding the gyro will give the true azimuth. This also implies that the gyro needs to be accurately calibrated, since the whole front-end positioning system depends upon it.

When using laser it is recommended that the gyro be calibrated daily — see Chapter 10.

The range and bearing function of the unit can be checked alongside during mobilization, at the same time as the reference target is mounted.

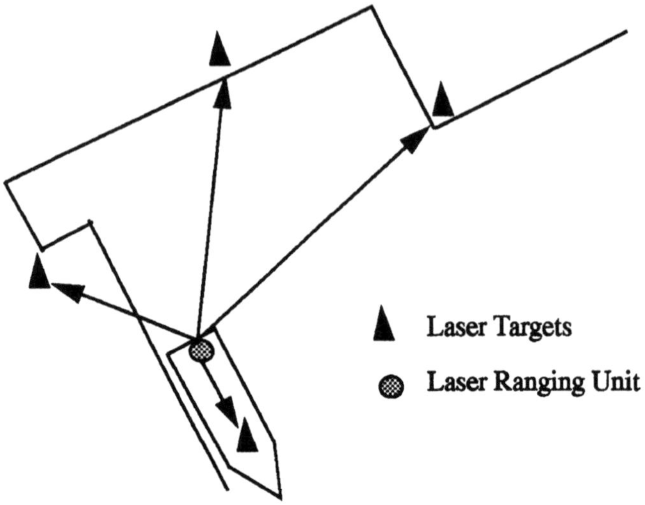

Figure 8.4 Laser calibration.

Figure 8.4 shows the vessel lying alongside the wharf. The reference target is set up on the vessel, and three other targets are set out around the wharf area, at ranges simulating the user ranges for the project. These targets are placed on previously coordinated points, and the LRU itself can then be coordinated by theodolite and electronic distance measurement (EDM). The laser is referenced (assuming that the gyro has been properly calibrated) and a number of ranges and bearings read to each target (the laser will scan automatically, swinging through an arc of about 180 degrees). Using the observed ranges and bearings and the known coordinates of the LRU, the positions of the targets on the wharf can be computed, thereby checking the system.

8.2.3 Operational considerations

The laser system is excellent for front-end positioning provided the following points are understood:

- Lasers will remain operational in far higher sea states than acoustics, and are not restricted by aeration and cavitation in the water. However, they will not work in fog or heavy rain because the beam becomes diffracted.

- The laser produces two lines of position for each target, in other words a range and a bearing. It therefore does not satisfy the criterion of redundancy on its own. But if two laser ranging units are used, one on each gun boom for instance, then there are two ranges and two bearings for each target, giving the redundancy required.

The laser will give results accurate to about 0.1 degree and 0.1 m, which when combined will give the target's position to within about 0.5 m. With an over-determined system the accuracy may be even better than this. But it is important to distinguish between the target's position and the position of the object under the water — the source or the streamer head. This is where acoustics have a distinct advantage, because the acoustic pods can be clamped direct to the cable. Accuracy with the laser depends upon the towing arrangements:

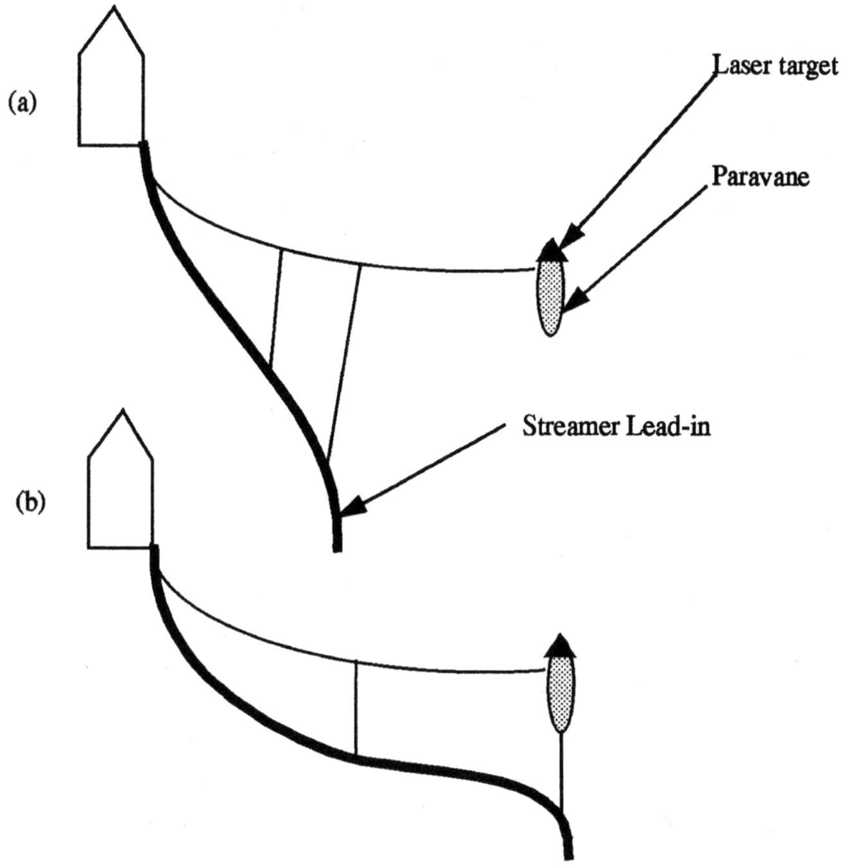

Figure 8.5 Laser targets.

In Figure 8.5(a) the streamer has been deployed from two wires supported on the paravane wire, whereas in Figure 8.5(b) the strops are attached to the back of the paravane so that the streamer leads directly astern. So if the second configuration is used, it is possible to be much more certain of where the streamer heads are. The guns present less of a problem because the gun string is suspended from the Norwegian buoys or sausage float. The only real difficulty here is deciding the correct layback for the guns.

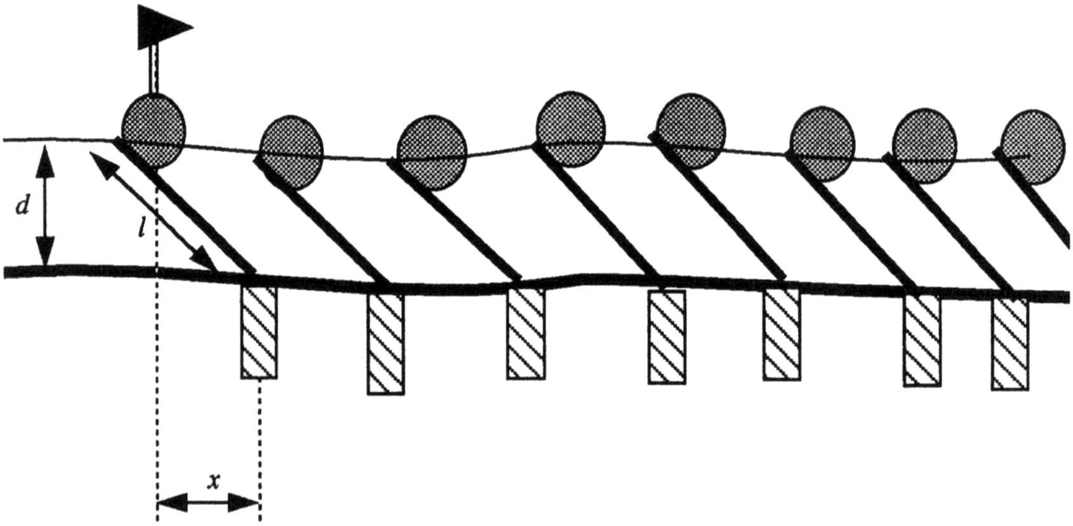

Figure 8.6 Gun string cross-section.

Figure 8.6 shows the cross-section through a gun string, with the string suspended from a line of Norwegian buoys. To find the distance x between the optical centre of the laser reflector and the centre of the first gun the following method is used. The distance d is known because we know the depth of the guns below the surface. If the length l of the towing strop is measured, the distance x is given below by:

$$x = \sqrt{l^2 - d^2} - k \qquad 8.1$$

where k is the (constant) horizontal distance between the point of attachment of the strop of the float and the target.

The weight of the guns and the constant speed of the vessel will mean this distance can vary by only a few centimetres. In fact it is felt that the laser method will give a more accurate result than acoustics for the guns, because of the difficulties in obtaining acoustic ranges through the vessel's wash.

8.2.4 Summary

The Mars laser system is now considered to be a very reliable front-end positioning method, particularly if two laser ranging units are used together. It does not suffer from re-radiation or false reflections to any great extent, and the recent introduction of a new encoder has made the system even more reliable. The only electronic equipment is on the vessel, so there is never a problem with servicing. Occasionally in very rough weather the targets may break off and need replacing, but in these cases there are nearly always other reasons to pull in the gear, such as leakage, tailbuoy malfunction, etc.

The very fact that laser has the ability to produce azimuths for the in-sea gear makes it a necessity in the author's opinion.

8.3 Acoustics

8.3.1 General

Acoustic systems have the most popular following for both front-end and down-cable positioning, and there are 2 main manufacturers — Sonardyne and Digicourse.

There are a number of different types of acoustic device:

- Transducers — these transmit an interrogation signal on one frequency and receive a reply on another frequency.
- Transponders — these receive an interrogation from a transducer and then transmit a reply on a second frequency, exactly the opposite of a transducer.
- Pingers — these transmit continually on a single frequency and are free running.
- Hydrophones — normally attached to the hull and only capable of receiving signals.
- Responders — these are triggered by a hard-wired control signal and then transmit an interrogation signal for receipt by a transducer or hydrophone.

The in-spread systems use a combination of transducers and transponders. The use of intelligent transponders allows each transponder in a network to obtain ranges from each of the other transponders in its vicinity. The ranges measured are almost invariably two-way ranges, although they are usually displayed as one-way ranges. All of the systems have some method of range gating to eliminate unwanted reflections and multipath ranges, and if a range is suspect will often output that range with a negative sign as a flag. The output formats are varied, and some (such as Western Geophysical's seismic sonar target system (SSTS)) are difficult to interpret by third parties.

The accuracy of acoustic systems depends on a number of factors. These will now be looked at together with the practical methods of reducing error, rather than embarking on a detailed description of each system.

8.3.2 Frequency

The ambient noise in the sea is generally below 5kHz, whereas that generated by ship machinery may be somewhat higher. Acoustic systems for seismic use operate in the range 10—200kHz. The higher the frequency used, the greater will be the accuracy.

But unfortunately, the higher the frequency, the shorter the maximum obtainable range because of attenuation and absorption, the smaller will be the packet of energy emitted in the first place.

Figure 8.7 plots frequency against range, showing that a frequency of 100kHz will yield a maximum range of about 500m. The diagram also shows that an accuracy of about 0.25m can be expected at a range of 350m, the sort of range over which the system will be operated. The theoretical accuracy and obtainable range are much better than shown on the graph, but it shows what is obtainable in practice.

8.3.3 Velocity of sound

Just as the propagation of radio waves is affected by temperature, pressure and relative humidity, so the propagation of sound waves in sea water is affected by temperature, pressure (density) and salinity. Pressure varies quite uniformly with depth, salinity has no great influence on the velocity but temperature varies greatly and exerts a big influence on the velocity of propagation of sound waves. Sound velocity increases with increasing temperature, and a typical variation of velocity and temperature in deep water is shown in Figure 8.8.

It can be seen that the greatest variation in temperature and velocity occur in the first 20m of depth. This gradient is even more pronounced if the sea has been calm for some time and the weather is hot, when a layer of hot water can build up on the surface. Conversely, if the sea is rough more mixing occurs in the surface layers and the gradient is less pronounced.

The systems are operating in the top 10m of water and therefore the sound velocity must be measured regularly to ensure that the ranges do not contain a systematic velocity error. Most vessels carry a velocity meter and it is a very simple task to obtain a measurement. If the vessel is shooting and cannot stop in the water, the meter can be deployed using the workboat.

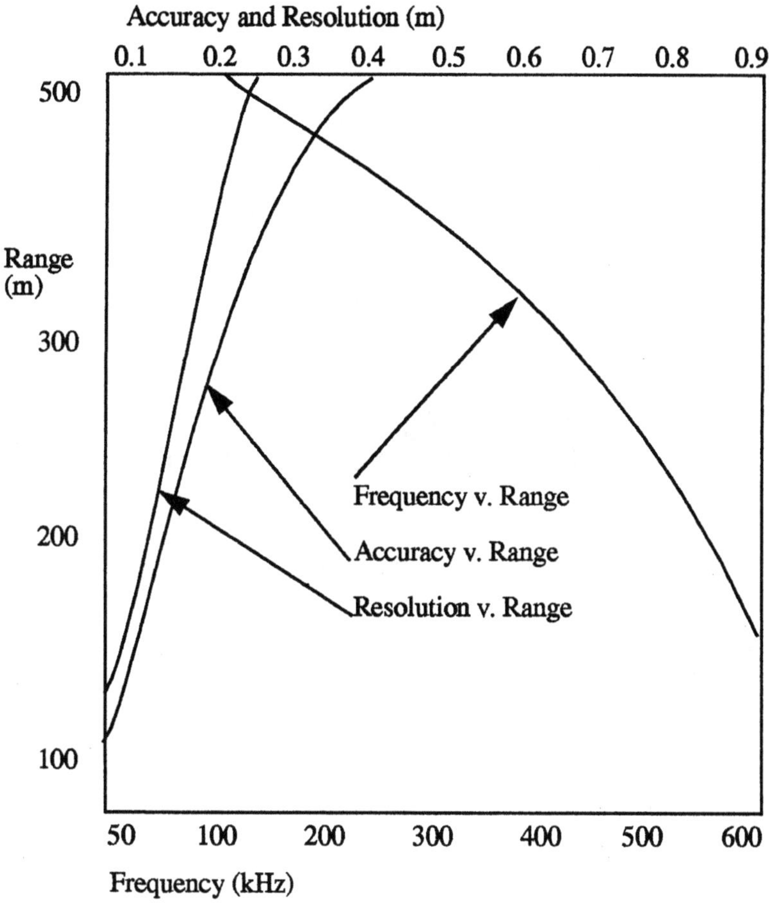

Figure 8.7 Frequency against range.

The average velocity on the North West Shelf of Western Australia is about 1520m/s. An error of 10m/s in the value of the velocity when used with a two-way range of 400m would make 2.63m difference to that range, or 1.32m to the one-way range, which is quite substantial.

The output from acoustic systems will usually be in milliseconds. Often the velocity can be entered into the control unit so that the system will output in metres, but it is recommended that the velocity is entered in the navigation software so that is can easily be traced later.

The equation to get from milliseconds (one-way range) to metres is:

$$dist = range \times velocity \times 10^{-3} \qquad (8.2)$$

For example, if the range is 62.37ms and the velocity is 1537.2m/s, the distance is 95.87m.

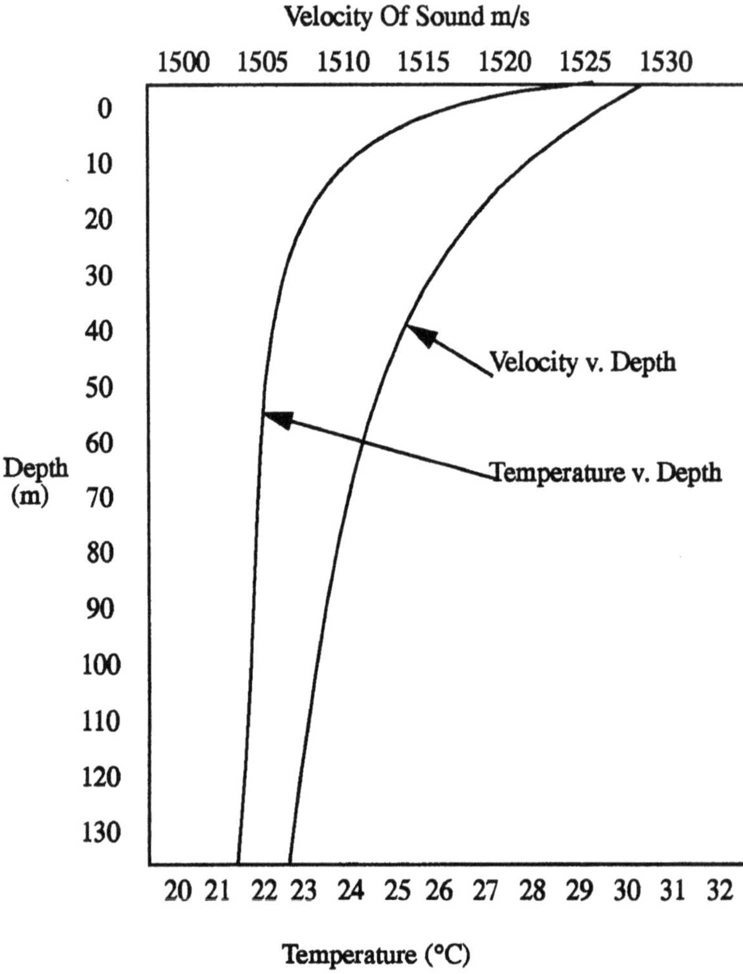

Figure 8.8 Thermocline.

8.3.4 Multipath and reflections

Like radio navigation systems, acoustic systems suffer from multipath. The problem is often noticed most in the front-end network, the area which is perhaps most important since it is used for defining for streamer head and source locations.

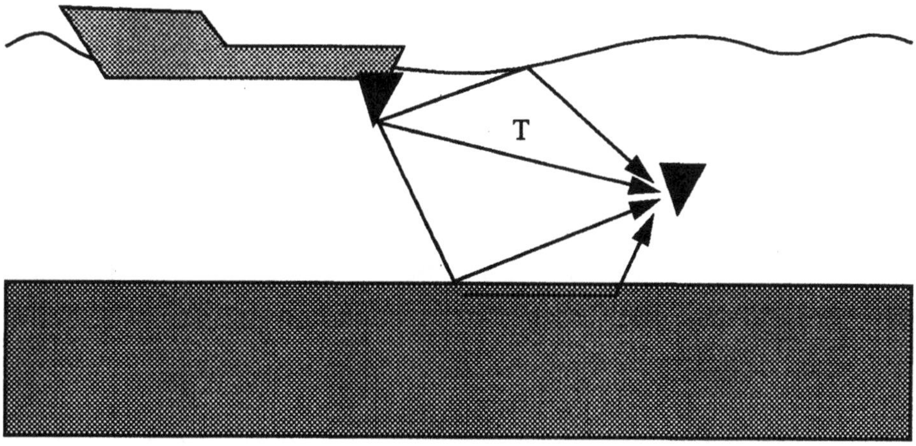

Figure 8.9 Acoustic multipath.

Figure 8.9 shows a one-way range from a hull-mounted transducer to a transponder about 200m astern of the vessel. There are three possible paths — the direct path, the surface reflection, the seabed reflection and the seabed refraction.

For the 200m range and a common unit depth of 6m, the surface reflection path is only 1m longer than the direct path. This is too small for a range gate to reject, and if turbulence is present from the screws in the area around T, in calm weather the surface reflection may well be the first ray to arrive at the transponder. If the sea surface is rough the chances are that the ray hitting the surface will be completely diffracted and absorbed.

If the screws are cavitating, or if the ship is causing aeration in the water, the direct signal may not arrive at the vessel at all. It only takes an air bubble 2mm in diameter to stop a 100kHz transmission. In this case the bottom reflection or refraction will be the first to arrive. If the water is deep then the range gate should reject this reflection, but if in the example the water depth was 30m, then the bottom reflection would only be 5.7m longer than the direct path and the range gate would need to be set to a very narrow value to reject the reflection.

8.3.5 In–line offsets

Consider the dual streamer spread in Figure 8.10. The sources have been omitted for simplicity, and acoustic ranges are observed from boom-mounted transducers to transponders on the streamer heads at X and Y. A cross-range is also observed between X and Y.

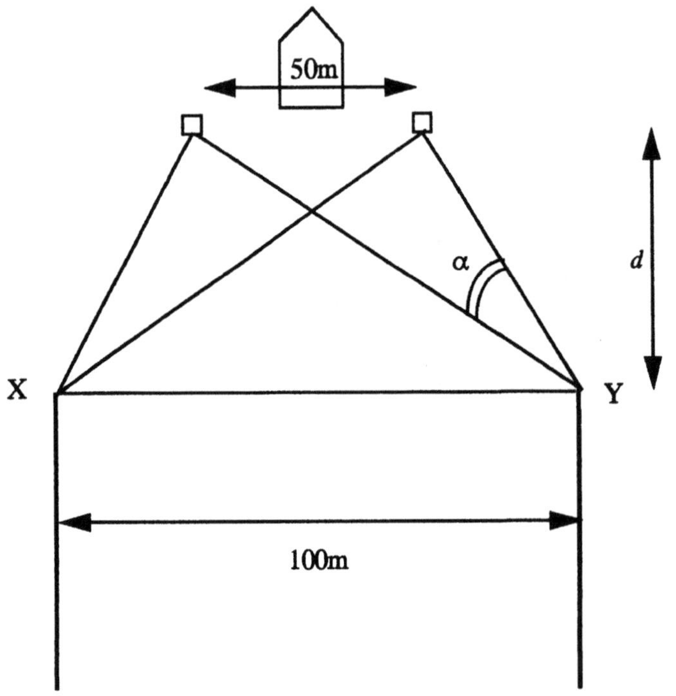

Figure 8.10 Front-end acoustic net.

It can immediately be seen that the angles subtended by the boom ranges at X and Y are very narrow. Just how narrow these angles are, and just how much the positions of X and Y can move in a cross-line direction if not constrained by the observed cross-range is shown in Table 8.1.

Table 8.1 Cross-line error

Distance from booms to X and Y	Angle α (deg)	Amount by which X and Y move cross-line if one range in error by 2m (m)
50m	30°	3m
100m	22°	5m
150m	17°	8m

Table 8.1 Cross-line error

Distance from booms to X and Y	Angle α (deg)	**Amount by which X and Y move** cross-line if one range in error by 2m (m)
200m	13°	11m
250m	11°	14m
300m	9°	18m
350m	8°	22m

The table shows that, at an in-line offset of 100m from the booms to the streamer heads, the subtended angle at the streamer head is only 22°. Also, if one of the ranges is in error by 2m, and the cross-range is not observed, the streamer head will move laterally by about 5 m. With an in-line offset of 300m, the movement grows to a staggering 18m for a range error of only 2m. Only with an in-line offset of 50m does the subtended angle satisfy the normal 30 — 150° acceptance rule.

The following points should therefore be noted:

- The inline offsets must be kept as short as possible to keep the subtended angle as large as possible and minimise cross-range error.
- Azimuth control should be introduced by using laser as well as acoustics for the front-end positioning.
- Cross-ranges in the front network are unreliable because they are trying to range through the vessel's wake.
- A higher level of redundancy is required with acoustics than is required with laser.

The last of these points, the question of redundancy, is worth pursuing. With a laser system using two LRUs there are two ranges and two bearings to each target, and therefore two levels of redundancy. The cost involved to achieve this result is for two LRUs only. There is no in-sea gear and the targets are relatively inexpensive. To achieve the same level of redundancy with acoustics in the example it would take two boom transducers, two hull transducers to define the booms' baseline and four streamer transponders to ensure that there were a reasonable number of cross-ranges. This is eight units altogether, all of which are in the water and expensive.

It is recommended that the correct mix of systems would be laser at the front-end and a ladder of acoustic units staggered at 500m intervals all the way along the streamer.

8.3.6 Calibration

There are several ways of calibrating acoustic systems. Because the ranges are two-way ranges, there will be a delay introduced at the far unit as the signal is received and then re-transmitted. Acoustic devices must be calibrated to find this delay, and the sound velocity must also be measured. Sound velocity measurement has already been discussed in section 8.8.3. The system can be calibrated in harbour as follows.

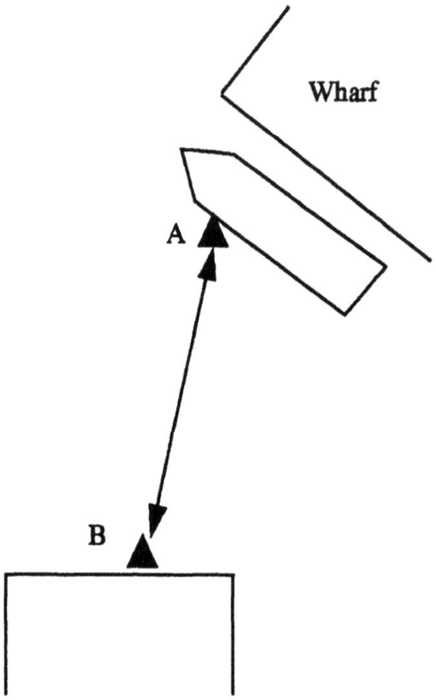

Figure 8.11 Acoustic calibration.

In Figure 8.11 the vessel is moored tightly to the wharf and transducer A is suspended over the outboard side of the hull. Transponder B is suspended over the side of the wharf at a distance of about 100m from A. The distance between A and B can be measured using EDM. The method assumes adequate water depth, no current, and little or no harbour traffic. The range AB is measured about 20 times and the mean compared with the known distance. A simpler variation is to suspend the units over the bow and stern of the vessel and measure the distance between them. If there is not enough water in the harbour, or it is found that the method is impossible because of reflections from the wharves, the vessel can proceed to anchor in deeper water, perform another sound velocity measurement and suspend the units from their booms. In any of these cases, the misclosure should be repeated for all transducers and transponders.

Once underway and on the prospect, the velocity should be measured again, and thence weekly during operations. An underway check can be performed in conjunction with the tailbuoy calibration outlined in section 10.5.2.

8.4 Compasses

8.4.1 General

The two main types of digital compass units used in seismic surveying are Syntron and Digicourse. Compasses are housed in watertight tubes clamped to the cable, and make use of the standard cable coils to send signals back to the vessel. They are battery driven, battery life being from 2 to 10 weeks, depending on type.

The cable clamp arrangement allows the compass to rotate freely around its longitudinal axis only. Most compass units are now combined compass and depth birds, and are fitted with wings to drive the cable up and down in the water to maintain depth.

Although operation of the main types of compass is substantially different, their output is almost identical in the form of a binary stream of unit identifiers, azimuths and depths. The compasses are magnetic compasses, and a few definitions should be observed before going further.

The earth can be visualized as a huge bar magnet, with the north magnetic pole currently near 76°N, 100°W and the south magnetic pole near 66°S,139°E. The positions of the poles change gradually, and erratically. Studies of mid-oceanic ridges show that the poles have actually reversed several times during the earth's life.

The relationship between magnetic north and true north is shown below:

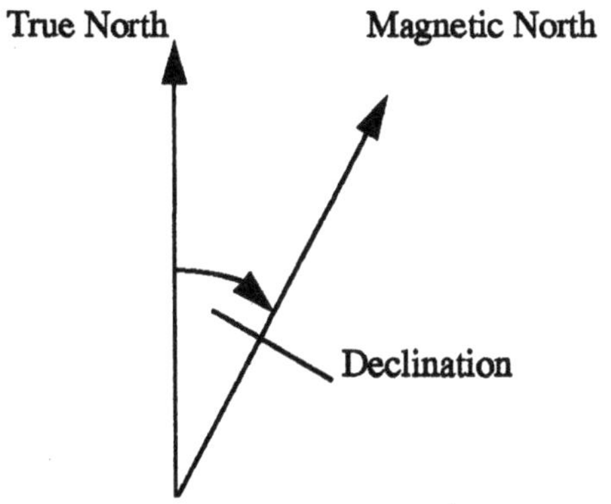

Figure 8.12 Declination.

The difference between true and magnetic north is called ***declination*** or ***variation.***

Declination is positive if the magnetic meridian is east of true north, which is the case in Figure 8.12. Declination around Australia varies from about 15 ° W on the south west coast to about 12 °E on the east coast. Its rate of change varies but is generally a few minutes of arc annually.

The mnemonic to get from magnetic to true azimuths is the CADET rule.

Compass to True ADd East.

The earth's magnetic field can be divided into horizontal and vertical components. As the compasses are constrained to lie roughly horizontal, they are acted on only by the horizontal field, which is strongest at the equator and zero at the magnetic poles. Consequently working with magnetic compasses on streamers in the Antarctic Territory would be rather difficult.

The value of declination has a considerable effect on our streamer shaping.

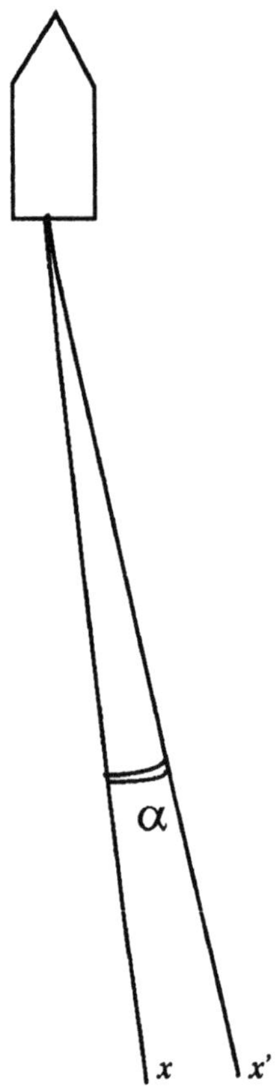

Figure 8.13 Declination error.

In Figure 8.13 the effect of an error in the declination is to rotate the streamer by the angle α. If the streamer is 4000m long, an error of 1 degree will move the far end of the streamer by 70m cross-line. Of course, if there is absolute positioning of some sort on the tailbuoy, it will be realized very quickly that something is wrong, but if there is no live tailbuoy or if the tailbuoy positioning is down, the error will be hard to detect.

When a magnetic compass is installed in a vessel, it experiences changes in the magnetic field due to the magnetic signature of the ship. The effect is known as ***deviation.*** This is counteracted by housing the compass in a binnacle and using soft iron spheres and rods to apply corrections for different ship headings. There may be a similar effect when placing the compasses on a cable. Furthermore, the effect may be different when the cable is run in different directions relative to the earth's field.

The compasses are also subject to errors induced by magnetic storms, which are connected with periods of high sunspot activity, The effect of this activity serves to show that although most of the earth's magnetic field is due to internal effects (convective currents in the outer core acting as giant coils), a substantial fraction of the total field is also due to external forces.

8.4.2 Calibration

Digital compasses need calibrating to check their functionality. The calibration should be carried out ashore in an area of low total field gradient. On the NW shelf of Australia the total magnetic field as measured with a magnetometer is about 53000 gammas. A baseline about 100m long should be marked out with pegs, and the total field strength measured every 2m along the baseline. The total variation along the baseline should not exceed 10 gammas. The field strength is also measured every 2m on a grid extending 4m around each peg, as shown in Figure 8.14.

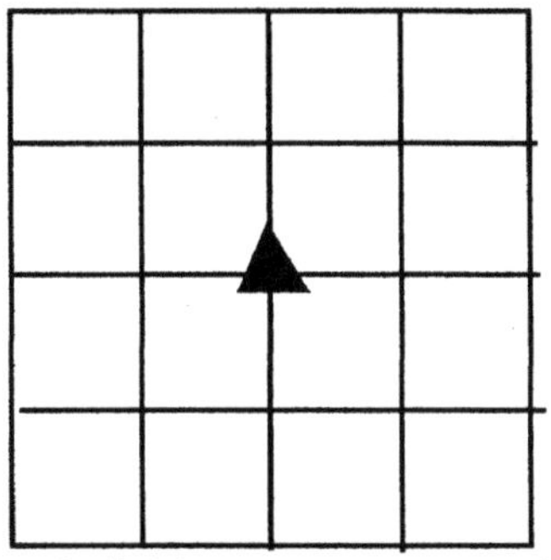

Total magnetic field measurements made at the intersections of the grid

Figure 8.14 Compass calibration.

The ground also needs to be level. It is interesting to note that Dampier and Karratha in Western Australia are entirely unsuitable as compass calibration sites because of the large ore content of the rock. The total magnetic field across Dampier football pitch changes by about 100 gammas every 10m! The nearest suitable venue to Dampier is Onslow.

Once a baseline has been established, one peg is occupied by a theodolite and the true azimuth of the baseline is established using observations of the sun or a star.

A flux-gate compass is mounted on the theodolite telescope and is calibrated by 'nulling' the compass in the east and west directions and determining the error for a single face pointing. With the compass mounted on its telescope the theodolite is able to read magnetic azimuths accurate to few seconds of arc. Meanwhile a jig for mounting the digital compasses is set up over the other end of the baseline. The jig consists of a cradle in which each compass bird can be clamped. It is able to be rolled and pitched through about 40 degrees. The compass's streamer mounting pins are aligned with the theodolite's vertical hair, after which the horizontal circle is read on the theodolite. This reading is used as the reference azimuth for the compass.

A coil is clamped to the compass's own coil with a rubber band so that the compass's readings can be down-loaded to a PC. The compasses themselves can be programmed with a smoothing function, which is usually set to about 10 seconds when deployed on a cable to help counteract wave action. The smoothing should be set to zero for the calibration

The compass is rotated through 180 degrees about 20 times, each time recording the compass value and ensuring that the cradle remains firmly clamped. Then the compasses are rolled and pitched up to about 20 degrees to simulate the environment in which they will be used, each time reading the compass values. In fact a pitch of about 5 degrees is probably adequate — if a compass were actually inclined at 5 degrees in the water, the vertical separation between adjacent compasses spaces 300m apart would be 26m! The roll parameters are more important because the compasses are free to rotate about the streamer axis.

The sets of forward and reverse readings are meaned, their standard deviations computed, and assessed by comparison against the flux-gate reading. The compasses are generally failed if there is more than 0.5 degree difference with the flux-gate, or if the compass fails when rolled and pitched.

This test can only check the general functionality of the compass. It does not test the compass in a dynamic environment, nor does it test the smoothing function. This must be checked in the field during the project.

The best way of checking the compasses in the field is to find the misclosure in the cross-cable direction when comparing compass-derived tailbuoy positions with those derived from the absolute tailbuoy positioning system (i.e. Syledis, DGPS, etc.). This is essentially the same procedure as that described in section 10.5.2.

8.4.3 Operational considerations

The following points should be noted when operating digital compasses on the streamers:

(a) If using multiple streamers, it is sometimes possible for the navigation computer to mix up the streamers, or get the compass order wrong. Some contractors label the streamers starboard to port, others port to starboard. Similarly, some contractors label the first compass at the front of streamer 1 and others label the first compass at the back of streamer 3. The only way to check is to compare the data on the compass controller itself against the data displayed by the navigation computer.

(b) Compasses should be spaced evenly down the cable, about 300m apart.

(c) If the cable operating depth is less than about 8m, the compasses will respond quite severely to swell. With the cables operated at 6m changes in heading of up to 10 degrees per shot have been noted in a single compass in a swell of 2m.

(d) As the survey progresses, the performances of each compass should be compiled into two spreadsheets, one for lines running in the primary direction, and one for reciprocal lines. Any changes in the magnetic declination (or other systematic error) will then be picked up in the mean residuals for each compass.

(e) It is recommended that the following parameters be computed for each compass down each line:

- Mean raw value
- Mean residual
- Standard deviation of the mean residual
- Delta standard deviation of the raw values.

(f) The following parameters can be used to reject a line based on compass performance:

- If the distance between valid compasses is greater than 600m
- If either of the front two compasses are out of specification or dead
- If the mean residual of a compass is greater than 2m per 100m between compasses (equating to 1 degree per 100m).

The mean value of a compass differing from the mean of all compasses by a certain amount is not thought to be a valid criterion, since the streamer may be permanently curved for a whole line.

Of course, if an integrated navigation system is used in conjunction with a Kalman filter, the filter will be able reject bad compasses automatically.

9 High-resolution survey

9.1 Introduction

High-resolution surveys are carried out prior to a rig moving on to location and spudding a new well. Such surveys are required for insurance purposes, to locate pockets of gas close to the surface which might cause a blow-out before the casing has been set and the blow-out preventer installed. There is also a requirement to check the type of seabed on which the rig's drilling template is going to be installed, the seabed slope and any obstructions which might be a hazard to mooring the rig on that location or to the drilling operation itself. In some cases current meters and wave buoys may also be installed, together with an automatic tide gauge.

The survey requirements are somewhat different to those required for standard deep 2D and 3D operations, in that the positioning accuracy must be high and repeatable, the depth of water must be known accurately to ascertain the seabed slope, the range of the tide must be known and the strength of tidal streams and currents must be measured.

It would be true to say that around the Australian coast many site surveys have been undertaken with very few technical specifications set in place, and also with no digital work involved. They have tended to be a grid of lines run at 100m or 200m intervals over a 3km square grid using an echo-sounder, side scan sonar and perhaps a pinger.

9.2 Tides

9.2.1 Movement of the sun and moon

For tidal theory we make the earth the centre of the universe. The sun appears to travel eastwards along the ecliptic, which is inclined to the celestial equator at an angle of $23°27'$, as shown in Figure 9.1.

The vernal equinox is also known as the First Point of Aries, and is the point from which Right Ascensions of heavenly bodies are measured in one of the many coordinate systems available. The autumnal equinox is also called the First Point of Libra. The sun reaches its maximum declination of $23°27'$ at the summer and winter solstices (note that spring, summer, autumn and winter in this diagram refer to the northern hemisphere seasons, which is the usual convention).

The ecliptic itself is not circular, but elliptical. The **furthest** point of the ecliptic from the sun is called **aphelion**, and the **nearest** point **perihelion.**

The equinoxes are not actually stationary; they precess westwards in the direction of the thick arrow at a rate of one complete revolution every 26000 years.

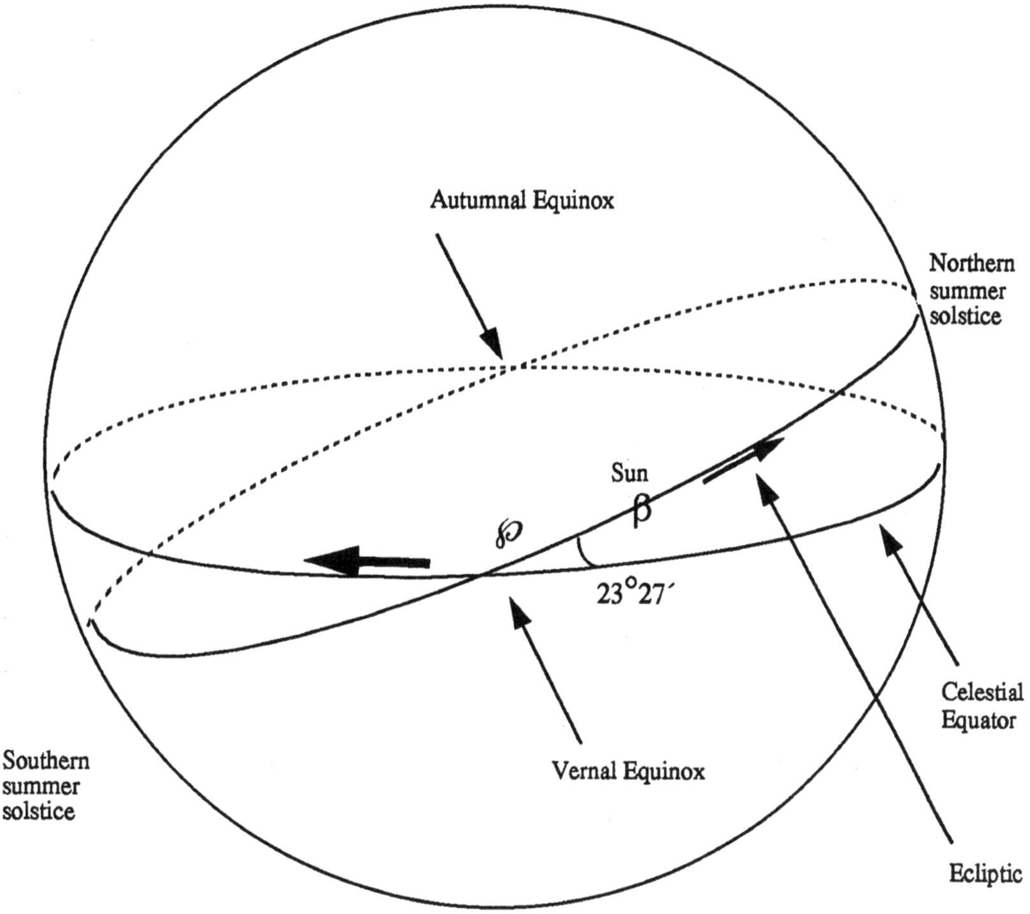

Figure 9.1 Apparent motion of sun.

The moon also moves eastwards around the earth (apparently). Its orbit is inclined at 5°09′ to the plane of the ecliptic (i.e. about 28°36′ to the celestial equator).

The moon crosses the ecliptic at **ascending and descending nodes**. The length of its orbit between consecutive ascending nodes is 27.2122 mean solar days. The nodes regress westwards with a period of 18.61 years. The apparent motion of the moon is seen in Figure 9.2.

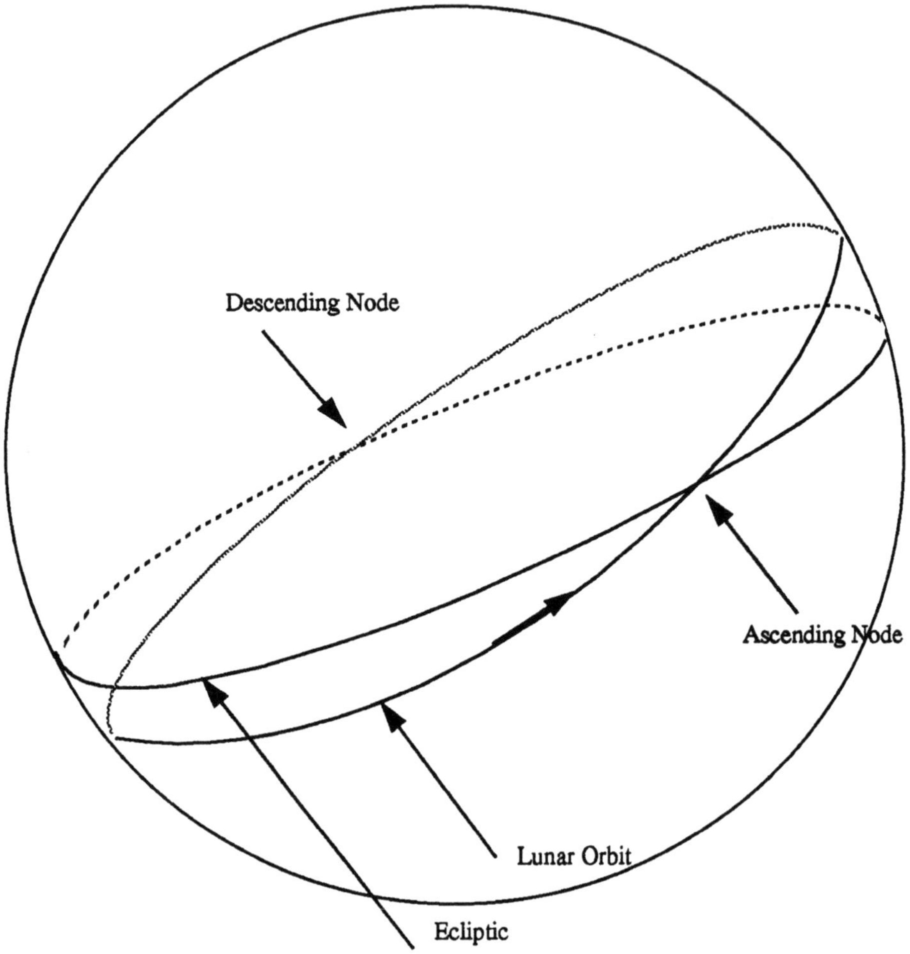

Figure 9.2 Apparent motion of moon.

The orbit is again elliptical, and the moon's **furthest** point is called **apogee** and the **nearest** point **perigee.**

9.2.2 Tide-generating forces

The tide-raising force is explained by considering the following diagram:

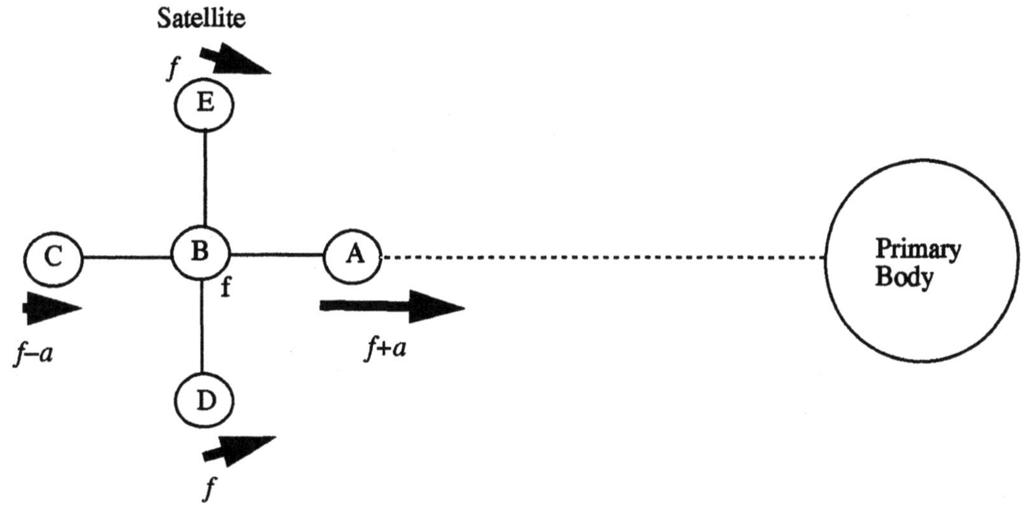

Figure 9.3 Tide-generating forces.

Figure 9.3 shows a satellite in orbit around a primary body. The satellite is represented in five different positions A, B, C, D and E. The attractive force between the satellite and the primary body will be greatest when the satellite is at A, least when at C, and more or less the same at B, D and E. If the earth is thought of as being completely covered in water, Figure 9.4 can be interpreted as follows:

The water tends to be pulled out towards the moon, creating a band of low water between the north and south poles. The explanation for there being another high water on the side away from the moon is that the earth is being pulled towards the moon more than the water on that side.

The tide-generating forces are due to both the sun and the moon, the ratio being approximately 5:11 in favour of the moon.

When the moon and the sun are in line there are higher high tides and lower low tides, a situation known as **springs**. When the sun and the moon are at right angles to each other the forces are less the tides are called **neaps.**

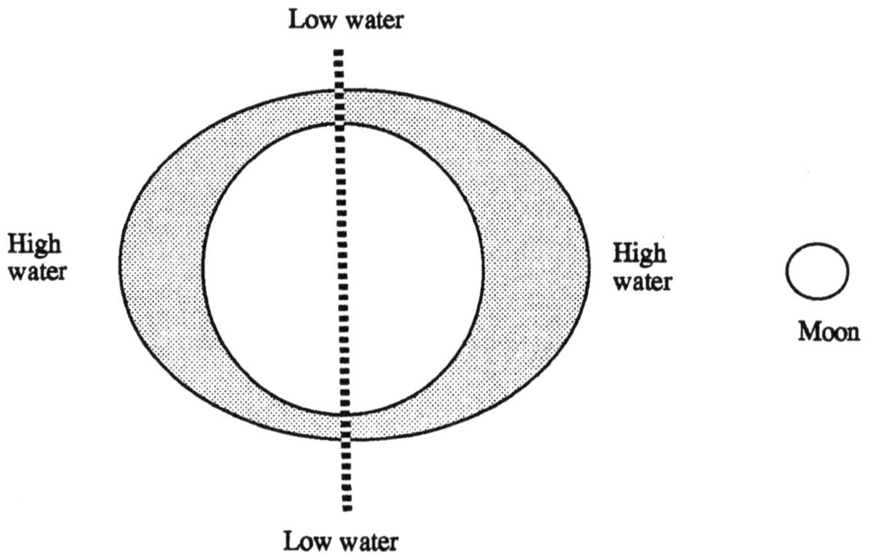

Figure 9.4 Tidal cycle.

Obviously the sun and moon will be in line when the moon is full and when it is new, and the most extreme of these cases is at new moon when the moon and the sun are both on the same side of the earth. The biggest spring tides occur when there is a new moon coincident with a vernal or autumnal equinox.

In Figure 9.5 high waters are raised at X and Y, and a band of low water encircles the earth between A and A'. A point along the parallel of latitude AB will have one HW per day - a **diurnal tide.** A point on the equator will have two HWs per day — a **semi-diurnal tide**. Melbourne will have a semi-diurnal tide with diurnal inequalities.

The various combinations of diurnal and semi-diurnal tides are shown in Figure 9.6. At this stage it should be mentioned that the above scenario is what would happen in a world completely surrounded by water. In practice this is not quite the case. To start with, the spring tides do not occur right on new and full moon, they lag behind these events by about 2 days. Secondly, the position of land masses causes great fluctuations in the tidal cycle.

A good case is Bass Strait between Tasmania and Australia, where the main tidal wave progresses from east to west. A secondary tidal wave goes into the strait from the western end, having been around Tasmania.

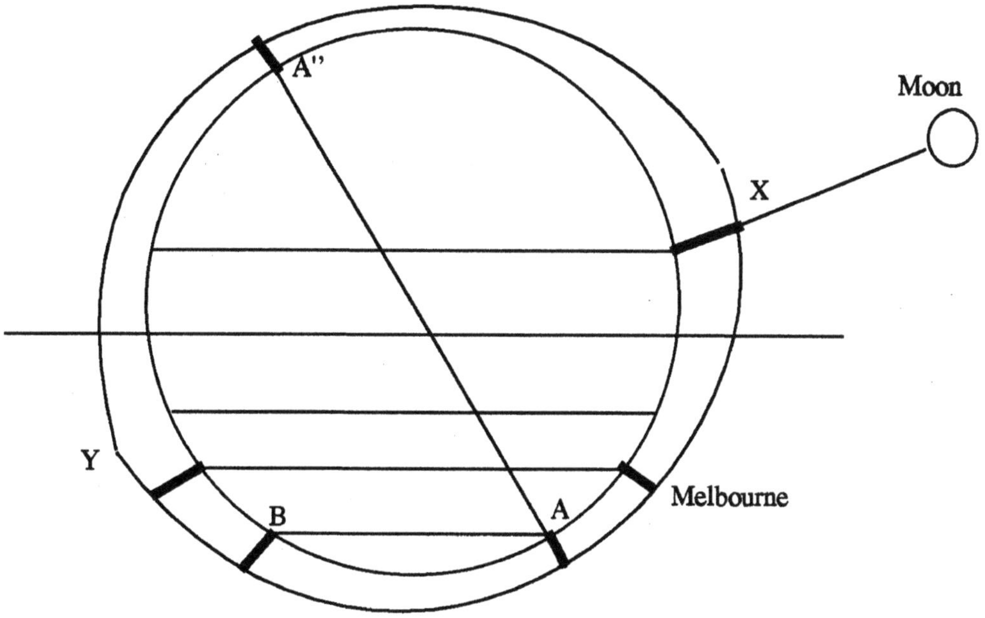

Figure 9.5 Tidal bands.

At Georgetown on the north coast of Tasmania the tidal range is sometimes greater at neaps than it is at springs, because of the unusual flow.

Where there are funnel-shaped estuaries or archipelagos of islands the tidal wave becomes compressed, gains height and speeds up. In the Severn estuary in the UK the tide rises so quickly and to such a height that it creates a bore — a wall of water several feet high on which it is possible to surf some way up the river. On the NW coast of Australia at Derby the tidal range at springs is about 10.5m, whereas at Perth it is only 0.6m. Perth is on a straight coastline whereas Derby lies at the head of King Sound.

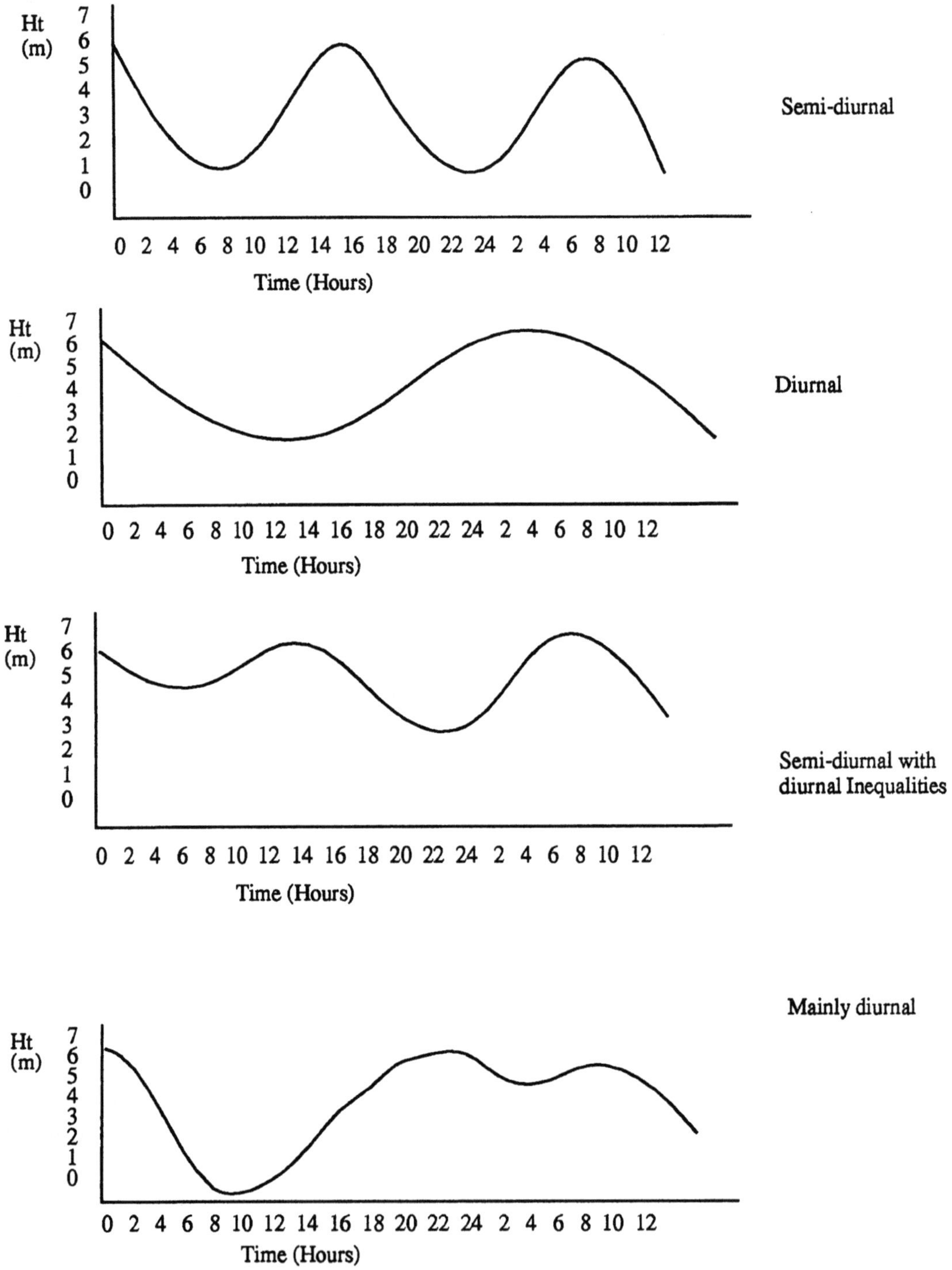

Figure 9.6 Diurnal and semi-diurnal tides.

9.2.3 Datums

One of the most important aspects of reducing observed depths for tide is to establish the sounding datum. The datum is the level above or below which all depths are recorded on the chart, and is not usually the same as the land-based datum. The most usual datum is the lowest astronomical tide (LAT), but there are others which can be used. LAT is used because it is the lowest level to which the tide is predicted to fall at a particular place, and if all soundings on a chart are made relative to this level the mariner will nearly always be on the 'safe' side as the height of tide at any time will always be some distance above LAT. Soundings on the chart will then always be deeper in reality because of the height of tide. Figure 9.7 illustrates the relationship of the various levels.

In areas such as Bass Strait the tides are quite well known and can be predicted accurately. The same is true of the North West Shelf area around North Rankin 'A' platform, for which tide tables are published. These predictions are made by observing the tide for a period of at least one lunar cycle (29 days). The observations can then be subjected to harmonic analysis. Any wave form which repeats itself can be represented by the sum of a number of pure sine or cosine waves whose wavelengths and frequencies bear a simple harmonic relationship to each other. Harmonic analysis is thus used to predict the tides from the month's observations.

The Australian National Tide Tables (ANTT) are published annually. The main tables contain the values of high and low waters at standard ports for every day of the year. They also contain a large number of secondary ports for which differences from the nearest standard port are given in terms of height and time at neaps and springs. These data are fine for predicting high and low water, but are not that helpful for reducing observed depths for tide. To do this the tidal height must be known at every epoch along the lines that have been surveyed, probably at 30 second intervals. Then the tidal height can be subtracted from the (time-stamped) digital echo-sounder depth to obtain the depth relative to chart datum. To enable this to be done the ANTT gives a listing of the main harmonic constants, seasonal corrections and shallow water corrections and a simple computer program listing which enables us to predict the height at any time.

So the ANTT give us good, reliable information for most of Australia's coastline. What about surveys hundreds of kilometres offshore? And what about tidal streams and currents? If there is a significant vertical range of tide, there will also be a correspondingly large horizontal flow associated with that range, and the strength of the tidal stream will be important to take into account when drilling. It may be necessary to deploy an automatic tide gauge and current meters in the area prior to the survey taking place.

Figure 9.8shows a string of current meters deployed. Automatic tide gauges can be deployed in the same way — they rest on the seabed and are usually pressure sensors. It is very important to plan the survey such that the current meters and tide gauges can be recovered prior to the start of the survey — otherwise equipment is bound to get snagged on them.

It is also possible to get direct-reading current meters and tide gauges and have the data telemetered directly to the vessel in real-time. In this case the soundings can be reduced for tide online to an arbitrary datum, and the guage must be located well outside the survey ground.

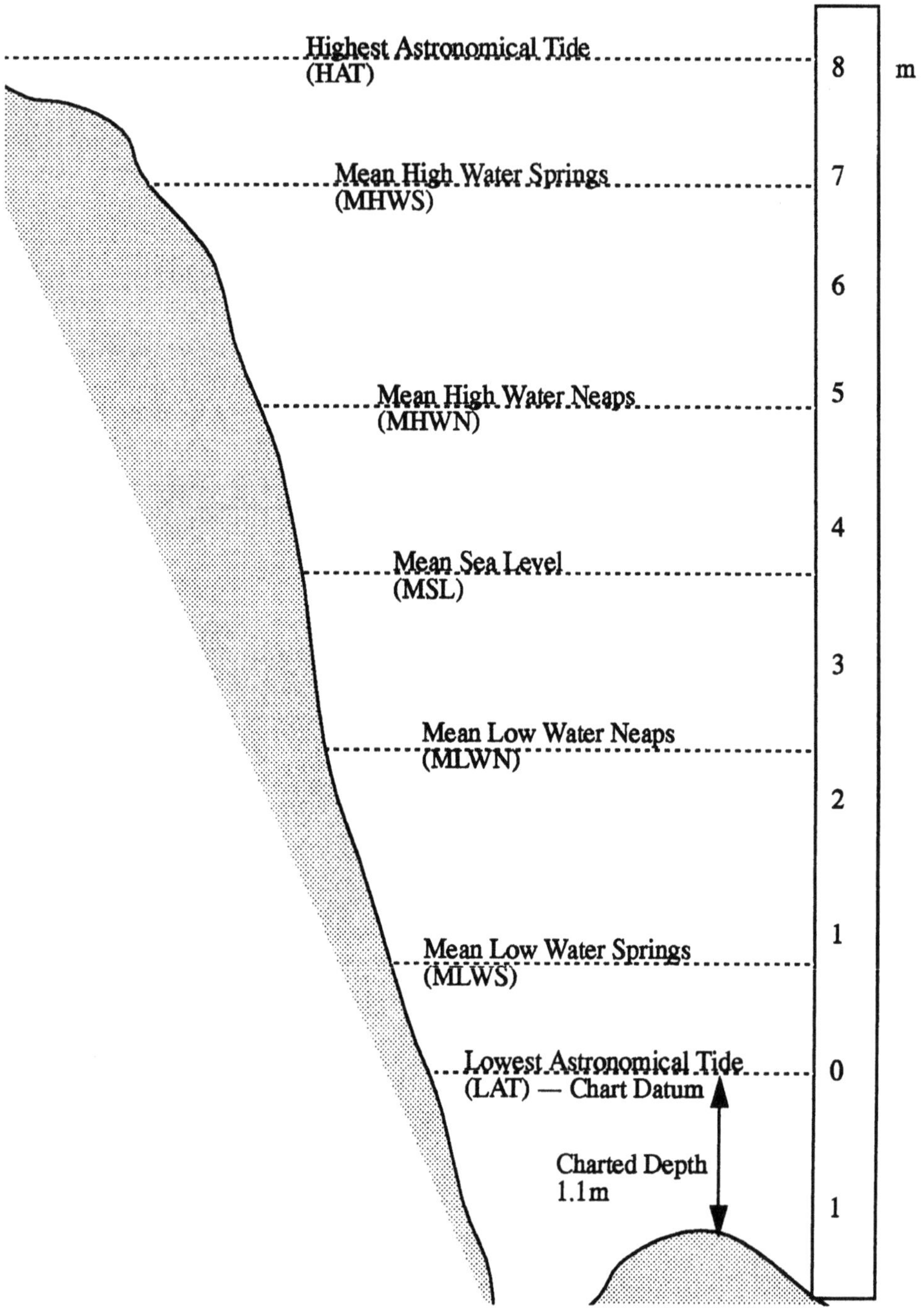

Figure 9.7 Tidal relationships.

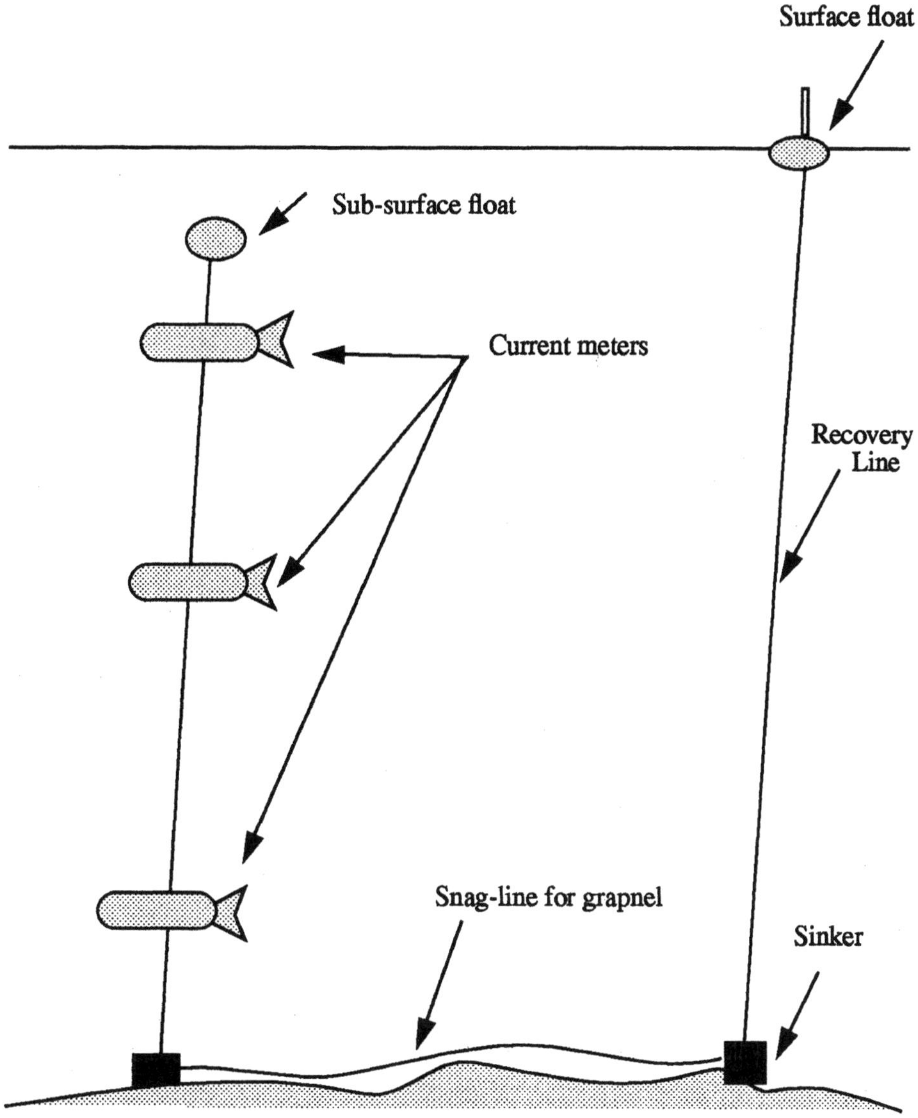

Figure 9.8 Current meter deployment.

9.3 Echo-sounders and pingers

9.3.1 Echo-sounders

It is beyond the scope of this book to go into the theory of operation of echo-sounders. Suffice it to say that all echo-sounders consist of the following fundamental components:

(a) A time base, controlling the emission of the outgoing pulse, and measuring the time taken for the pulse to return after reflection from an object.

(b) A power source to activate the transducers.

(c) Transmitting and receiving transducers (usually a single transducer transmits and receives).

(d) An amplifier to boost the weak return signal.

(e) A means of recording the arrival time of the return pulse and its strength.

The resolution of echo-sounders depends upon their operating frequency and pulse length. The Krupp Atlas Deso 20 sounder has been in use for many years and is probably the most popular survey echo-sounder. It has two transducers operating on separate frequencies of 30kHz and 210kHz. The resolution of the higher frequency is about 0.01m. The 210kHz beam has a 9deg beamwidth and the 30kHz a 16 deg beamwidth.

9.3.2 Calibration

To calibrate an echo-sounder a bar check should be carried out. The bar itself is made of a rigid material such as angle-iron or tubular aluminium, and should be made long enough to pass under the hull at the transducer's location as shown in Figure 9.9.

The bar's lowering lines are made of wire rope or chain so that they do not stretch, and marked off at intervals corresponding to 5m vertical depth intervals, taking into account the difference between the bar's length and the distance between the lowering points on deck.

The bar is lowered to the shallowest 5m mark that will fit under the transducer, and the transmission mark on the echo sounder is set so that the correct depth of the bar is read out. The bar is then lowered to its fullest extent (usually 35m will suffice) and the echo sounder's motor speed is adjusted to read the exact depth again. The bar is then raised at 5m intervals, all the while checking that the correct depths are being read. The check at the shallow depth sets the transmission mark, and the deeper ones adjust for the velocity of sound.

On the Deso 20 a 1m long test transducer is supplied which can perform the check. It consists of a small transducer set exactly 1m from a reflecting metal plate. The instrument is lowered through the water to various depths, checking at each that the reading of 1m is maintained.

A velocity meter should be used to check that the speed of sound derived from the bar check is correct. The bar check should ideally be carried out on site, but often the swell makes the task very difficult. It is usually more practical to do the transmission check in harbour and then use the velocity probe on site to determine the sound velocity.

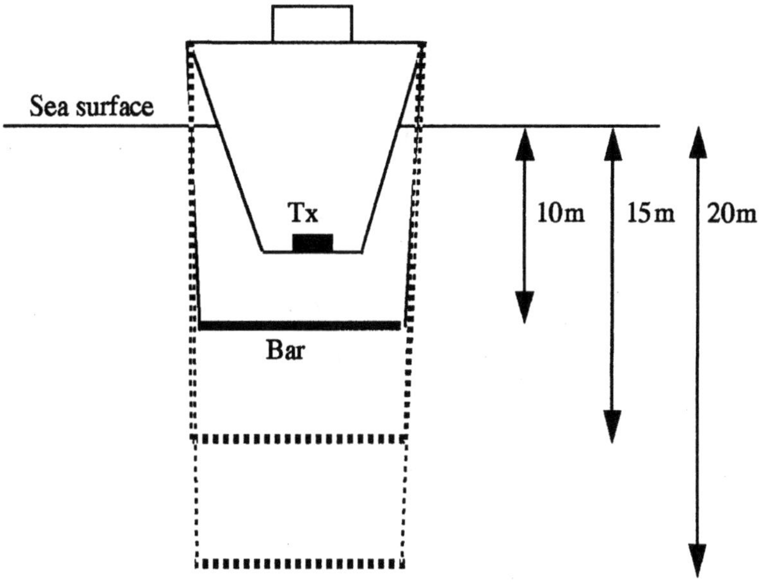

Figure 9.9 Bar check

When observations are carried out with the echo sounder, the digitized depths should be compared with those on the echo trace at regular intervals to check their validity.

The results obtained using echo-sounders should be viewed with care. In shallow water (up to about 15m deep), the sounder will generally give the shallowest depths observed, but this is sometimes not the case in deep water.

When the water is deep it can be seen that the footprint of the echo-sounder expands, and that an obstruction on the edge of the footprint will not be reported as its true depth. In very deep water (over about 400m deep) the echo trace becomes a series of hyperbolic echos from which it is very difficult to pick out the true depths of any objects on the seabed.

This is why trying to find a wellhead with just an echosounder in deep water is almost impossible unless one runs a grid of lines about 10m apart so that the wellhead will eventually appear in the centre of a footprint. Another way of achieving the same object would be to use a multi-transducer or Fansweep echo-sounder.

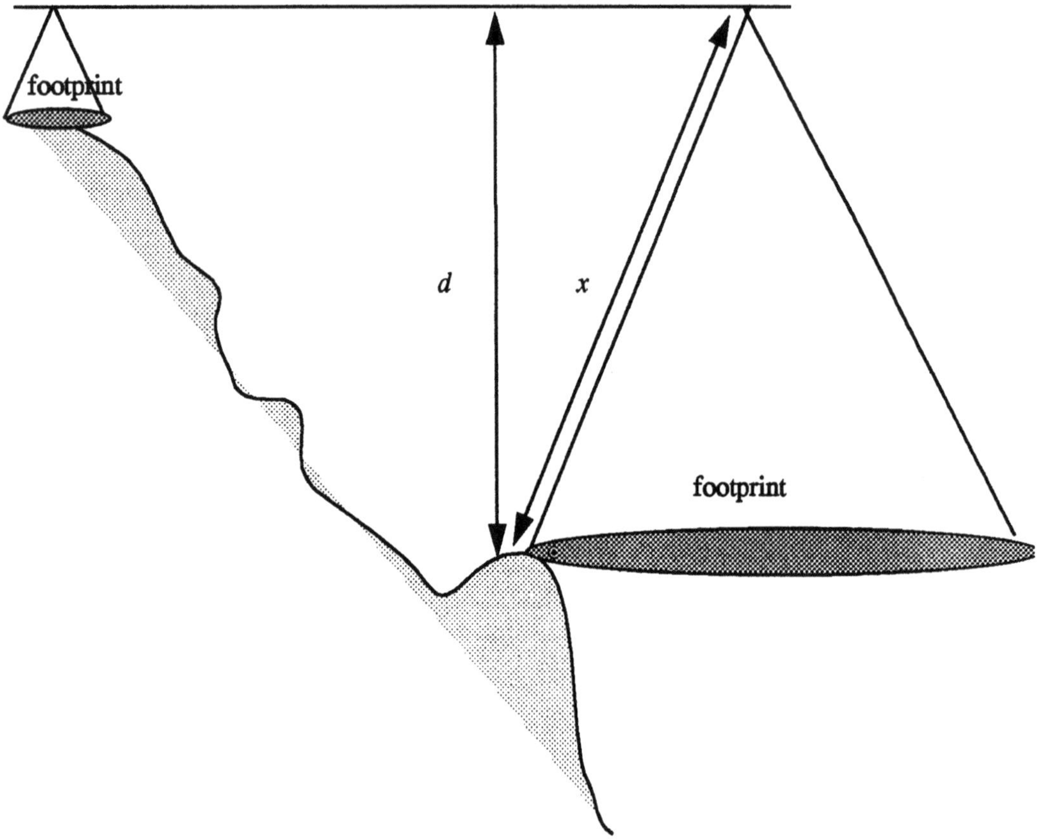

Figure 9.10 Echo-sounder footprints.

The use of a dual-frequency sounder is highly recommended, as the lower frequency can often penetrate a soft seabed to a depth of a metre or more, giving the interpreter a good idea of the hardness of the bottom and of any silted areas.

Specifications for echo-sounders should meet the following minimum requirements:

Depth — 0 to the maximum depth on site plus 100m.

Units — metres.

Frequency — 33kHz and 210kHz (or similar).

Frequency recording — simultaneous.

Paper speed — adjustable.

Speed of sound setting — adjustable and displayed on the record, checked and logged daily.

Transducer draught setting — adjustable, checked and logged daily.

Event mark — manual and external input.

Output — analogue paper record and digital output to survey computer.

9.3.3 Pingers

Pingers are high-resolution sub-bottom profilers.

Applications for pingers include the determination of surface sediment thickness and the location of buried pipelines. Their transducers are typically piezoelectric, operating at frequencies of 3 — 10kHz and mounted in a surface-towed fish or mounted in a bracket over the side of the vessel. The records obtained are analogue profiles and they will penetrate sandy sediment to a depth of a few metres only.

An instrument in common use is the ORE model 3000.

Pingers should provide the following requirements:

Frequency selection — low frequency (3 — 7kHz) and medium frequency (11 — 20kHz).

Pulse length — 0.1 — 0.5ms, adjustable.

Output power — 2 — 10kW, adjustable.

Repetition rate — 1 — 100pps, adjustable.

Graphic recorder — high resolution digital, thermal paper with the following capabilities:

Sweep speeds up to 1/32s, adjustable.

Ability to display up to 2 channels (i.e.different frequencies).

Provide scale lines.

variable chart speed (100 — 200lpi).

TVG control — time variable gain.

Calibration of pingers is carried out by monitoring the pinger pulse in port with the source at shallow depth and with a suitable calibrated hydrophone and storage oscilloscope at a distance of 2m below the centre of the transducer. The pinger source should be shown to produce a stable outgoing signal meeting the clients specifications. These specifications will include the pulse length (usually less than 1.5ms for the low frequencies and 1.0ms for the high frequency) and the peak-to-peak amplitude of the pulse train.

9.4 Sonar

In the previous section we discussed the echo-sounder and its limitations. Another tool used in high-resolution surveys for defining the seabed is the sidescan sonar. The sidescan sonar is basically a pair of sideways-looking echo-sounders mounted in a single towed body, and they have been in used since the late 1960s. The latest instruments can store data digitally and produce a rectified image of the seabed taking account of the vessel's position, course and speed over the ground.

The sidescan sonar is towed astern or abeam of the vessel, and may be equipped with a short baseline or ultrashort baseline acoustic positioning system. The sonar needs to be towed about 10 — 15m above the seabed to produce a reasonable image, and a 100kHz system will look sideways from the towed body to a range of about 200m. This will mean that lines will need to be run no less than 150m apart to ensure that adjacent lines are completely insonified.

The sidescan sonar record is essentially qualitative, although it is possible to make some measurements from the trace to a low standard of accuracy. It is possible to measure the height of objects above the surrounding seabed by measuring the length of the 'shadow' behind the object, although if the object is directly below the 'fish' there will be no shadow. This is why adjacent footprints need to overlap to give a complete picture of the seabed.

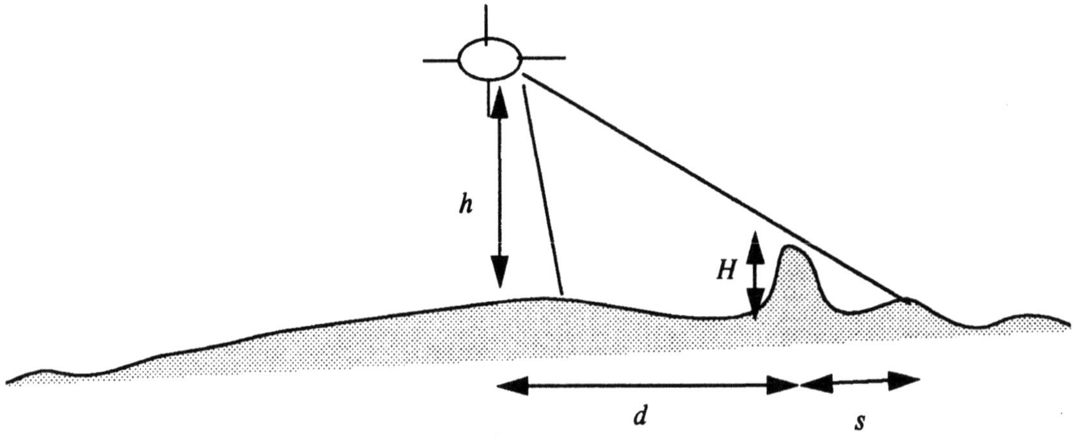

Figure 9.11 Sidescan sonar equations.

In Figure 9.11 an obstruction of height H is picked up on the trace. The area behind the obstruction does not receive any of the sound wave and is therefore represented by an echo-less shadow of length s. The distances s, d and h can be measured from the trace and so the height H can be determined by similar triangles.

The positioning of the towfish is all important in this type of operation. Usually there is no form of positioning sensor on the towfish, and it will need to be streamed to a distance of about 500m astern of the vessel in a depth of 150m. All the surveyor can do to position the towfish in these circumstances is to use the drift angle between the gyro heading and the course made good to determine feather. Often this does not work well because the currents near the seabed are completely different from those at the surface. The best method is to use some form of acoustic positioning system.

The following specifications should be the minimum requirement for sidescan sonars:

Towing characteristics — to ensure that a minimal layback from the vessel is maintained, preferably without using a depressor.

Height of the towfish — the towfish height and the vessel speed need to be adjusted to ensure full coverage of the survey area.

Course alterations — kept to a minimum to avoid yawing of the towfish.

Towing cable adjustments — made gradually in order to maintain towfish height.

Line spacing and range setting — specified so as to provide a 100% overlap on all lines. Change in range setting needs to be accompanied by a corresponding change in firing rate or triggering.

Transducer tilt angle — variable and set to maximum scope for the range setting in use, taking into account depth of water and line spacing.

Gain settings —constantly monitored to ensure that maximum definition is maintained.

Adjacent lines — run in opposite directions to minimize layback errors on targets.

Significant contacts — should be boxed in with four lines at high firing rate and large range scale.

9.5 Digital equipment

Where digital equipment is to be used for site surveys, a short streamer (usually 600m) and single airgun string are normally defined. It is not usual to use compasses in the streamer, although perhaps this should be considered as otherwise the processing is reliant on feather angle determined by radar. Radar ranges can be very accurate, but radar bearings are notoriously inaccurate (+/– 5 degrees).

The following specifications would be typical for high resolution digital work:

Near trace offset — 20m.

Streamer length —600m.

Group interval — 6.25m.

Group length — 6.25m.

Depth controllers — 3 (head, mid, tail).

Source depth — 3m.

Gun pressure — 2000psi.

Gun volume — 350 cu inch

Streamer depth — 3m.

Channels — 96.

Move-up rate — 6.25m.

Sampling rate — 1ms.

Record length — 1.5s.

Low cut filter — 8Hz/18dB.

High cut filter — 250Hz/72dB.

Fold — 48.

Reference signal — 20 microvolts.

FTB to gun phone — 55ms.

Line spacing — 25m.

This type of survey can easily be controlled with a DFSV instrument or similar, using a small vessel.

9.6 Acoustic systems

9.6.1 General description

Some of the sensors mentioned above, and in particular the sidescan sonar and pinger, can be positioned using long-baseline acoustics. This system gained favour with site survey work during the early days of DGPS because it provided accurate positioning during the periods when satellite coverage was insufficient to continue.

Operation of the system is very simple — an array of transponders is laid on the seabed just outside the corners of the site (assuming a 2km square site — the maximum range of these systems is about 3km). The network may be left down after the site survey to provide positioning for the incoming rig, and the siting of the transponders needs to be planned so that they are some distance away from the planned position of the rig's anchors.

The transponders are equipped with acoustic releases and have floats attached to them so that when the release is operated they come back to the surface and can be recovered.

Sonardyne is the best-known manufacturer of these systems, and its equipment is well-known in the industry for its robustness and reliability.

9.6.2 Calibration

The network needs to be calibrated, and this can be achieved in a number of ways:

Box-ins.

The vertical separation between the vessel transducer and the transponders is determined from echo-sounder readings and the depth read-out from each transponder.

The position of the transponder is fixed as it is deployed, and then the vessel steams around the transponder in a circle of radius three times the water depth, observing about 50 — 100 ranges to the transponder while simultaneously fixing the ship with the primary navigation system. The vessel's position should be computed and displayed online together with the raw acoustic ranges to ensure that these requirements are met, and the transponder's position can then be computed using a simple least squares adjustment.

While the vessel is steaming these 'box-in' circles, its own transducer is usually housed in a towfish streamed alongside the vessel (not astern to avoid the wake) from a davit at about 10m depth.

Array calibration

The Sonardyne transponders (Compatts) are intelligent and can be made to measure the baseline distances between each pair. The measurements are sent to the vessel, which maintains station in the centre of the network. At least 20 measurements of each baseline should be made. These measurements are normally very reliable, having standard deviations of less than 0.5 m.

Having measured all the baselines (in both directions), the network may then be adjusted by least squares.

Dynamic calibration

The network can be calibrated by steaming a cloverleaf track around all of the transponders, simultaneously observing the vessel's position and all available ranges to all transponders. The calibration is computed by a sequential least squares method, which gradually builds up the optimal estimates of position for each transponder. This system of calibration works very well with the right software, and can be checked by boxing-in any two transponders in the method described above.

Verification

Once calibrated, the integrity of the system can be verified by the vessel steaming two tracks through the centre of the site and at right angles to each other.

10 Calibration

10.1 Introduction

10.1.1 General

The importance of calibration cannot be over-emphasized. There is absolutely no point in investing in expensive systems if those systems are not used to their full potential. Geophysicists need to know the accuracy of the positions they are obtaining for their acquired data, and the degree of reliability that they can expect,

All systems are subject to the influences of the propagation of radio waves, pattern geometry, ground conductivity and climatic conditions. They are also subject to errors made in offset calculation, base station coordinate errors, incorrect software parameters and incorrect datum parameters.

Most of the potential sources of positioning disasters can be averted by good calibration.

10.1.2 The objective

The objective is to set down certain procedures which will test a system meaningfully and, by minimizing or recording the effects of the environmental conditions pertaining at the time of the calibration, make an assessment of the true capability of the system itself. It must be established whether or not a system is operating to its capacity and also to the specifications laid down in the contract.

Once the initial calibration has been completed satisfactorily, it must be ensured that regular calibrations are performed throughout the survey, so that the integrity and relative accuracy of data points within that survey are maintained. This is particularly true of 3D surveys where lines are shot in blocks. Once the prime coverage in a block has been finished, it may be some weeks before lines on either end of that block are shot, and before infill within the block is acquired. Much can happen to the positioning system during that time, and it is necessary to be absolutely certain that all systems are operating to the same levels of absolute and relative accuracy and reliability.

10.1.3 Static calibration

Although methods vary greatly for individual systems, some form of static and dynamic trial can usually be conducted. The static trial tests the functioning of the main system components in the field prior to deployment. It is important that these trials be conducted by the contractor and supervised by the client. By doing this both client and contractor begin to establish a relationship with the system and its peculiarities, and will know what to watch out for during the project.

For most radio positioning systems all that is required is a measured baseline of about 2km in length, preferably over a sea path, though a land path is also suitable provided the land is flat with no obstructions, no grazing rays and not susceptible to interference from power lines, etc. A distance of 2km is too short for propagation errors to have a marked effect, and too long to risk damage to equipment or personnel through exposure to high-power transmissions.

Such a static trial will enable equipment delays to be quantified and the general functionality of the system to be tested.

10.1.4 Dynamic trials

Dynamic trials are intended to determine the accuracy of the system when installed on the vessel in the final configuration. Now the system will be used in conjunction with the contractor's and QC's navigation software, so that the datum, delay, C–O and velocity correction parameters can be checked.

It is necessary to check for repeatability and reliability in different parts of the prospect, and a secondary system must be available where positioning needs to be very accurate, as for 3D projects.

10.1.5 Systems

We will address each system individually, as the calibration techniques for each can be quite diverse.

10.2 Syledis

10.2.1 Static calibration

Static calibration for Syledis falls into two categories — short baseline and long baseline.

10.2.2 Short baseline calibration

A baseline of about 2km is set up such that the path is free from obstructions and electrical interference. Good sites are along quiet roads, along beaches or across a coastal bay. The baseline can be between known points, or temporary marks can be set up. In either case, the baseline should be measured with an infrared or optical EDM system to check its length.

Prior to calibration an inventory is made of all equipment, making a note of all serial numbers and marking cables with coloured tape. It should be determined which beacons are going to which shore stations prior to calibration, and one of the mobiles should be set up as the primary mobile with other mobiles labelled as spares. A typical set-up is shown in Table 10.1 and Table 10.2

Table 10.1 Static calibration - general

Client:	An oil company
Project:	Prospectus Novus 3D Marine Seismic 1993 - P0193
Location:	Onslow, W.A.

Table 10.1 Static calibration - general cont.

Equipment:	Syledis SB5	
Mobile Station	Onslow Local Baseline North	Ht: 15.78m (AHD)
Remote Station	Onslow Local Baseline South	Ht: 19.32m (AHD)
True Distance	2135.42m	
Date:	25 June 1993	
Weather:	Fine. Dry Bulb 28.3°C Wet Bulb 26.5°C Pressure 1017.3HPa Wind SE 18 km/h Cloud 3/8 Cu	

The equipment is warmed through for at least 20 minutes prior to observations being taken, and all internal delays and corrections are set to zero. This is important because it must be ensured that all delays and corrections are set to zero so that all delays and corrections used during the project are entered into the navigation software, not into the hardware. ***Software set-up errors are salvageable: hardware errors aren't.***

Table 10.2 Static calibration - systems

Mobile:	Prime					
Serial No.	1021					
Cable:						
Serial No.	132					
Colour	White					
Length	30m					
Antenna:	Omni					
Serial No.	432					
Colour	White					
Beacon:	Alpha	Bravo	Charlie			
Serial No.	4256	4908	6987			
Colour	Blue	Green	Red			
Cable:						
Serial No.	156	406	407			
Colour	Blue	Green	Red			
Length	15m	15m	30m			
Antenna:	Yagi	Yagi	Yagi			
Serial No.	786	739	876			
Colour	Blue	Green	Red			

Table 10.2 Static calibration systems cont.

Start Time	1047					
End Time	1058					
Obs:						
1	2298.3					
2	7.5					
3	7.9					
4	8.4					
5	9.1					
6	7.9					
7	6.9					
8	7.5					
9	7.9					
10	8.4					
11	9.1					
12	7.9					
13	6.9					
14	8.3					
15	8.4					
16	7.6					
17	7.5					
18	7.9					
19	8.4					
20	9.1					
Mean	2298.0					
SD	0.7					
Delay	162.6					

After warming through, at least 20 observations are made for each mobile — beacon pair. Each mobile and beacon is calibrated using its own antenna and cable (including the spares), and an extra calibration is made for each spare set of antennae and cables by swapping them over on at least one mobile and beacon. Only by doing this can we calculate the delays for the spare antennae and cables when used with any mobile or beacon. If this cross-calibration is not carried out and a station or mobile goes down, we have to stop the survey to recalibrate equipment. Typical results are shown in Table 10.3.

Table 10.3 Short baseline results

Parameter	Alpha	Beta	Gamma	Delta
Mobile: Prime				
Known range (m)	2671.3	2671.3	2671.3	2671.3
Raw observed range (m)	2839.1	2764.1	2913.7	2863.8

Table 10.3 Short baseline results cont.

Parameter	Alpha	Beta	Gamma	Delta
Computed delay (m)	167.8	92.8	242.4	192.5
Mobile: Secondary				
etc. ...				

The observed delays can now be checked against those from the bench calibration which will have been carried out in office by the operators. This bench calibration consists of setting up the equipment in office using attenuators and systematically measuring the delays in each item of equipment. The short-baseline values should fit the bench calibration very closely, and any marked differences need investigating.

Note that the standard deviation of the observations in Table 10.2 has been computed. Large standard deviations indicate poor reliability, and the whole observation set should be repeated.

If a suitable computer and software are available, a much larger number of observations can be taken, with an increased measure of reliability.

10.2.3 Dynamic calibration

Dynamic calibration takes the form set out in the following paragraphs.

10.2.4 Long baseline calibration

Now the instrumental delays and corrections are known, it must be determined what corrections are necessary for propagation effects. Remembering that Syledis is a UHF, ducting system, any changes in propagation velocity can be interpreted as a scale factor to be applied to the chain.

In Figure 10.1 the four beacons A. B. C and D are taken to their final locations and set up in their correct configuration ready for the survey. Care is taken here that antennae and cables are matched with the correct beacon, that directional antennae are pointed in the right direction and that the permanent power supplies are stable and charging correctly.

The mobiles are taken to a convenient coordinated station near the prospect. Very often there will be no such convenient mark: in these cases the mobiles could be taken to station A and set up there, temporarily taking beacon A to station B. Whichever route is taken it must be ensured that the ranges from the mobiles to the stations approach the ranges that will be experienced whilst surveying, and that the ray paths do not travel over differing ground conductivities.

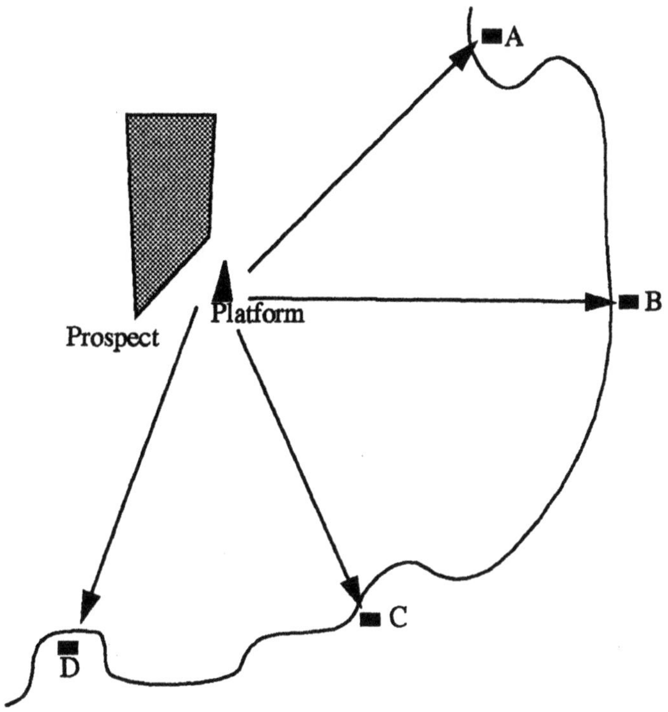

Figure 10.1 Dynamic Syledis calibration.

The mobiles are switched on and warmed through thoroughly; once again it must be ensured that all internal delays and corrections are removed. At least 20 observations are taken in the same way as for the short baseline calibration. It is very important to take meteorological measurements at about 30 minute intervals for temperature, pressure and relative humidity, so that the changing refractive index can be computed and monitored. The results will be similar to Table 10.4

Table 10.4 Long baseline results

Parameter	Alpha	Beta	Gamma	Delta
At Cal-1:				
(1) True distance (m)	67341.7	75219.4	50753.9	87326.1
(2) Raw obs. distance (m)	67513.2	75317.3	50998.4	87527.9
(3) Delay (from short baseline)	167.8	92.8	242.4	194.5
(4) Corr. obs. = (2)–(3)	67345.4	75224.5	50756.0	87333.4
(5) Misclosure = (4)–(1)	3.7	5.1	2.1	7.3

Table 10.4 Long baseline results

Parameter	Alpha	Beta	Gamma	Delta
(6) Scale correction = (5)/(4)	0.0000549	0.0000678	0.0000414	0.0000836
(7) Mean scale correction	0.0000619			

It can be seen that as the range to each station increases, so does the misclosure. This clearly demonstrates the scalar nature of the errors. It could be argued that a different scale correction should be applied to each station, but it would be found that if the long baseline calibration were performed on another day, or at a different time of day, the results would be a little different. In fact the refractive index would change all the time in sympathy with temperature, pressure and relative humidity changes, and it makes more sense to apply an overall scale factor. If a commitment is made to making an independent scale factor for each beacon, then the meteorological measurements should be taken daily and the scale factor adjusted daily.

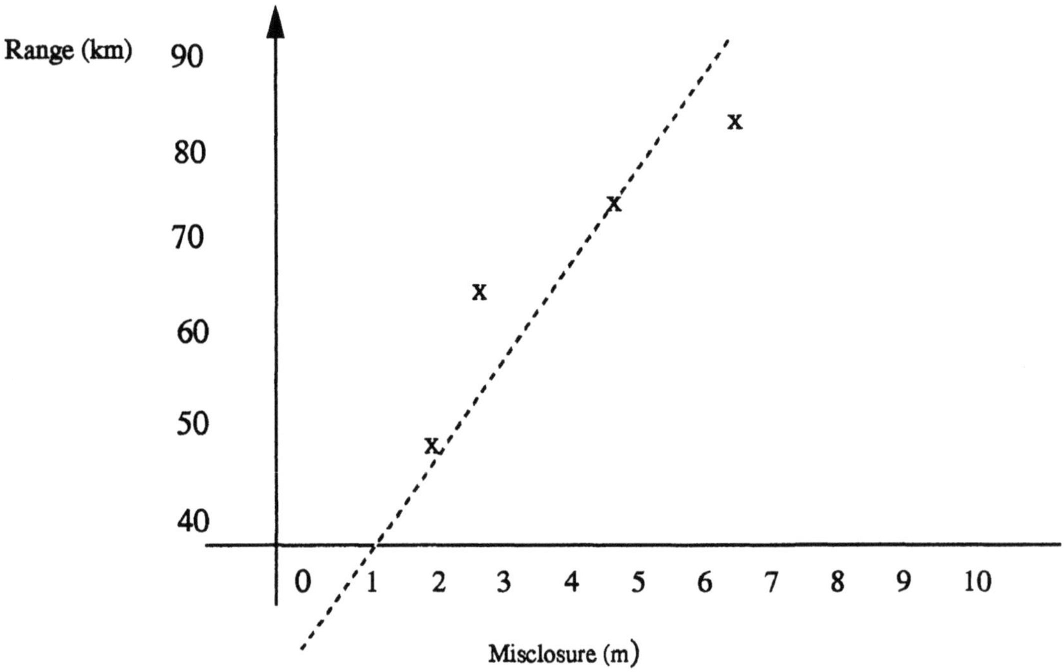

Figure 10.2 Misclosure.

Figure 10.2 shows a plot of misclosure against distance for the calibration. A best-fit line has been drawn through the data, because the introduction of a scale factor should produce a linear fit.

10.2.5 Alongside calibration

The prime mobile can now be installed on the vessel, together with its assigned antenna and cable. If the wharf at the mobilization port is within the chain coverage, the following trial should now be performed.

The vessel is moored very tightly to the wharf, to restrict surging and change of heading to an absolute minimum. All crane and similar movements should be stopped during the trial to reduce re-radiation and reflection effects to a minimum.

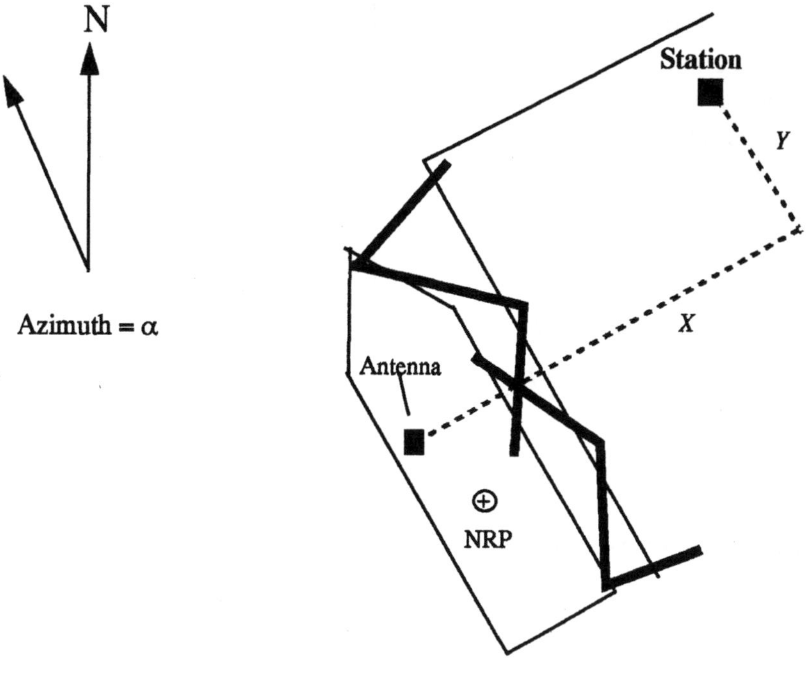

Figure 10.3 Alongside calibration.

Figure 10.3 illustrates the situation. There is a coordinated station on the wharf, whose coordinates must be known in terms of the survey datum. The offsets X and Y at right angles and parallel to the ship's head are measured by tape, and the gyro is calibrated (see later notes).The mobile is interfaced to the main navigation computer and all delays, propagation scale factors and other parameters are set as if the survey was about to be commenced. The system is cycled at a rate of about 10 seconds, logging the antenna position, navigation reference position (NRP) and the gyro heading at each cycle. The trial should continue for about 30 minutes, and should then be repeated with the spare mobile, antenna and cable. The mean antenna, NRP and observed range residuals should be computed, together with their standard deviations. A sample computation is shown in Table 10.5.

Table 10.5 Alongside calibration results

Parameter	Values	
Station coords.(E,N)	341678.2	7768521.9
X, Y offsets station to antenna (m)	–21.6	–12.2
Vessel heading (deg)	346.9	
Computed antenna coords.(E, N)	341659.9	776805.1
Observed antenna coords.(E, N)	341661.2	7768503.9
X, Y offsets antenna to NRP (m)	+2.1	–5.3
Computed NRP coords. (E, N)	341663.1	7768500.1
Observed NRP coords. (E, N)	341664.4	7768498.9
Range 1 residual (mean, sd)	–1.2	0.1
Range 2 residual (mean, sd)	+0.7	0.2
Range 3 residual (mean, sd)	-0.2	0.1
Range 4 residual (mean, sd)	+2.0	0.6
Range 5 residual (mean, sd)	-0.8	0.4
Range 6 residual (mean, sd)	+0.3	0.2

The offset computation is carried out as follows:

$$\begin{bmatrix} E_1 \\ N_1 \end{bmatrix} = \begin{bmatrix} \cos\alpha & \sin\alpha \\ -\sin\alpha & \cos\alpha \end{bmatrix} \begin{bmatrix} X \\ Y \end{bmatrix} + \begin{bmatrix} E \\ N \end{bmatrix} \tag{10.1}$$

The results show that the observed NRP is about 1.4m from the computed position, a result well within the expected accuracy of the system (5m). None of the range residuals are large and their standard deviations show good reliability. In practice it will be hard to achieve a result as good as this.

This trial does more than confirm the delays and scale factor obtained from the short and long baseline calibrations. It also confirms that the navigation software is using the right datum and that the offset computations are the right way round. Many errors from surveys can be traced back to one or both of these being incorrect, but an alongside trial of this type will almost certainly eliminate such errors. The trial also tests the integrity of the interface between the navigation system and navigation computer.

10.2.6 Harbour trials

Harbour trials can be conducted when the vessel sails for the prospect, but are not recommended as a very accurate method. Pairs of beacons or other shore marks are selected to form transits along which the vessel steams, as in Figure 10.4.

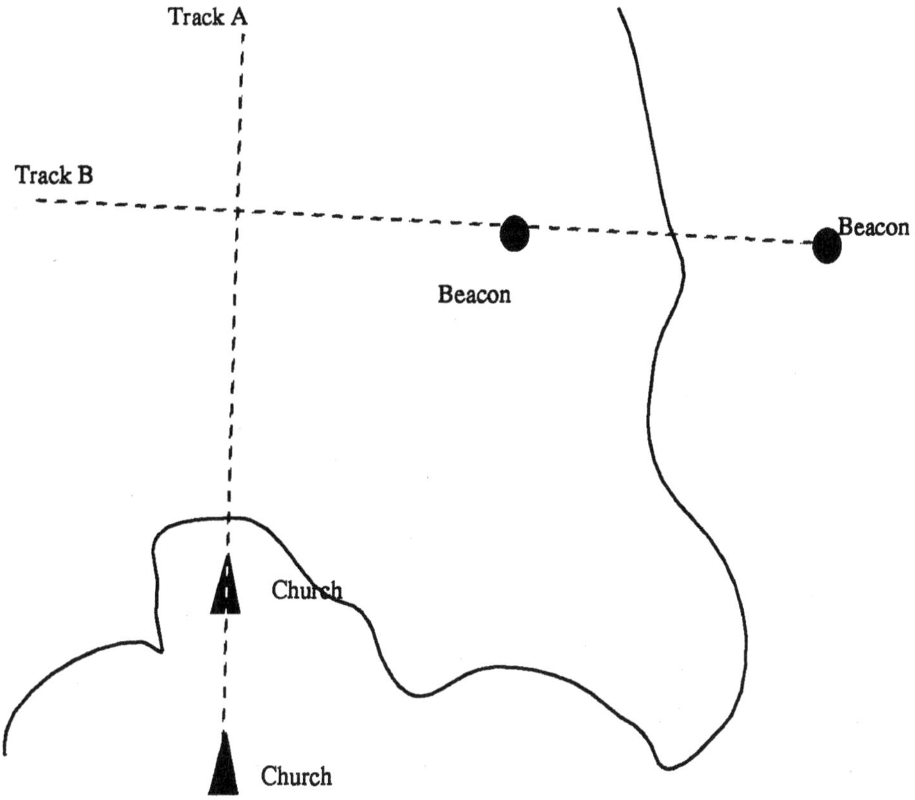

Figure 10.4 Transits.

The two transits are selected so that they cut as closely as possible at right angles, preferably aligned north — south and east — west. Each of the marks forming the transits must be coordinated, so that the navigation computer can be programmed with the planned tracks. The vessel steams the tracks indicated in both direction, towards and away from the marks, with the helmsman keeping the vessel on track by eye. The navigation system logs the distance off track and plots the results up.

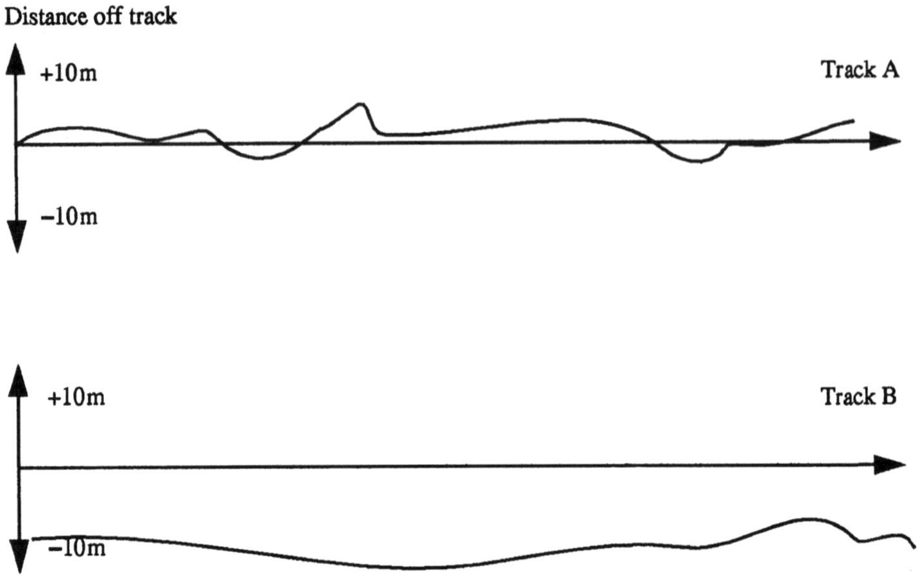

Figure 10.5 Systematic track error.

In Figure 10.5 the two tracks are shown. Track A has a mean offset close to zero, but track B's mean offset demonstrates a systematic error of −10m. Because of the ways the transits are orientated this systematic error is attributable wholly to northings, and is investigated with that in mind.

The trial's success is dependent upon the helmsman's ability to stay on the transits, and for that reason is not always successful. However, the trial does enable the system's dynamic capability to be quantified under reasonably controlled conditions.

10.2.7 Baseline crossings

As the vessel proceeds to sea it should cross as many baselines as possible. Baselines are the imaginary lines joining pairs of stations, and the idea of the trial is that at the instant the vessel crosses the baseline the sum of the observed ranges to the two stations should equal the distance between the stations. If the vessel cuts the baselines near their centre, any misclosure will be attributable to both stations in equal parts, but if the baselines are cut within (say) 10km from one station, then there will be negligible propagation correction for the near station and all of the misclosure will be attributable to the far station, assuming that the instrumental delays have been assigned correctly. The vessel should cross the baseline at right angles for two reasons:

- The vessel is travelling tangential to the range circles from both stations, minimizing the rate of change in range.
- The plot of the sum of the observed ranges against time approximates a parabola, and a best fit can be made using a second-order polynomial. This is the method most software suites use, and if the vessel steers an erratic course the fit will not be as good.

A plot of the sum of the observed ranges against time is shown in Figure 10.6:

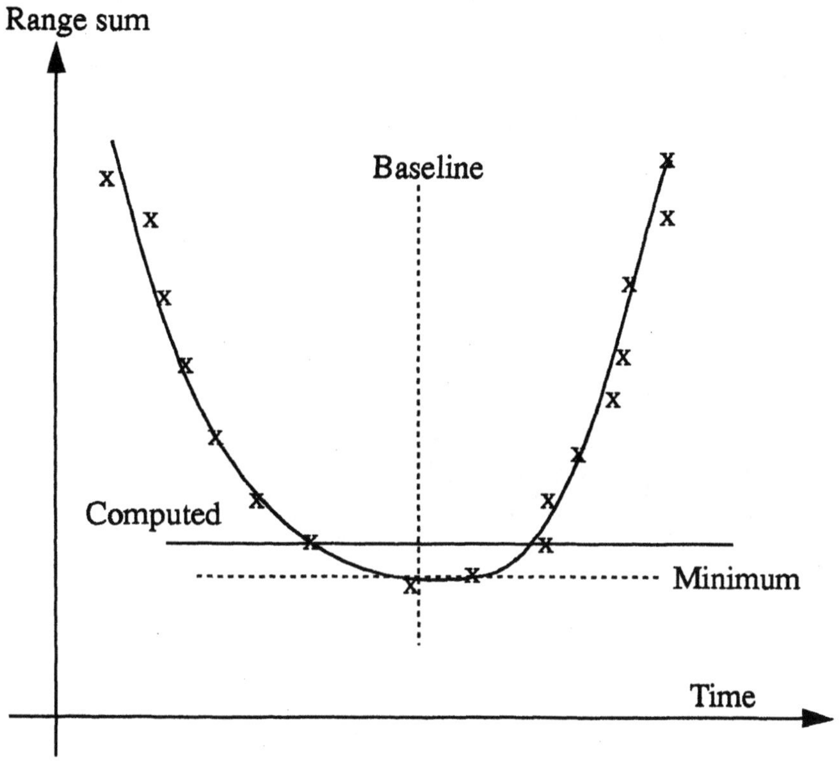

Figure 10.6 Baseline crossing.

The best-fit curve to the data should approximate a parabola, and the curve's minimum should coincide with the baseline itself and be within 1 or 2 metres of the computed baseline length. Any misclosure is usually attributable to propagation errors, unless there is a gross delay error in the system. Most software suites are capable of computing the curve and also of computing the range sum at the point of inflexion.

Baselines should always be crossed in both directions, to help eliminate systematic errors.

10.2.8 Offshore trials

All that remains now is to test the system on site. A good way of doing this is to steam two boxes through the prospect in opposite directions, logging the Syledis positions against those of the secondary system (which may be DGPS).

The trial should be conducted in calm weather, not at dawn or dusk (avoiding ionospheric problems) and during a good observing period for the secondary system (i.e. at least five SV's and regular correction updates if using DGPS). The results will indicate any systematic errors that are due to the vessel steaming in a particular direction, and logging continuously during each leg (about 5 km long) will reveal how well the system is performing in terms of residuals and variances.

Figure 10.7 illustrates this technique.

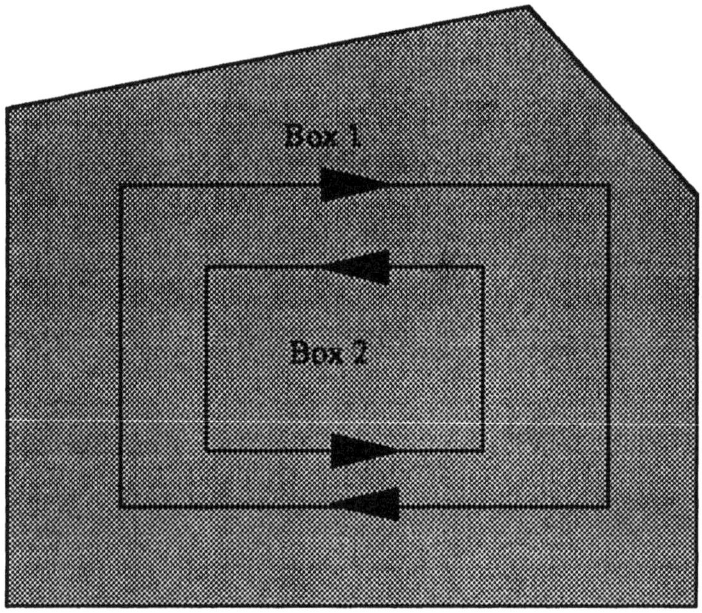

Figure 10.7 Offshore trials.

10.3 Argo

10.3.1 Static calibration

Argo contains instrumental delays in the same way as Syledis — they are in terms of lanes, not metres, and are known as 'partials' because they are corrected using partial lane corrections. Ideally we would perform the same short baseline calibration as with Syledis, but this isn't practical because of the antenna size (30m), ground net size (radius up to 80m) and power requirements. Therefore the chain is set up ashore and the mobile unit is installed on the vessel.

Prior to sailing field strength trials can be conducted on the vessel and the shore stations to determine if all parts of the system are transmitting at their optimum levels.

10.3.2 Dynamic calibration

If the mobilization port is within chain coverage and reasonably free of obstructions, the trials outlined in sections 10.2.5 and 10.2.6 can be carried out, but these will only give approximate values before the partials are known. The trials may be of more use as a confidence check during crew changes if the vessel returns to port.

10.3.3 Partial calibration

To calibrate Argo we need a secondary terrestrial radio-navigation system at least as accurate as Argo itself. Good candidates are Micro-Fix, Trisponder and Miniranger. The secondary system is calibrated using a short baseline calibration as described in sectio 10.2.2, and secondary remotes are installed about 10m up the Argo antenna towers, making sure that the Argo and secondary electrical centres are as close as possible and within 1m of each other. The calibration can be performed with the secondary system set on a satellite station, but the computation is a little more complicated and therefore more prone to errors. The secondary mobile is installed on the vessel, and the offsets between it and the Argo antenna and the NRP measured.

The vessel proceeds to the vicinity of each station in turn and steams reciprocal lines about 2km long at a distance of 10 — 15km from the station. The intersection of the tracks should be on the line between the stations and the prospect, and the lines are orientated in the direction of the primary sail line and at right angles to them. The set-up is illustrated in Figure 10.8.

The navigation computer is set to cycle at 10 second intervals, and on each cycle the Argo and secondary ranges are logged simultaneously. The difference between the two ranges will be equivalent to the partial lane correction in metres. An example is given in Table 10.6.

This computation is repeated for every cycle; most software available will do this automatically, and provide statistics for the partial corrections. An integrated system which can automatically deskew observations from multiple systems to the NRP can be used by giving the secondary system range a standard error of (say) 25m and the Argo range a standard error of (say) 5m. Then the partial correction in metres will be equal to the difference between the residuals for the two ranges.

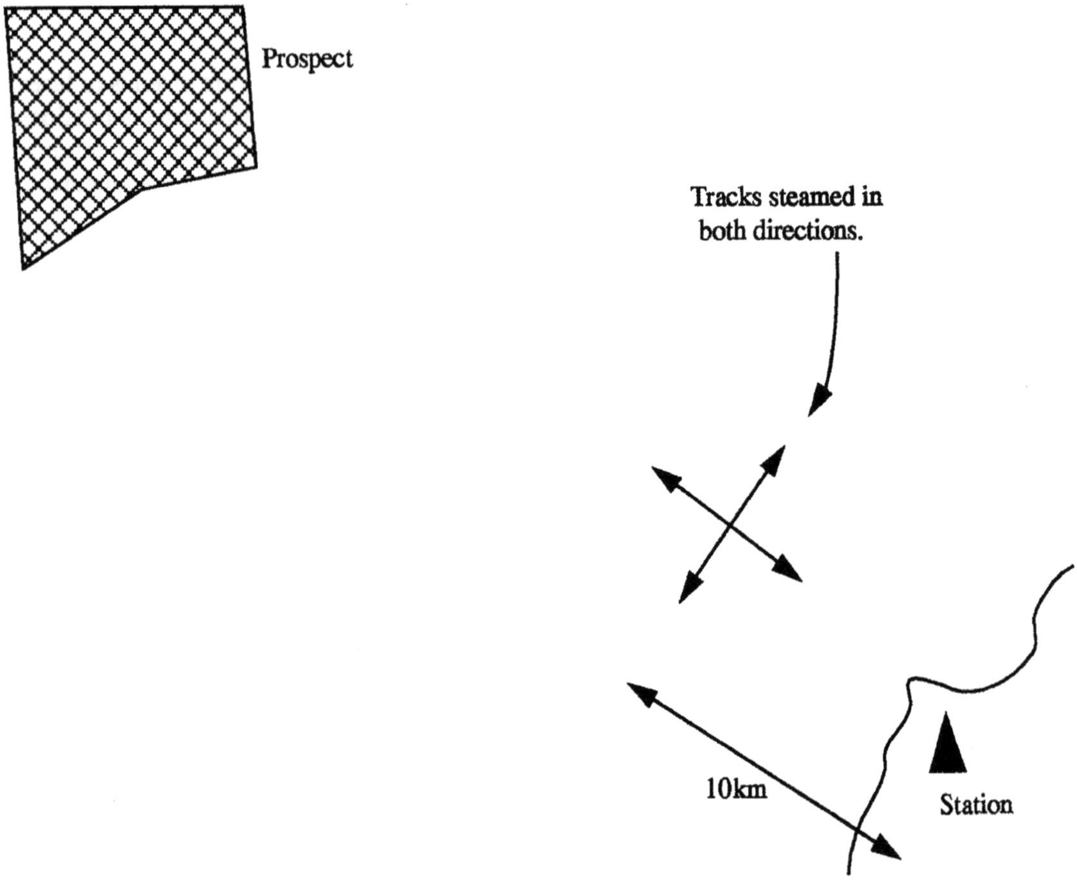

Figure 10.8 Argo calibration.

Table 10.6 Argo partial calibration

Parameter	Value
Argo velocity (km/s)	299792.5
Argo frequency (kHz)	1815.7
Argo lane width (m)	82.55562
Obs secondary range (m)	10423.6
Obs Argo lanes	126.32
Computed Argo range (m)	10428.4
Misclosure (m)	4.8
Partial Correction	0.06

The reasons for steaming lines in this configuration are as follows:

- The intersection point is on the line between the station and the prospect so that it lies on the propagation path to be used for the work. There will be no abnormalities due to an induced land path, for instance.
- The lines are run in the same direction as the primary sail lines because any phase error induced by the vessel running in that particular direction will become evident.

The resulting partial corrections from the four lines should be very close to each other (within 0.03 lane).

10.3.4 Baseline crossing

To calibrate the system for propagation scale factor, we again resort to using a secondary system. In this case the secondary remote is placed on a coordinated point near the prospect, and baselines are cut between it and the Argo stations.

The baselines are multi-system, with a secondary station at one end and an Argo station at the other, but other than that the technique is the same as in section 10.2.7. The baselines are crossed within 10km of the secondary station, so that there will be no propagation effect from that station. Then all the baseline error should be attributable to the Argo range and the propagation scale factor can be deduced using the same computations as in section 10.2.4.

10.3.5 On-site calibration

If there is no suitable structure of island offshore to conduct baseline crossings, the propagation factor can be deduced from an on-site calibration.

In Figure 10.10 the vessel steams two boxes inside the prospect. The box sides are about 5km long and the sides are aligned with the primary sail lines. The vessel steams the boxes with the navigation system cycling at about 10 seconds, and logging both Argo and secondary system positions.

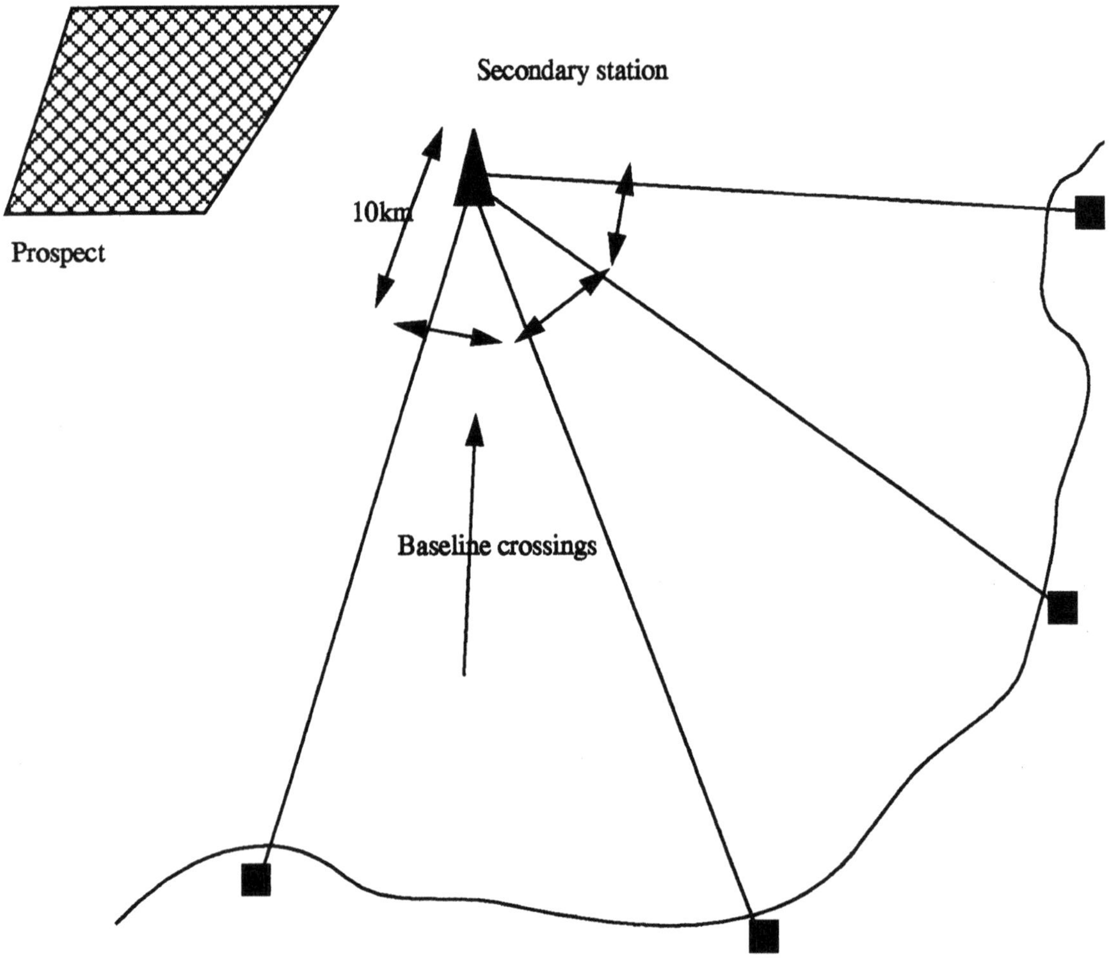

Figure 10.9 Baseline crossings.

Production of residual, variance and separation statistics between the two systems will point to any errors. Needless to say, the secondary system will need to be as accurate as Argo for this test to be successful; the secondary system most often used is DGPS, and great care must be taken that the DGPS is operated using optimum constellations, elevation masks, update rates and so on.

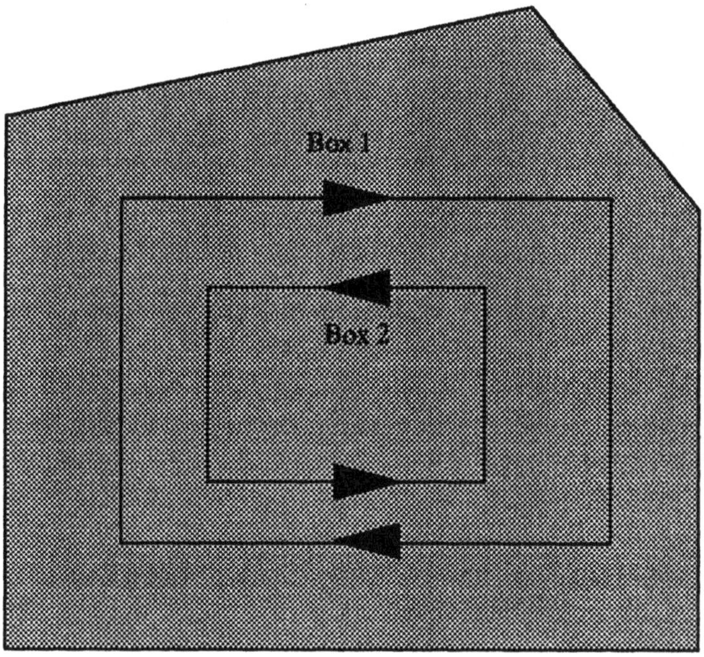

Figure 10.10 On-site Argo calibration.

10.4 DGPS

10.4.1 General

Although GPS is shrouded in mystery to non-surveyors, its operation is actually very simple and similar to a land-based ranging system. We obtain lines of position and adjust them by least squares in just the same way. The big difference is that the stations are moving, and the broadcast positions of those stations are deliberately degraded. However, we can overcome this by placing a receiver on a known station, computing the corrections between where the receiver thinks it is and where we know it to be, sending those corrections to the offshore mobile and using them in conjunction with the ranges observed offshore. The workings of the system are explained elsewhere in the course. So we can isolate the different sections of the system, and require to calibrate or check the following:

- Reference Receiver
- Datalink
- Vessel receiver
- Overall system

10.4.2 Reference receiver

The reference receiver is sited at a known coordinated station. The station itself should have been coordinated by GPS and also tied into the existing national geodetic network. The derivation of the reference station coordinates must be scrutinized and proved. The whole system revolves around these coordinates and **there is no redundancy.** It is recommended that the station be coordinated to at least second-order levels in all three dimensions.

The reference station coordinates can be checked by putting another two GPS receivers on nearby primary control points and making phase measurements at all three sites simultaneously. Proprietary software can then be used to check the reference station coordinates and also serve a useful purpose in detecting the amount of multipath and cycle slip experienced during the measurements.

A multipath audit should be conducted. This procedure takes at least 48 hours' observation time and consists of a post-processing study of all pseudo-range and phase measurements made during the 48 hour period. It is recommended that standard deviations on code and carrier measurements should not exceed 3.0m and 0.05 (at the 1-sigma level) respectively. During the trial a monitor receiver should be set up less than 10km away in open space to minimize its own multipath and observations taken concurrently.

The reference station antenna should be placed at least 2km above any horizontal surface, including the ground, and at least 30m away from any vertical structure, specifically MF and LF datalink masts.

10.4.3 Datalink

The datalink is proved once the vessel receiver is installed. The most common link now is by communications satellite, as in the SkyFix and MN-8 systems. The rate of correction update should be better than 5 seconds to stand any chance of breaking selective availability jitter, and the correction transmitted should be pseudo-range corrections.

10.4.4 Vessel receiver

The receiver can be tested alongside the wharf prior to departure, as in section 10.2.5. For 2D surveys the system will probably be used without any secondary system, so this check is crucial. It should be carried out under the following guidelines:

- Ensure the vessel is moored tightly to the wharf.
- Check the gyro heading at regular intervals.
- Stop all wharf and upper deck movements during the trial.
- Elevation mask 15 degrees.
- Over-determined positioning using at least five satellites.
- No height aiding.
- At least one change of constellation during the trial.
- Correction update rate better than 5 seconds.
- Convert position to the survey datum.

The use of an over determined solution cannot be over-emphasised. Most current navigation software only accepts latitudes and longitudes from the differential receiver, and it is easy to lose sight of the fact that we need to solve for position in three dimensions and also a time unknown.This makes four unknowns so we need five LOPs to ensure redundancy. If we only use 4 satellites it is exactly the same as using two-range Syledis.

The height can be restrained at a preset value (this is called height aiding), but for calibration purposes this can only be seen as a potential source of systematic error. A four satellite, height-aided fix is not really over-determined, but does at least possess one degree of redundancy.

For offshore trials, the technique set out in 10.3.5 can be used if there is a secondary system available.

It is very strongly recommended that the alongside trial be conducted every time the vessel returns to port, for example during crew changes.

10.5 Tailbuoys

10.5.1 Static calibration

The static calibration for any tailbuoy receiver is exactly the same as for the vessel receiver. The mobiles and beacons can be set up for short baselines, long baselines, alongside trials, harbour trials and baseline crossings (depending on the system), as discussed on the previous pages.

10.5.2 Dynamic calibration

To perform an offshore trial the following method can be used.

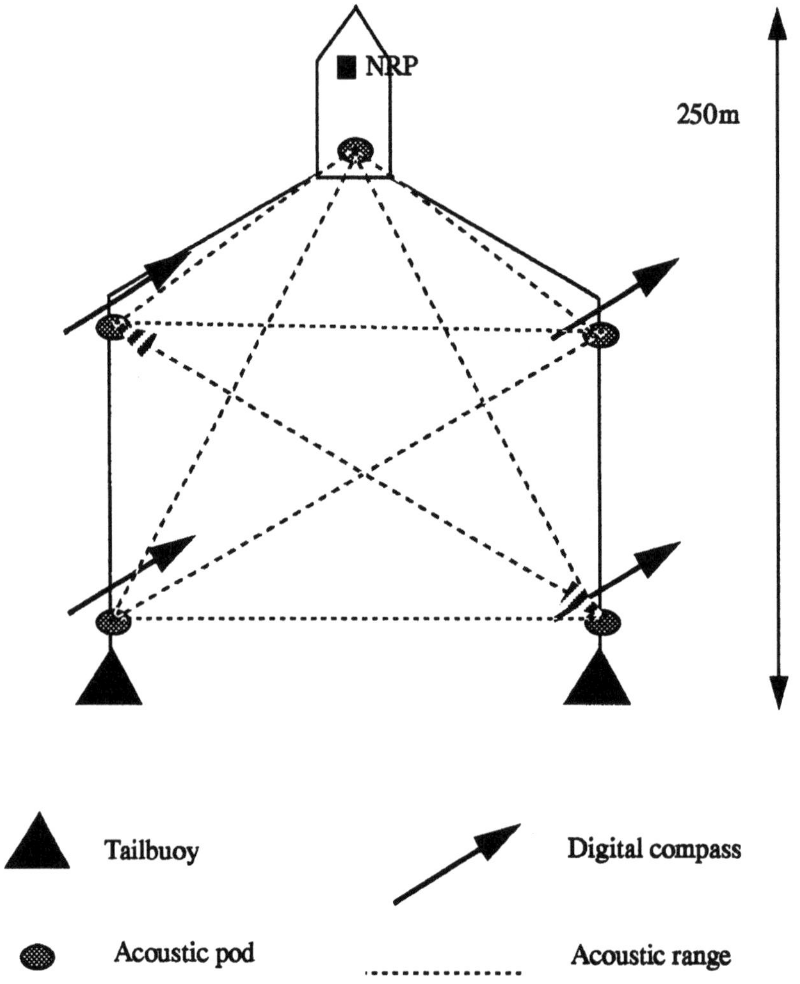

Figure 10.11 Tailbuoy calibration.

In Figure 10.11 the streamers have been deployed to a distance of about 250m astern of the NRP. Each streamer is fitted with at least two acoustic pods (or laser targets) and digital compasses. The acoustic pods and compasses will have already been calibrated prior to deployment, and the best estimate for magnetic declination should be used for the trial.

The vessel again steams two boxes in the prospect, selecting a calm-weather window with weak tidal streams, in conjunction with good observing conditions for whichever system is being used. The difference between the tailbuoy trial and that for the vessel receivers is that there is almost always no secondary absolute system on the tailbuoys.

There is no need to use any more than a second-order polynomial at the very most to compute the streamer shapes — since the tailbuoys are only 250m astern of the NRP a straight line fit would probably suffice. In either case there is more than enough redundancy in the system to position the streamers.

In an integrated system we can monitor the tailbuoy, compass and acoustic residuals to measure the performance of the system. For an older system we can compute the streamer positions using the acoustics and compasses and then calculate the 'rotation angle' to fit the streamers to the tailbuoys. In both cases the fit should be excellent at such short range.

The trial can be repeated with the streamers at their fully deployed length, but it becomes harder to judge whether any misclosure is due to an incorrect magnetic declination, incorrect gyro angle, incorrect convergence or whether there is something wrong with the tailbuoy. The clue here is that any rotational parameter such as convergence or magnetic declination tends to increase with distance from the ship, whereas an offset or other systematic error tends to remain static down the streamers.

10.6 Gyro

10.6.1 Static calibration

The survey gyro is a very important piece of equipment on board the vessel, particularly in 3D surveys. the static calibration should be carried out with the assistance of a qualified land surveyor, since the use of a theodolite is necessary.

The calibration requires two coordinated stations on the wharf, as demonstrated in Figure 10.12. The theodolite is set up to the two stations in turn, and angles observed to marks set out on the ship's centre line at the bow and stern of the vessel. If EDM equipment is available, then distances should be measured to reinforce the solution.

The method depends upon the vessel being tightly moored to the wharf so that there can be no change of heading during the measurements and the mean heading is compared with the computer centre-line heading to determine the gyro correction. the grid convergence will have to be taken into account when converting grid bearings to the true azimuth of the gyro.

10.6.2 At sea

Once at sea the gyro is often forgotten about. But it should be checked daily on the bridge as standard navigation practice, and this can be done by using an azimuth circle and observing the sun's azimuth at sunrise and sunset. The true azimuth can be looked up in Nories Tables by entering the day of the year and the latitude and this can then be compared with the observed reading. Of course, the azimuth circle will generally only be usable on the ship's gyro, so the ship's and survey gyros will need to be correlated at the time of observation.

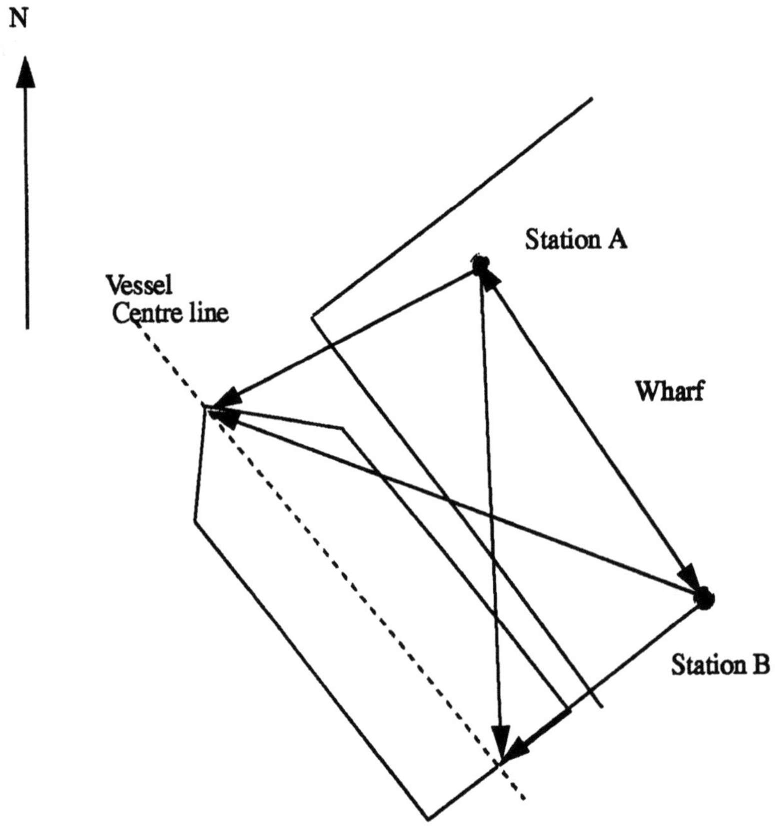

Figure 10.12 Gyro calibration.

10.7 Relative systems

The calibration of relative systems (laser, acoustics and compasses) is dealt with in Chapter 8.

11 Planning the network

11.1 Introduction

In common with most activities, careful planning of the survey control for a seismic survey will pay dividends.

11.1.1 History

In the early days of seismic acquisition, navigation stations usually had to be surveyed from scratch. Often this meant extending control from a known geodetic station by triangulation.

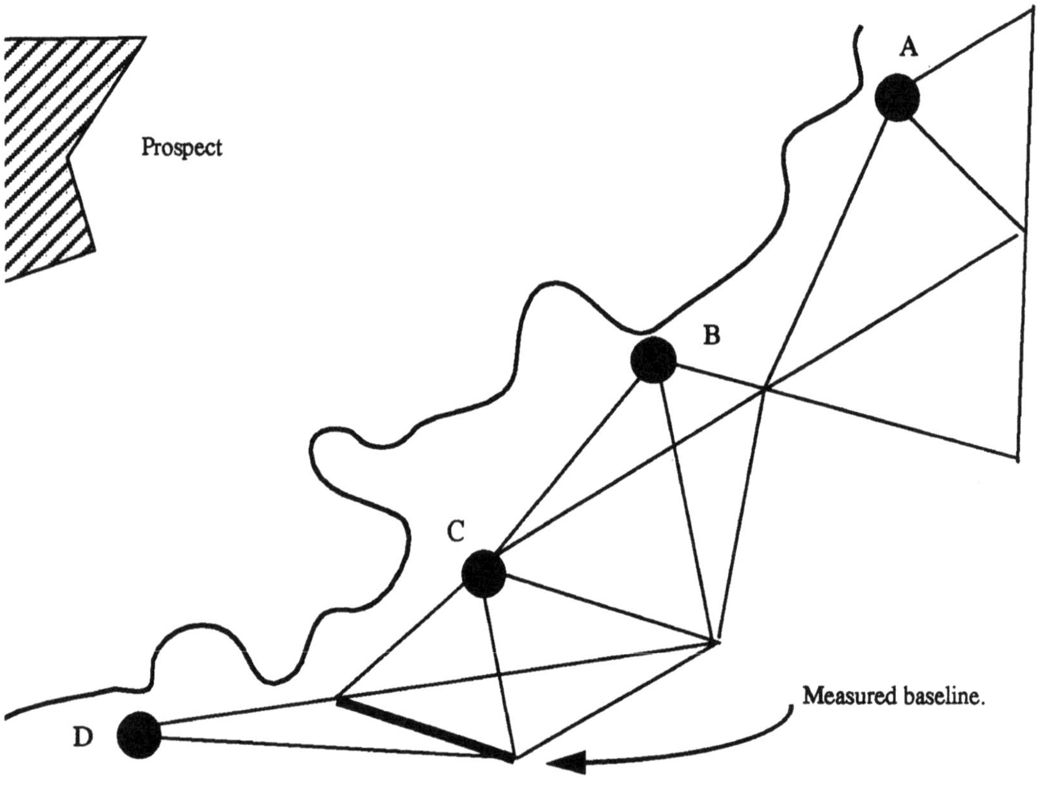

Figure 11.1 Triangulation.

In Figure 11.1 a system of triangulation has been used to set out stations A, B, C and D. We start with a measured base (the thick black line) and then observe well-conditioned triangles and quadrilaterals using a theodolite, gradually extending the network along the coastline. Needless to say, we are restricted to line-of-sight measurements and the task is laborious and expensive.

With the advent of EDM (electronic distance measurement) the task became much easier.

In Figure 11.2 we start at a known station X and finish at another known station Y. We also rely on auxiliary stations X' and Y' to orientate the survey and observe opening and closing angles. Then we use angular and distance measurements between the stations to form a traverse along the coast. This method can produce accurate results in much less time than the old triangulation method.

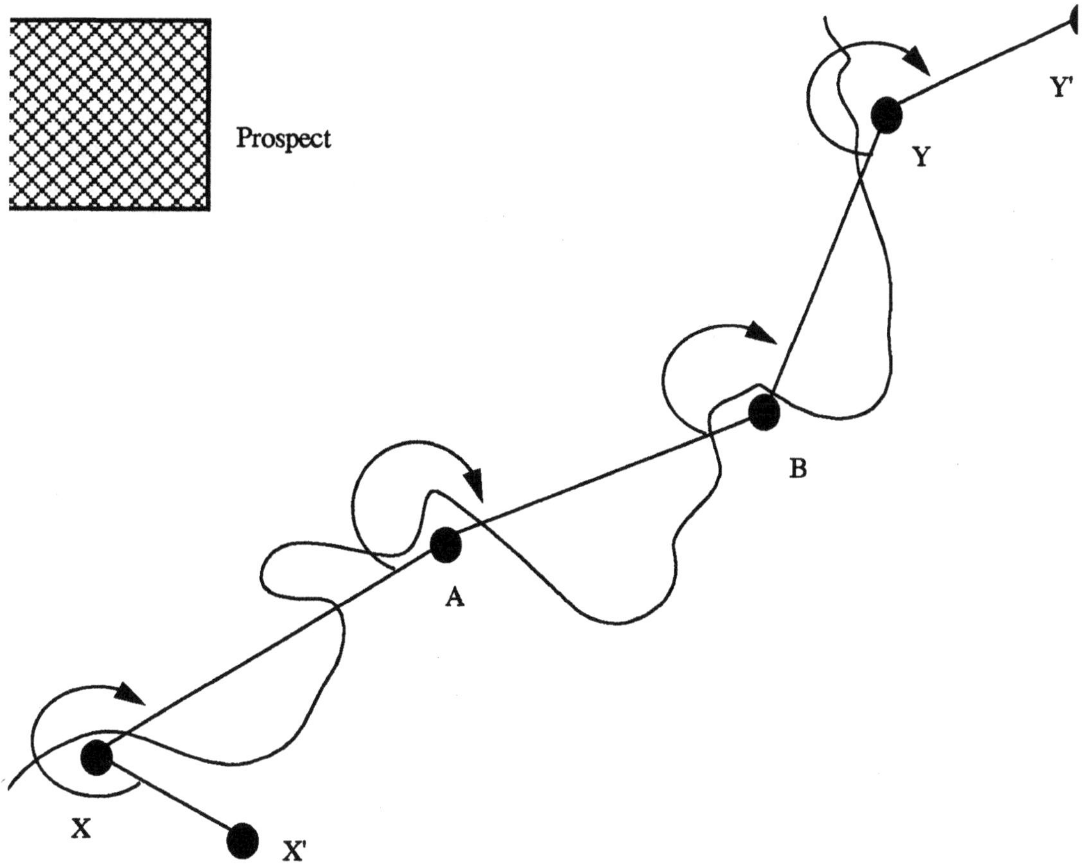

Figure 11.2 Traverse.

Both of these methods have now been superseded by GPS. Using GPS we can survey and interconnect pairs of stations using phase measurements in about one hour and to millimetre accuracy. No longer do we require line of sight between stations. Now that we have almost 24 hour coverage from GPS it is possible for a two-man team to survey a network of six stations or so in about a day.

However, there are some points worth mentioning when using GPS for control:

(a) If the station coordinates are left in the WGS84 datum, which is the datum for GPS, there may be several hundred metres difference between the station coordinates and those used on maps and charts for the area. So due care must be taken that the coordinate system is uniform.

(b) The second point is particularly relevant for surveys near the coast, near obstructions such as wrecks or in shallow water. All features on the nautical chart for the area will be on the datum for that area. So there is good reason for converting the WGS datum coordinates to the same datum. But there is a further consideration here. All features on the chart will have been surveyed **relative to shore marks**. Stations surveyed using GPS bear no relation whatsoever to shore marks. A good example is the Monte Bello islands off the NW coast of Western Australia. Geodetic stations on the islands were surveyed in using systems such as two-range Maxiran from stations on the mainland. The positions of all shoals and wrecks near the islands are relative to the geodetic stations set up on the islands. Of course, in recent years the shore marks themselves have been re-surveyed using GPS, but until the waters surrounding the islands are re-surveyed relative to the new marks, it is quite likely that a GPS position for a vessel suggesting that the vessel is several hundred metres from danger may in reality be much closer to disaster.

The lesson to be learned is that if a promising station is reconnoitred and subsequently surveyed in using GPS, it is prudent to connect that station to the existing shore control by terrestrial methods and investigate any discrepancies. This action will at least give us more confidence in the new station.

11.2 The problem

11.2.1 Main requirement

The main requirement of the control net is to provide positioning to the required accuracy at minimal cost and optimum efficiency. Although DGPS is being used for primary control in 2D surveys, it has not yet been fully accepted for 3D work, and the use of terrestrial radio-positioning systems will continue for the foreseeable future.

The most sophisticated acquisition equipment and techniques can be used for a project, but they will be rendered next to useless if the positioning is not up to the required standard. One of the costliest aspects of the whole job can be the establishment and manning of the radio-navigation system.

There are a number of points which need to be considered when setting up a network; these include, but are not limited to:

(a) Accuracy obtainable in the prospect.

(b) Intersection angles between 30 — 150 degrees.

(c) Cost of establishment.

(d) Cost of manning and maintenance.

(e) Site accessibility.

(f) Site security.

(g) Probability of flooding.

(h) Probability of lightning strikes.

(i) Station height.

(j) Power requirements.

(k) Site ownership.

(l) Logistics of resupply.

(m) Land paths.

(n) Suitability for earth mats — ground conductivity.

(o) Ease of calibration — baseline crossings.

(p) Multipath and re-radiation problems.

(q) Storm protection.

(r) Evacuation procedures.

Our problem is how to optimize all these variables.

11.2.2 Historical solutions

Even now, with all the power of computers and artificial intelligence at our command, the selection of survey control is pretty much a hit and miss affair. The survey planner has to meet some or all of the requirements listed in section 11.2.1 above, and he has few tools to help him.

The first requirement is accuracy. There are software programs available now where the user can input a number of proposed station locations, together with estimates of precision in terms of standard errors, and the programs will generate curves representing contours of expected positioning accuracy, much as in Figure 11.3.

In some of these programs the user can interactively choose sets of stations and generate contours for each set. The contours themselves are produced by a least squares computation based on the assumed standard errors.

So there are tools to help decide the accuracy achievable from different geometrical configurations, but none to help us solve the remaining problems in our list.

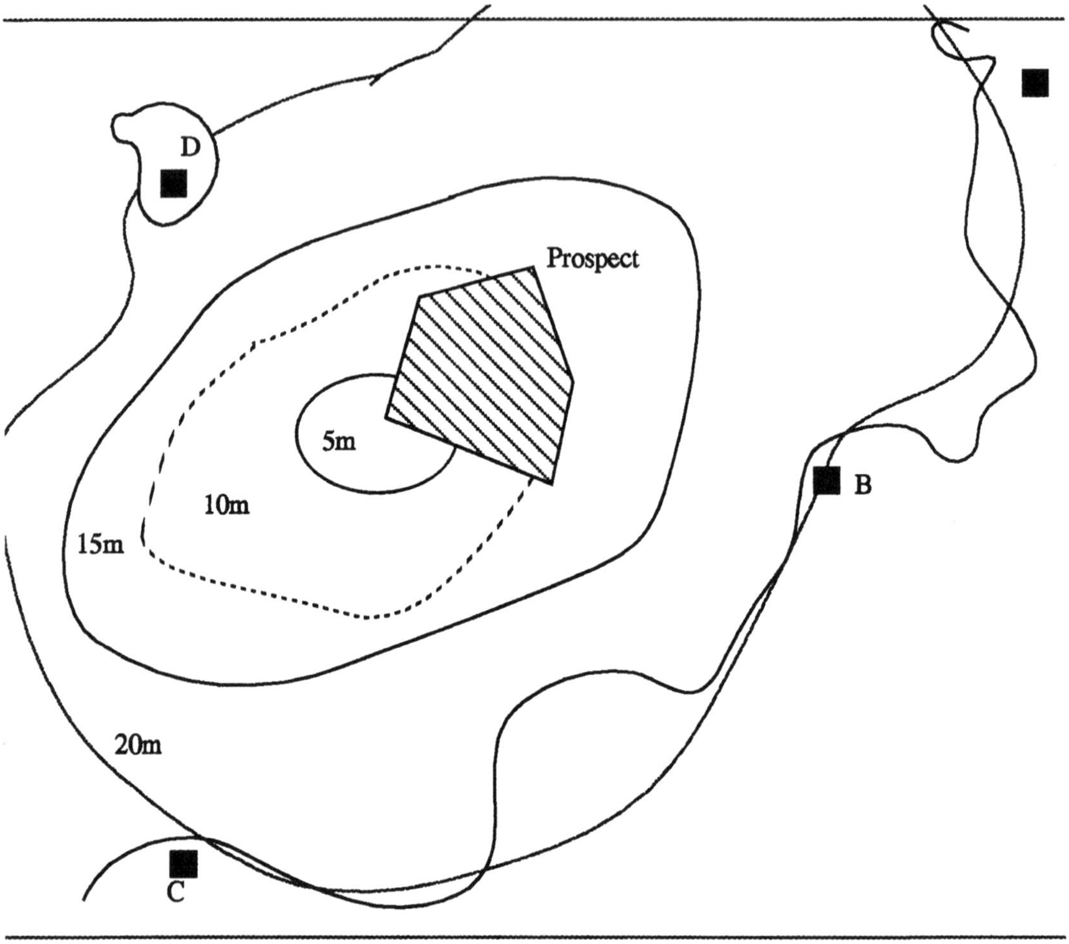

Figure 11.3 Accuracy.

11.3 The solution

11.3.1 The state function

Firstly we must decide on an initial state. In fact there may be a number of initial states — one for each proposed system. Each initial state can be constructed as follows:

- INITIAL STATE 1 — MICRO-FIX SYSTEM

Required vessel accy (1 sd) 5 m

Required tailbuoy accy (1 sd) 5 m

Number of stations 4

Station Locations:

Name	Easting(m)	Northing(m)	Height (m)	SE (m)
1. Alpha	345789	7731207	533	5
2. Beta	357312	7724128	353	5
3. Gamma	361782	7728716	273	5
4. Delta	372341	7742815	613	5

From the initial state we can compute the following data:

Inter-station distances (m)		Angles of cut (deg)
Alpha-Beta	13523	53-168
Alpha-Gamma	16185	48-67
Alpha-Delta	28978	91-117
Beta-Gamma	6405	72-95
Beta-Delta	23980	35-54
Gamma-Delta	17614	49-81

Distances to Prospect:

	Maximum (m)	Minimum (m)
Alpha	12650	25436
Beta	13675	41239
Gamma	6789	17892
Delta	24318	37962

Accuracy figures at 1 sd level for various configurations:

	Maximum (m)	Minimum (m)
1234	4	3
123	5	4
124	5	3
134	6	3
234	5	4

From the above data we have established that using all four stations meets our accuracy criteria. We have also established that if one station goes down, any combination of the remaining three will also meet the criteria. (Not strictly true because combination 134 yields a maximum accuracy of 6 m, but this is only just over our limit of 5 m and the probability of only station 2 going down probably justifies the configuration). Also we can see that the angles of cut fall between the usually accepted criteria of 30 — 150 degrees and that all of the ranges fall within the maximum capability of the system.

Now we have established that the accuracy is acceptable we need to go through our remaining criteria to ensure that these are met. For the time being we will leave out the cost — this will be introduced as a separate function later.

(a) Accuracy — already computed and acceptable.

(b) Intersection angles — already computed and acceptable.

(e) Accessibility. This can often be determined by reference to maps or by searches at the Titles office for ownership and contacting the owner. If the station is a geodetic station, then the Department of Land Administration's Geodetic Branch can be consulted: they will sell details of coordinates, azimuths and distances to nearby marks and ownership/accessibility details. Failing this (for instance if the site is a new one on Crown land), a reconnaissance may be necessary.

The quality of accessibility required depends upon the type of installation. For instance with the Micro-Fix system accessibility by 4WD or boat would be enough, but for a Hyperfix station where we need to erect a substantial tower, provide generators and provide some form of camp for personnel we may need to think about using a large helicopter or truck.

(f) Security. Generally this is not a problem around Australia. The biggest problems arise when operating in areas such as the Dampier Archipelago where there are a lot of people - the heavy duty batteries used for powering Micro-Fix and Syledis stations are ideal for small boats, and they tend to 'walk'. Some form of security, such as chains and padlocks, is necessary in this case. Electric fences may be needed to keep out animals.

(g) Flooding. Common sense will dictate that we do not install stations in an area liable to flooding. Usually the stations are on the highest ground in the area anyway, but we also need to check that accesses to the station don't get flooded.

(h) Lightning strikes. If the station is sited at the Top End during the summer the camp area, generators and fuel should be kept well away from the transmitter and antenna (they should anyway!). This is the one case where consideration may need to be given to siting the station away from the highest point on the ground.

(i) Height. Generally select the highest point, the above comments not withstanding. If occupying a geodetic station, it will usually be on the highest point for visibility to other stations. In our example, height would not be a problem because the maximum ranges to the prospect are only 24km, but height certainly becomes an issue when trying to operate a system near its limits. Although the maximum range for Micro-Fix is quoted as 80km, there is no way that this can be achieved if the remotes are sited at sea level.

(j) Power requirements. Our chosen system (Micro-Fix) uses 16 — 30v dc power. The usual method is to use pairs of 12v batteries connected in parallel, with each battery being connected in series to its partner. Therefore at least 4 and up to 8 × 120amp/hr batteries should be used in this way. The system draws 12W operating and 3W in stand-by mode, so 8 batteries should last about three days without recharging. It is usual to provide a charging system - this can be one of the following:

(i) If mains power is available it should be used, with a battery back-up.

(ii) In areas such as the NW Shelf use solar panels - these need to be cleaned weekly because of dust and bird strikes.

(iii) SE Australia and the Top End in summer would work well with gas or diesel generators.

(iv) Wind generators can be used in the southern half of Australia during winter.

(k) Site ownership. The ownership of the land on which the station is to be sited is all-important. In cases where the station is on privately owned land or on pastoral leases it will usually be found that the owner is very helpful. He will be able to point out the best access routes and can offer advice with regard to shelter, evacuation and so on. If the station is to be unmanned, requiring only a weekly check on power supplies and general serviceability, the owner will usually oblige. Conversely, if the proposed station is on native land, it is prudent to check out the station some months in advance. A visit to the station will almost certainly be required, accompanied by a tribal elder who will assess whether or not the site is sacred and so on.

(l) Logistics of resupply. There is no problem with Micro-Fix because the stations can remain unmanned. We need a back-up plan in case of station failure; one person ashore will be nominated as chain manager and will be responsible for chain maintenance. He will need to visit the whole chain on a regular basis to provide essential maintenance — checking batteries, cleaning panels, and so on. The quickest (and probably most expensive) way for him to get around will be by helicopter. If a station goes down there must be still be enough redundancy to obtain an over-determined position - this is true if the other three stations are still working. But there must be a contingency plan to get the chain manager out to a station quickly to effect repairs. There must be a radio link between the vessel and the chain manager and there should be at least two scheduled radio checks per day. For manned systems such as Argo there will be two personnel per station and they must also have at least two radio schedules per day. They will also need resupplying at regular intervals.

(m) Land paths. These are crucial considerations for the lower-frequency systems such as Argo, but it is always prudent to site any station as near to the coast as possible to remove any possibility of phase lag. Also, by siting right on the coast we reduce the temperature and relative humidity differentials between the station and the vessel, and reduce the possibility of our path crossing over buried power lines which could cause interference.

(n) Earth mats and ground conductivity. Sea water is the best conductor, and loose sand is one of the worst. The better the conductivity, the less attenuation the signal will suffer and the better the station will be. At the same time, we cannot set the station up in the water, and in the case of Argo and Geoloc (for instance) we need a radius of about 80m around the antenna tower to set up an earth mat. The mat consists of upwards of 60 copper wire radials which need to be in contact with the ground and be capable of being staked down to prevent movement. The impossibility of laying earth mats has made many potential sites useless, and this is where good reconnaissance will pay dividends.

(o) Calibration and baseline crossings. Few chains have been put in where the ease of calibration has been a priority, but a little forethought can mean that the baselines can all pass quite close to the prospect. If they don't, the vessel has to pick up gear and travel some distance to cross them, which puts pressure on both client and contractor and often means that they get put on the back burner.

(p) Multipath and re-radiation. Multipath is caused by reflection of the signal on nearby reflective surfaces, and re-radiation is caused by excitation of nearby structures acting as antennae. The solution is to install the station away from reflective surfaces, man-made structures, TV antennae, etc.

(q) Storm protection. Lightning strikes have already been mentioned, but consideration should also be given to how exposed a station is to high winds, sea level surges and so on.

(r) Evacuation procedures. All personnel manning stations should have first aid knowledge and know what to do in the event of snake bite, spider bite, fractures, etc. There must also be a comprehensive evacuation plan in place for injury, cyclones, etc.

Assuming that all the above criteria can be met, we need to implement a cost function next.

11.3.2 The cost function

The total cost computed for the state is the sum of the capital cost of the chain, the daily running cost of the chain for the duration of the work, and an emergency fund. The client should be in a position to specify in the tender documents which stations need to be occupied and which systems are acceptable for the project. It is then a question of specifying schedules for the work — these make up the total cost function. There are a number of ways of presenting such specifications:

(a) Lump sum for everything. If this method is chosen, the schedules break down the job into (for example) mobilization, demobilization and total running cost for the system. The method is attractive for the client because no matter how long the job takes the budget for the navigation system is fixed. It is most unattractive for the contractor, however, because if the job goes on longer than expected (which is often the case), or if there are unexpected expenses involved, his fingers may get badly burned.

(b) Day rate for everything. In this case daily rates and stand-by rates are specified for every aspect of the work, including mobilization and demobilization. This is the most favourable system for the contractor, because his income is guaranteed whatever happens. But the client can suffer, particularly at the start of a project, if the contractor brings in equipment which is below standard and the mobilization takes longer than expected.

(c) A combination of lump sums and day rates. This is a compromise to suit both contractor and client and generally works best. However, currently tendering documents are often put in place without a good plan on the client side. In section 11.3.1 above we have specified a state function which allows us to cost every part of the system and which is flexible enough to allow for the addition or subtraction of stations and personnel from that state.

A sample schedule of prices may look like Table 11.1.

Table 11.1 only specifies a skeleton schedule. When included in the tender documents each file should be expanded to ensure that there can be no confusion over what is included in each rate and what isn't. The day rates for equipment do not start until after the entire system has been accepted as operational, in other words on completion of the dynamic trials. Note that there are operational rates and stand-by rates for equipment.

The next task is to compute a transition function.

Table 11.1 Schedule of prices

Item	Quantity	Unit	Rate ($)	Cost ($)
1. MOBILIZATION				
Transport personnel and equipment to site	1	Lump sum		
Permanently mark station	4	Lump sum		
Install station	4	Lump sum		
Short baseline calibration	1	Lump sum		
Long baseline calibration	1	Lump sum		
Vessel installation	1	Lump sum		
Dynamic trials	1	Lump sum		
2. DEMOBILIZATION				
Transport personnel and equipment off site	1	Lump sum		
Unrig station and clear site	4	Lump sum		
3. STATION MANNING				
Chain manager	1	Day rate		
Station Personnel	8	Day rate		
Victuals	8	Day rate		
4. VESSEL MANNING				
Operator	2	Day rate		
Victuals	2	Day rate		
5. OPERATIONAL				
Station	4	Sby/day		
		Op/day		
Vessel	1	Sby/day		
		Op/day		
Spare equipment (specify)	1	Sby/day		
6. LOGISTICS				
Shore accommodation	1	Day rate		
Shore transport	1	Day rate		
Helicopter support	1	Round trip		
Consumables	1	Day rate		

11.3.3 The transition function

The transition function allows us to move from one state to another using the cost function from section 11.3.2 above. We specify in the tender documents that we may wish to increase or decrease the number of stations, manning levels and so on, and when the tenders arrive back we can use the schedule of prices to adjust the total cost accordingly.

Some of the variables we may want to consider here are:

(a) Add another station that has already been coordinated — to give greater redundancy.

(b) Add another station that has yet to be coordinated — again to give greater redundancy.

(c) Swap an existing station for a new one to improve logistics and reduce costs.

11.3.4 Artificial intelligence

At the beginning of this chapter we mentioned the power of computers now available to us. It would be perfectly feasible to use computers to do much of the planning of survey control for us, since the state, cost and transition functions can be defined heuristically.

12 Processing and quality control

12.1 Introduction

12.1.1 The current state of affairs

Many contractors can now produce fully processed navigation and seismic data on board the vessel. If not processed in real time, the data will only be about 24 hours old when reaching the processed stage.

However, to reach the 'processed on board' stage there are a number of hurdles to be overcome, not least of which is the proper quality control of the data.

Certainly, the concept of processing on board holds a number of advantages over the current practice of sending data ashore:

- The processing is performed by the personnel involved in shooting the survey. They have the most intimate knowledge of the work, and it can be strongly argued that they are the only personnel qualified to process the project that they are involved in.
- If any errors are found in the data, it is not too late to correct them — lines can be re-shot and parameters can be re-set for new lines because the vessel is still on the prospect. This is not the case when data is ashore and the vessel has been demobilised.
- It is suggested that the time required to process will be vastly reduced. On board a ship shifts and watches are enforced rigidly, and personnel are already at their place of work. The pressure of keeping the processing time to within 24 hours of the data being acquired to ensure there is no backlog will keep the work progressing quickly.
- The tendency to leave errors uncorrected and improperly documented so that 'they can be sorted out in processing' will be obviated.

It may be more convenient to send the data ashore on a daily basis (via Inmarsat for example) to be processed in pseudo-real time ashore. Under these circumstances some of the above criteria can still be met; but it remains the author's view that data processing will be most efficient on the vessel.

12.1.2 Why are data processed ashore?

The main reason for data being processed ashore is a combination of history and computing power. Processing is still carried out on mainframe computers in many cases, and it has been traditional to have these computers set up in office environments ashore. Processing is carried out by teams of processing staff who are not usually qualified surveyors.

The second reason is that the processing algorithms used involve a pre-processing stage in which groups of data are reviewed and despiked. This despiking process is often done by taking 'average values' for a parameter and using a gate based upon that average. Any observation falling outside the gate will be rejected. We have seen in Chapter 5 that this practice is dangerous without proper statistical testing, and that such testing can only be carried out easily if the whole spread is treated as an integrated system.

Therefore it seems that the logical way to process navigation data is to use an integrated system which is capable of filtering and rejecting data using least squares, in real time if possible. The Kalman filter is not the only way of doing this. There are other filters such as the Bayes Filter which will give the same result as the Kalman filter. However, the Kalman filter is probably more suited to surveying techniques with respect to ease of processing.

The combined use of powerful workstations and integrated software should mean that all navigation data can be processed offshore.

12.2 New computer technology

12.2.1 RISC

RISC is an acronym for Reduced Instruction Set Computer, and is a style of computer architecture emphasizing simplicity and efficiency. RISC designs begin with a necessary and sufficient instruction set, and a few simple operations account for almost all computations.

The advantage of RISC is the inherent speed of a simple design and the ease of implementing and debugging a simple design. Currently, RISC machines are about five times faster than comparable traditional architectures such as PCs and the old HP systems, and are easier to implement.

RISC is to hardware what the UNIX operating system is to software. The UNIX system proves that operating systems can be both simple and useful (cf. DOS) and hardware studies suggest the same conclusion. As technology reduces the cost of processing and memory, complex instruction sets become a liability for performance. The designers of RISC machines strive for hardware simplicity, with close cooperation between machine architecture and compiler design.

Portability between platforms has been the key to the success of UNIX, and the same is true of RISC architecture. Everybody in the commercial world has their own idea of which system is best, and software developers need to ensure that their products can be recompiled and installed from platform to platform with the minimum time and cost.

There are a number of characteristics which are typical of RISC machines:

- Most instructions are executed in a single cycle.
- There is little or no microcode. Microcode adds a level of complexity and raises the number of cycles per instruction.
- All computation instructions involve registers — memory accesses are made with only load and store instructions.
- All instructions are the same length (32 bits).

- The instruction set allows for the processing of several instructions at the same time (pipelining).

- RISC machines have high-performance memory — at least 32 general purpose registers and large cache memories.

- Only those features that really improve performance are implemented in hardware. Software contains sequences of simple instructions for executing complex instructions themselves, which improves system efficiency.

The real keys to enhanced performance are single-cycle execution and keeping the cycle time as short as possible.

Microcode is made up of low-level hardware instructions that implement the higher-level instructions required by a particular architecture. This code used to be burnt into ROMs, but this became prohibitively expensive because fixing a bug meant replacing the ROM chip. So microcode was then placed in read/write memory chips called control store RAMs. As integrated circuit improvements in the 1970s made micro-programming more efficient, instruction sets became more and more complicated.

The person who probably did more than anyone for RISC (in its early days) was Seymour Cray. He believed that complexity was bad, and in 1975 he said:

"Registers made the instruction very simple. That is somewhat unique. Most machines have rather elaborate instruction sets involving many more memory references in the instructions than the machines I have designed. Simplicity, I guess, is a way of saying it. I am all for simplicity. If it's very complicated, I can't understand it."

As semi-conductor memory began to replace ferrite-core memory in the early 1980s, a new set of simplified design criteria for computers emerged:

- Instructions should be simple unless there is a good reason to the contrary.

- Microcode is generally no faster than sequences of hardwired instructions, and the use of microcode makes modifications harder.

- Fixed format instructions and pipelining are more important than program size. Program size was always one of the most major considerations in old main-frames and PCs.

- Compiler technology should simplify instructions, not generate more complex instructions.

The speed advantage gained by RISC machines is all important in on-line navigation processing. We are trying to compute a complicated least squares solution, and the solving of a large matrix inversion in itself requires many thousands of iterations and, therefore, machine cycles. Using any given benchmark, the performance P of a computer is inversely proportional to the product of the benchmark's instruction count, I, the average number of clock cycles per instruction, C, and the inverse of the clock speed, S, as in the following formula:

$$P = \frac{1}{IC\left(\frac{1}{S}\right)} \tag{12.1}$$

The way this formula can be used in practice is as follows:

Suppose that an RISC machine and a conventional machine have the same clock speed, so that S is identical. The number of cycles per instruction, C, is from 1.3 — 1.7 for RISC machines but 4-10 for traditional ones. This would make the RISC machines three to six times faster than the traditional machine, except that the RISC has to perform about 20 — 40% more instructions because the traditional machine's instructions are more complex and powerful. This reduces the RISC's advantage to two — five times faster.

Compiled programs are larger on RISC machines than on traditional machines, partly because several (simple) instructions replace a single (complex) instruction, and partly because of decreased code density. All RISC instructions are 32 bits wide, whereas on traditional computers instructions may be narrower.

12.2.2 SPARC architecture

SPARC stands for Scalable Processor ARChitecture. The term scalable refers to the size of the smallest lines etched on a chip. As lines get smaller, chips get faster, but some chip designs do not shrink very well — they do not scale properly — because the architecture is too complicated. The term architecture in computers refers to an abstract structure with a fixed set of machine instructions. Because SPARC architecture is very simple, it scales well. This means that SPARC systems will get faster and faster as better chip-making techniques are perfected.

The SPARC instructions fall into five basic categories:

- Load and store instructions (the only way to access memory). These instructions use a register and a constant, or two registers to calculate the memory address involved.
- Arithmetic/logical/shift instructions. These instructions compute a result that is a function of two source operands and put the result in a register.
- Floating point operations, working concurrently with integer operations.
- Control transfer instructions, which include jumps traps, calls and branches. They are usually delayed until after execution of the next instruction, so that the pipeline is not emptied every time a control transfer occurs.

- Read/write control register instructions, which include instructions to read and write the contents of various registers.

A unique feature of SPARC design is overlapping register windows. This can be compared with radial tyres on a car. As the tyre rotates, the zigzag tread grips a different portion of the road surface. The tread is analogous to overlapping register windows - results left in registers become operands for the next operation, obviating the need for extra load and store operations.

12.2.3 What is happening now?

All three main contractors (Geco-Prakla, Western Geophysical and PGS) are migrating, or already have migrated to SPARC systems. All are able to produce P2/91 files and P1/90 files in real time or pseudo-real time.

For companies trying to port from a DOS environment, there is a steep learning curve in getting to grips with the concepts of open systems and the graphical user interface GUI. In the DOS environment the program has complete control at all times, so the programmer knows what is going to happen at a particular instant. In an open system such as SPARC this is no longer the case. Several processes can be running at the same time on a machine, and it is very important for the programmer to realise that his application has to get along with all the others that may be running on the network and on his machine. No longer does his application have exclusive control of the CPU or the even the screen, so the strategy for program development is very different.

12.3 Quality control

12.3.1 Preconditions

One of the main reasons that navigation processing is carried out ashore, and one that we have deliberately left until now, is the problem of quality control (QC). The means of adequately controlling final processing of data offshore have generally escaped contractors and clients alike. UKOOA have a produced a document *(Guidelines for the Measurement of Quality and Basic Quality Control of Offshore Positioning Data. 1992)* which sets out to implement a Quality System for positioning, and the notes that follow are taken mostly from that publication. The publication itself was intended to be integrated into a technical specification for offshore positioning, and part inclusion here should be seen as an attempt to reinforce the idea that adequate quality control is crucial to the success of a survey.

The preconditions applied to QC are as follows:

(a) Simultaneity — observations that are processed in one batch (i.e. a shotpoint) should refer to the same instant in time. In practice this is rarely achievable and for that reason the time mismatch (skew) should be kept as small as possible. The maximum mismatch between the event time and the actual measurement time, multiplied by the maximum rate of change likely to occur in the relevant observation, should not be more than one-fifth of the standard deviation of that observation.

(b) Raw Data — No pre-processing, such as smoothing and filtering, except that carried out in the receiver or measurement device outside operator control should be applied to the observations before logging. All operator-controllable filtering in positioning equipment should be disabled.

(c) Redundancy — the survey network should contain sufficient redundant observations to enable biases to be detected to an adequate level. A well-controlled network will have at least 30% redundant observations. The minimum percentage of redundant observations should be agreed between client and contractor prior to the start of the survey. The redundancy should be evenly spread, such that the network contains no uncontrolled observations. However, operational requirements may force exceptions to be negotiated between supplier and client in special situations.Examples of solutions with less 30% redundant observations are:

- A GPS fix from five pseudo-ranges (25% only).
- DGPS positioning, where pseudo-range corrections are transmitted from only one shore reference station. This effectively gives rise to uncontrolled observations.

In accepting such cases it should be realized that positioning configurations without sufficient redundancy are undesirable in principle as lack of redundancy impairs QC or even prevents it altogether.

12.3.2 Quality, accuracy and precision

Quality here is defined as the accuracy of the positioning data satisfying the tolerances specified by the client.

The objectives of QC in this context are:

- Control of the propagation of the **random** errors of the observations into the coordinates of the point(s) to be positioned — the **precision.**
- Identification of biases in the data, whether systematic, periodic or incidental (**reliability**).

The second objective is generally the more important, as undetected biases can have far greater consequences than greater **random** errors expressed in **higher standard deviations.**

Precision should be specified on the one standard deviation (one sigma) level.

Tolerances should be specified on the 99% probability level.

12.3.3 Position calculation

The computation of position should be performed by means of weighted least squares.

Hybrid fixing, i.e. the integrated adjustment of data from more than one positioning system, is encouraged as it enables QC to be more effective.

Ideally, complex survey networks should adjusted in one step, **integrating all available positioning data** for any one event. However, a fully integrated adjustment must be supported by a sophisticated QC method offering proper error detection capabilities. Should a fully integrated adjustment be used, a detailed description of the QC method should be supplied as part of the positioning quality plan.

The weighting of observations should be the **inverse of their covariance matrix**. The elements of the covariance matrix should be derived from realistic and accurate predictive estimates of the standard deviations of all observations and the correlation coefficients between the observations

12.3.4 Positioning quality plan

The contractor should submit a positioning quality plan to the client prior to the start of the survey.

The objective of the positioning quality plan is to provide a basis for agreeing the quality requirements for the survey and to provide the evidence that the positioning systems and survey tolerances as proposed by the contractor are able to meet the positioning tolerances.

The plan should include:

- A description of all positioning systems and **nodes** proposed for the survey.
- A diagram or description of the proposed **survey network**, clearly showing the relative positions of all nodes and giving their offsets with respect to the navigation reference point (NRP).
- A list of all proposed **observations** with their respective **standard deviations** and **correlation coefficients** between the observations, believed to be an accurate and realistic prediction of the **random** behaviour of the observations. Alternatively, a listing of the **covariance matrix** could be presented.
- A description of the **functional model** and the adjustment method.
- A description of all data quality measures and controls relevant to the positioning process. A detailed description of the **associated QC method** should be provided.
- A calculation of the **predicted position precision**, achievable with the proposed systems, providing proof that the quality requirements for the survey can be met anywhere in the survey area.
- The predicted position precision referred to above should be computed for any points such as **NRP, common mid points, streamer receiver groups, sources, etc.** as directed by the client. The **standard error ellipses** of these points are computed in a sufficient number of representative locations in the survey area. In all cases the semi-major axis of the stand-

ard error ellipse multiplied by 3.03 should not be greater than the position tolerance specified in the contract for any of the specified points.

12.3.5 Data acceptance and rejection

Data acceptance and rejection criteria should be applied to data sets. A data set in this context is defined as the minimum set of time contiguous raw positioning data and derived coordinates, that is large enough to yield a statistically meaningful sample of quality measurements. The data set size should be agreed between client and contractor, and a value of 200 events (or shots) is suggested.

The same data may form part of a large number of data sets which is achieved by moving a window with a width of 200 events along the time contiguous data.

This mechanism is intended to capture and reject consistently poor positioning, while allowing the occasional outlier or spike in the raw data to be accepted on the basis that the spike or outlier can be (and should be) removed in on-board processing.

Data, forming part of data sets satisfying all the **data quality criteria** defined in section 12.3.6 should be considered of adequate quality and should be accepted. Conversely, **only** data that do not satisfy the requirements should be rejected.

12.3.6 Data quality criteria

A data set is considered to satisfy the quality criteria of the survey if all of the following conditions are satisfied:

- At least 99% of all events constituting the data set have a **standard deviation of unit weight not greater than its tolerance**, while the effect of any spikes or outliers have been removed by processing, e.g. filtering, smoothing or interpolation techniques.

- The semi-major axis of the standard error ellipse of each of the specified points multiplied by 3.03 is not greater than the **position tolerance** in any position fix which is part of the data set.

- The number of **redundant observations** in any position fix which is part of the data set is not less than the agreed minimum redundancy, expressed as a percentage unless it has already been agreed that a configuration with an inherent lack of redundancy is to be accepted.

12.4 Integrated quality control

12.4.1 Introducing the seismic volume

Almost all of the book has been devoted to navigation. However, in terms of seismic acquisition navigation is not an end in itself. The following paragraphs outline the manner in which quality control of the navigation volume can and should be integrated into the seismic volume. Acknowledgement is made here to Jeff Hume of Energy Innovations Inc , Houston. Jeff's forward thinking is revolutionizing quality control of seismic projects at sea, on land and in the transition zone environments. The CENSUS™ software is referred to in several places in the text. This system, from Input/Output Inc, Houston, offers a true total quality management approach (TQM) to 3D surveys, allowing quality attribute information computed during survey planning and design to be stored and compared with quality attribute information computed during acquisition. The system allows integrated quality control of both navigation and seismic data and results in a coordinate- based database in which individual mid point or reflection point quality is known, allowing edits of 'bad shots' and 'bad traces' to be complete by the end of the field project. CENSUS™ is used to examine the spatial variation of quality of all measured and computed quality attributes and offers the field geophysicist tools to examine the offset stack response and bin smear of the data actually in the bin, after removal of data on a selected quality attribute basis (see Chapter 13 for an explanation of these terms).

The final database is used to QC 3D seismic processing by ensuring the field based decisions and recommendations are carried out by the processing centre and, in some cases, where the output format is supported by the processing centre, it can be used to directly drive trace selection and other steps in seismic processing.

12.4.2 3D project quality management

The basis of any quality management requirement of a 3D project is the assurance that acquisition and processing of the 3D volume is performed in the **absence of geophysically significant errors.** In addition it is desirable to have an environment in which auditable quality attributes of the design, acquisition and processing of the data are available to the client in a manner in which cost effective decisions can be made.

Survey design, acquisition and processing must be considered as a single entity that is fundamentally integrated. Cost-effective acquisition cannot be done without a thorough understanding of the processing choices available. Survey design needs to consider the practical limitations of the acquisition method and the effect of the processing sequence.

For some time the typical client 3D QC approach has been to utilize independent systems that essentially duplicate the computations made by the contractor, offering little added value. With the advent of powerful RISC workstations and the ability to maintain a true trace level quality database in CENSUS™, a new generation of quality-based decision making tools is now available to clients,offering verification and interpretation of quality attribute data provided by the contractor. These new techniques include:

- Pre-survey planning, survey design and simulation tools for marine, bottom cable and land methods to analyse contractors' proposals and build survey design databases populated with modelled quality attributes for comparative use in the field.
- An interactive seismic analysis environment (Bin Analytical Tools) in which client project geophysicists and interpreters can make geophysically based decisions on survey designs and whether to shoot infill or make up shots, apply flex binning or leave a hole in the data during the acquisition phase. These tools include 'offset stack response' and ' bin smear analysis'.
- Ray traced modelling and CENSUS™ target illumination and 'quick look' DMO (Dip Move Out) shift based testing, where assessment of coverage for the theoretical survey design or the actual surface sampling acquired in the field is achieved based on a range of effective reflection points rather than mid-points of data for selected targets.
- A comprehensive set of linear and spatial comparative tools in which real-time positioning performance can be compared with final P1/90 reprocessed coordinates.
- Spatially variant flex tools that map 'trace distance from bin centre', allowing the effect of flex binning in terms of the moving of mid-point data to be better understood.
- An interactive quality based binning environment in which duplicate trace selection is performed, not on a 'first one in the bin' basis, but on the basis of client-specified quality attribute selections. Quality based binning allows selection of low-noise over high-noise duplicate traces, selection of traces positioned near the bin centre over traces positioned far from the bin centre or selection of traces based on any other measurable quality attribute.
- Output of a seismic processing QC or script file based on trace selections and flex criteria established in the field, offering great savings in turnaround time and much enhanced continuity between acquisition and processing.
- Modular database construction, which allows clients to analyse and interact with the CENSUS™ display databases in their office and use bin analytical tools to verify field recommendations at very low cost.
- Archived field trace level database using final *XY* coordinates in which all survey design and acquisition quality attributes are stored provides a permanent, auditable and easily accessed source of reliable indicators of potential error problems or sensitivity of the data volume, to which processors and interpreters can refer.

In the CENSUS™ environment, TQM becomes truly total as all phases and aspects of the quality of the project are fully monitored. As with all quality management philosophies, the CENSUS™ approach requires a commitment by the client to a change in the traditional contractual approach to the project by allowing seismic contractors more responsibility for their own quality control, while insisting on access to verifiable quality attributes and raw and processed data in a timely manner.

Traditional QC and production of auditable quality attributes should be the responsibility of the seismic contractor. Audit and spot check verification of the data, interpretation of the attribute information in the context of the project objectives and cost versus data quality risk decision making must clearly remain the responsibility of the client.

Real cost savings can only be achieved if the client, geophysical consulting company and contractor all understand the common objectives of all aspects of the project.

12.4.3 TQM methods and requirements

The quality management methods proposed below can be very valuable and cost effective when used simply as traditional quality control approaches applied to the field operation. The cost savings and continuity of quality assurance is multiplied significantly when these methods are applied to a 3D project as part of a TQM strategy using an integrated approach in a properly designed contractual environment.

The basis of the approach is that a 3D project is a single entity requiring a continuous, seamless, client-referenced TQM process for maximum efficiency.

Total quality management of 3D seismic data requires that quality information be computed and stored **for every trace for every shot on the survey**, and that trace selections be made not on the basis of 'first in the bin' but on the quality of the seismic trace, its proximity to the bin centre and whether all the survey systems were operating within specification when the shot associated with that seismic trace was fired. This is the process we call **quality based binning.**

No matter how sophisticated the quality management process that is applied, there is always an ongoing requirement for conventional supervisory quality control methods. We believe that most contractors now offer reasonable real-time or near real-time QC monitoring systems for survey, seismic instruments and source control, all of which deliver machine-readable and auditable measures of quality which can be recovered post-shot and incorporated into the CENSUS™ database.

During field operations, the contractor's real-time 'as steered' positioning data for source and receivers are collected and verified by first loading the data and associated quality attributes, (as provided by the contractor), into the CENSUS™ system where it can be 'replayed' to spot any obvious errors and then analysed in 'edit data base'. This facility of CENSUS™ is the most powerful set of survey data and attribute analysis tools on the market today and allows detailed graphical analysis of input attributes and computation of attribute correlation.

Navigation, seismic operations and seismic data quality attributes are merged with the final accepted P1/90 datasets so that quality based binning trace selections can be made based on 'final' source receiver coordinates and measures of shot and trace quality.

The type of attributes that can easily be monitored vary from contractor to contractor. In some cases key attributes from other contractor-supplied data need to be computed. Gross errors in phone or shot locations can be identified by comparing the seismic arrival time information with predicted arrival times based on the input coordinate data. In land operations, predicted uphole times can be compared with actual measured uphole times and the results plotted spatially and correlated with the driller's name! This type of data can also be used in 3D processing to aid efficiency of statics and other surface problems.

The CENSUS™ 'bin analytical tools' module, which uses representative normal move out (NMO) corrected common mid-point (CMP) gathers and target horizon wavelet information, in conjunction with final source/receiver position data, represents a real seismic response equivalent to that which is found in the processing centre. In addition, CENSUS™ offers a comprehensive set of trace selection utilities that duplicate any of the seismic processing contractor's selection/rejection schemes. As all of this information is stored as database attributes and the final contractor coordinate data is used, CENSUS™ can generate a fully corrected header mask or can drive 3D seismic processing.

The aim of the TQM process is to take into seismic processing, at a minimum, a verified location data set in which trace selection criteria have already been decided based on rules laid down by the client. This dataset is available at any time as a reference either during processing or, potentially, during early interpretation of the data in the event of problems arising.

The achievement of these results is possible with nearly all modern seismic vessels and field crews and the results obtained are complementary to and not competitive with or a duplication of the contractor's effort. As identification of errors is initially carried out exclusively with contractor-provided data, and verification is done against final data provided by the contractor, there is little room for argument. To ensure that this type of result can be obtained, the TQM process must be important to the client, the client must intend to use it at later stages of the project and the contract specification must require timely delivery of raw data, processed data and attributes from the contractor's systems to the TQM environment. Failure to implement these important "access to data" clauses, will reduce but not eliminate the substantial cost and time savings of the method.

13 Binning systems

13.1 Introduction

13.1.1 What is binning?

Binning at sea can be viewed purely as a quality control function. It involves computing all (or some) of the common mid points (CMPs) and populating the cells of a grid with those CMPs. The cells are known as bins, and the operation is called binning. This is the basic conecpt of binning — ensuring that the required number of CMPs fall in each bin. As we shall see later, there is much more to it than this.

Probably the largest problem confronting any binning system is the amount of data. Let us look a dual-source example in which there are three streamers and 300 groups in each streamer. Suppose that there are 1000 shotpoints in each sail line and that there are 200 sail lines in each prospect. If a bin measures 50m long and 25m wide, and the shotpoint interval is 25m, the number of bins is computed as follows:

Length of prospect in bins = 1000/2 = 500 bins.

No. of CMP lines per sail line = 6.

Width of prospect in bins = 6 × 200 = 1200 bins.

Therefore total number of bins = 600000.

Finally, let each bin be represented by 1 bit for each of the 300 traces. This is very simplistic, but will serve to illustrate the amount of data involved

Each bin contains 300 bits (1 for each trace)

= 38 bytes per bin.

Size of database required to hold the prospect

= 38 x 600000 bytes

= 23MB

Thus we need 23MB of storage just to hold this very simple database.

However, so far each trace is represented by a single bit in a bin - and this can only be a 0 or a 1. Therefore, so far we can only really state whether or not a trace has landed in a bin. But what if we wanted to determine where a trace was in relation to the edge of a bin? This information is needed to enable us to re-bin at a different bin size, or to overlap the bins, and the smallest way in which the information could be represented is an unsigned character, which requires a byte. This would immediately increase the size of the database to (8 x 23MB = 184MB).

Now, consider that the database needs to be manipulated by a computer. In the (once again very simplistic) loop of code which is needed to determine the correct bin for a particular trace, there may be (say) 1000 machine instructions. So to manipulate the data for a single line would take (1000 × 1000 × 900) = 900 million instructions for our 1000 shot line. Assuming one cycle per instruction this would take 9 seconds on a 100MHz machine that had no other concurrent processes running, even without thinking of display overhead, swap space and memory required.

Obviously it would take a substantial computer to hold and manipulate a database of 184 MB in memory, and so the database must be split into more usable segments.

The point of this discussion is that to bin data on a small computer in real time is no mean feat, and takes an enormous amount of forethought and planning, not to mention some pretty smart code to write a binning program to run in real time.

13.1.2 Data types

There are several ways in which the binning process can be approached. As discussed above, the data we actually need for the binning algorithm are the CMPs. If we know the source position and the group position, the CMP (trace) position is very simply computed:

$$X_{CMP} = \frac{(X_{SOURCE} + X_{GROUP})}{2} \tag{13.1}$$

$$Y_{CMP} = \frac{(Y_{SOURCE} + Y_{GROUP})}{2} \tag{13.2}$$

The source and group positions are all listed in the standard UKOOA P1/90 files, which we discussed in Chapter 12. But let us look at the size of one of these files in the context of our example. Each line of a P1/90 file is 80 bytes long. In the header part of the file there will be about 50 lines. For each shot, there will be two lines for the antenna and source positions, then the group positions are listed in lines containing three groups each. So for each shot there will be:

$$(2+900/3) \times 80 \text{ bytes} = 24160 \text{ bytes.}$$

For our 1000 shot line there will be

$$(50+1000 \times 24169) = 24.1\,\text{MB in the whole file.}$$

Using file compression techniques this file can be compressed to about 20% of its original size.

One way of reducing the input file size would be to store the source position and the streamer polynomial coefficients for each shot instead of all the group positions. This would reduce the input size file size to about 240 bytes per shot, or 240kB in total — only a hundredth the size of the original file. Then the binning algorithm has the problem of computing all the group positions, a far more complex and time consuming task than that in equation (13.1) and equation (13.2). In fact it turns out that this task is so time consuming it is not really a viable option.

The other requirement for a binning system is that the bins need to be aligned along the CMP or sail lines. Unfortunately, most surveys are not aligned with the projection grid (i.e. along grid east to grid north), and equally unfortunately the standard UKOOA processed data files require the source and group coordinates to be in terms of grid coordinates. So the binning algorithm has yet another task to do — rotating the coordinates into the right azimuth. This can be done simply but involves sines and cosines which are computer intensive.

13.2 Binning problems

There are a number of problems which may be encountered whilst shooting the data and trying to populate the binning grid, and these are listed in the following paragraphs.

13.2.1 Shooting in opposite directions

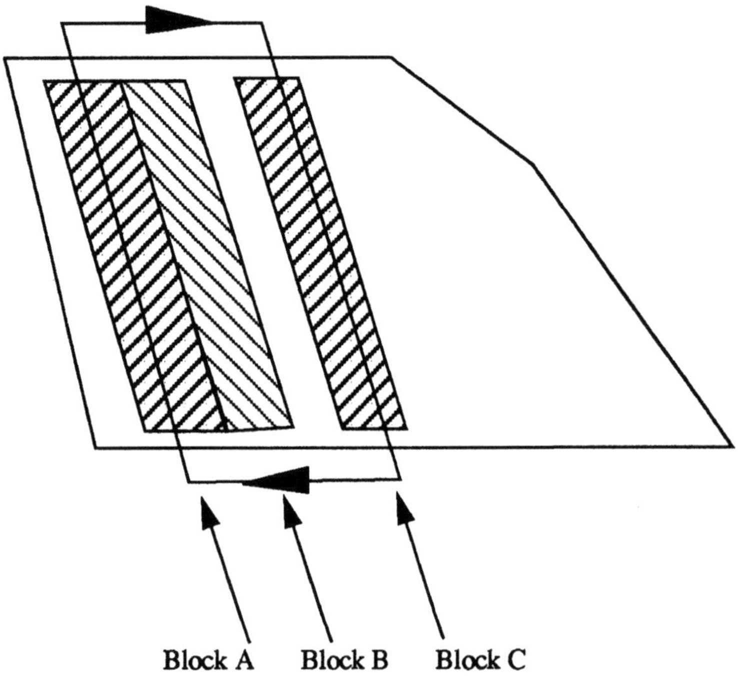

Figure 13.1 Binning seams.

In Figure 13.1 Blocks A and C are shot simultaneously as part of a 'racetrack'. The sample track indicates the vessel steaming northwards through Block A and then south through Block C. The blocks are shot as separate entities and it should be fairly easy to match up lines within those blocks. The problem comes along the seam between Block A and Block B, particularly if Block B is shot in the opposite direction. The data along the seam are subject to more sources of error because any inline errors in the data will effectively be doubled. Quite apart from this problem, there is the physical difficulty of trying to match data in perhaps four individual streamer segments.

13.2.2 Streamer feather

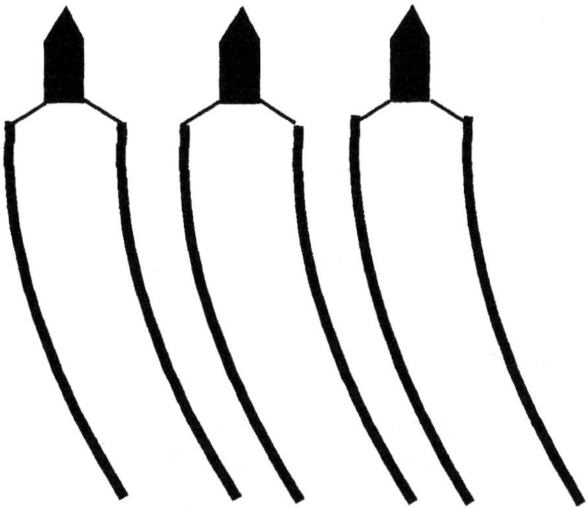

Figure 13.2 Effect of feather.

In Figure 13.2 three lines are shot in a northerly direction. Showing the underlying grid would unnecessarily complicate the diagram, but it can be seen that if the vessel does not steam the lines with the same feather a gap will be left, which is often the case.

In the North Sea, where the tidal flow is very predictable, it has been found most economical to steam just four lines a day, two in each direction, to make the best use of the tidal stream. If the data are poor along a line, that line may be scratched, but the vessel carries on steaming its intended track to be in the right place for the next line.

Unfortunately the tidal streams around Australia are somewhat less predictable, and the amount of resulting infill is often as much as 40% of the original prospect.

13.2.3 Solutions

This is where a flexible binning system can come into its own; if the binning system is able to vary the width of the bins (this is called flexing), effectively borrowing data from adjacent bins, the coverage may well improve considerably whilst still retaining data of suitable quality for the interpreters. This can result in shooting far less infill, but we have to be certain that what we see on the binning system is the truth.

13.2.4 Steering for coverage

In theory, a sail line is entered into the navigation system and the vessel steams up that line, perhaps filling up four or more CMP lines simultaneously. In practice, the vessel only uses the preplot sail line as a guide at the start of line. From then on the vessel steers for bin coverage, trying to match coverage on the current line with that from the adjacent line. The success of this practice depends upon a number of factors:

- The ability of the helmsman (or automatic steering system) to respond to required changes of course.
- The skill of the navigator in predicting the feather and the vessel's track and giving directions to the bridge.
- The quality and update rate of the binning display in giving the navigator the information he needs.

13.2.5 Data rejection

If a line is binned in real time but is subsequently rejected or part rejected from the coverage for any reason, we need the ability to be able to 'unbin' the data for that line.

This means that in our database we must also have knowledge as to which line each trace binned belongs to, otherwise we may end up removing traces from the wrong line. The retention of this knowledge in the database implies an even larger database; the alternative is to be able to rebin the whole (or relevant part of the) database very fast.

13.3 Specific requirements

13.3.1 Speed

As already stated, the size of database we are dealing with is very large, as are the input files. Manipulation of the database and the input files is going to require a lot of random access memory and CPU cycles, so we need to minimize both of these. One way of doing this is to use a SPARC based machine, which has a reduced number of cycles over other machines in the first place. Another way is to reduce the amount of work the binning algorithm has to do by using input files which already contain the CMPs. This can be done by the on-line navigation system - if it already produces a P1/90 file, then it is a small amount of extra work to produce a similar file containing CMP positions.

The next change we can make is to reduce the size of the database. Using current it is rather impractical to try and manipulate a database of some 720MB in real time. At the same time, it is not really necessary to view the whole database at once, but we do need to ensure that a reduced database is capable of binning all of the data on a current line even under conditions of extreme feather. As an example, consider a spread whose streamers are 40000m long, steaming a line under 15 degrees of feather. (It is worth pointing out here that theoretically the lines can be shot with any feather angle, and in any direction. It becomes purely a matter of economics in deciding whether shooting with high feather can give good coverage). Suppose we have a bin width of 25m in our example. The maximum distance from the desired line of bins will be at the far traces, i.e. at a distance of 2000m (approximately). The distance off will be (2000 × sin 15°) = 517 m. Therefore we must ensure that our database extends at least 20 bins either side of the current line to ensure that nothing is lost with a 15° feather.

If we use a display that extends 25 bins either side of the current line, we must also ensure that the coverage on the edge of the display is correct, which would mean binning lines "off the edge" of the coverage. Assuming that there are four CMP lines to each sail line, the number of sail lines that would need to be binned would be about 20. Since the number of required lines is now much smaller, we can almost certainly say that to re-bin this part of the prospect can be done from scratch, which immediately obviates the necessity to store offsets for each trace. We are therefore back to the state of using 1 bit per trace, and for 20 lines our database would be about 2.5 MB. This can very easily be manipulated by a SPARC machine without slowing down through using swap space.

Another way we could increase the speed is to bin only a proportion of the data. For instance, we could bin every tenth trace and assume that the database was then representative for all the other traces. This practice is not considered very ethical, nor is it safe; there may be systematic errors in the data which can easily be overlooked. Binning every trace is by far the most preferable method.

13.3.2 Segmentation and tapering

In general, the client will not require uniform distribution in the bins. It is now common to divide the streamer into three or four segments, although there is absolutely no justification for doing this. Bin flexing should be based on the target depth and should be spatially variable. The segments may be of unequal length, and the client then also specifies the coverage required for each segment as a percentage of traces. An example is given in Table 13.1:

Table 13.1 Streamer segmentation

Segment	Traces	Required percentage
1	1 — 75	90%
2	76 — 150	80%
3	151 — 225	70%
4	226 — 300	50%

This segmentation requires that we split our database into four parts and have the ability to display any segment at any time. The navigator may only be trying to get coverage for segments 2 and 3, and will need to switch repeatedly between the two displays to navigate the vessel.

As well as splitting the streamer into segments, we also need to be able to taper the binning. This effectively increases the bin size down the streamer, and ideally it should be such that the tapering occurs gradually throughout each segment length. For example, the nominal bin width may be 25 m, but we might wish to taper the bin width from 25 m at the front of segment 1 to 35 m at the back of segment 1. The technique involved is one of flexing the bins.

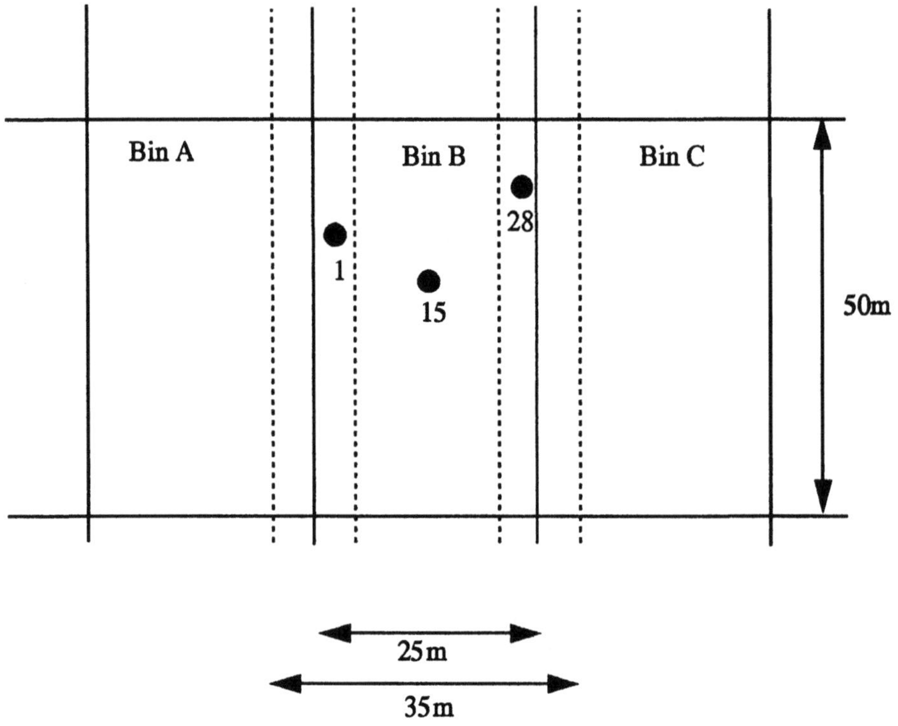

Figure 13.3 Bin flexing.

In Figure 13.3 we show three adjacent bins, A, B and C. The nominal bin width is 25 m and the required expansion is 35 m. Trace 15 falls near the centre of bin B, and so does not occur in bins A or C. But trace 1 falls near the edge of bin B, and is in the overlap area between bins A and B. Trace 1 therefore gets binned in both bins A and B. The same is true of trace 28 between bins B and C. In this way traces can be borrowed from adjacent bins to improve the coverage. Of course, the result of doing this from the interpreter's point of view is that resolution will be lost in processing.

13.4 Attribute analysis

So far we have looked at binning only the CMP positions. However, during the course of a line a huge amount of data is produced from a navigational aspect, including the following:

- DGPS PDOP

- DGPS ellipse ø
- DGPS age
- SVs in use
- DGPS correction latency
- Acoustic ellipse ø
- Acoustic C–Os
- Raw water depth
- Key node ellipse data
- Tidal correction
- Tailbuoy rotation
- Cable separations
- Compass bearings and ellipses
- Bird depths
- Time
- Array separations
- Seismic signal-to-noise ratio

This list is by no means exhaustive, but it represents some of the attributes that are capable of and should be logged as a matter of course. The traditional way of reviewing this data is to look at end of line statistical printouts and graphs, assess whether or not the data are good and move on to the next line. However, in a 3D survey the seismic data are a volume of data, not a two-dimensional cross section. Similarly, the navigation data are also a 3D volume, and they should be reviewed spatially, not linearly.

Essentially what we require is a geographical information system such as CENSUS™ that will allow us to view, manipulate and assess the data spatially.

14 Typical specifications

14.1 System specification

14.1.1 Introduction

This chapter gives examples of four different types of specification. The first is for a single system, the second is for job responsibilities, the third is for the whole project specification and the fourth is for techniques used with multiple integrated systems.

14.1.2 Radio positioning system — Argo

The first part of the specification lays out general rules for dealing with radio positioning systems, followed by descriptions of the specific system and calibration and operating procedures. Argo has been used as the example but it could as easily have been Syeledis, Hyperfix, Geoloc or a number of others.

14.1.3 General rules

(a) All radio positioning systems shall be operated according to the best surveying principles and standards — Best Practice.

(b) All systems are subject to:

the influences of the propagation of radio waves

pattern geometry

ground conductivity

climatic conditions

errors made in offset calculation

base station coordinate errors

incorrect software parameters

incorrect datum parameters

These potential sources of positioning problems shall be minimized by good calibration.

(c) All care must be taken to avoid any ambiguities in the data. Hybrid fixing, i.e. the integrated adjustment of data from more than one positioning system for the vessel, is encouraged as it reduces the opportunities for ambiguities to occur.

14.1.4 Calibration

(a) All systems must be calibrated at three stages of the survey project:

(i) Before the start of the survey (to create a benchmark)

(ii) At regular intervals throughout the survey (to ensure there is no movement during the survey)

(iii) On completion of the survey final calibrations shall be carried out (to ensure the system is still operating within specifications)

(It should be noted that there are occasions when a post-calibration is unnecessary. For example, where continuous comparisions are carried out between two systems the post-calibration is not required.)

(b) Calibration as specified for each system type must include any spare receivers and back up equipment and systems

(c) Trials shall be carried out both statically and dynamically

(d) Calibration shall test a system meaningfully to assess the true capabilities of the system, allowing for the atmospheric conditions at the time. It must be operating to specification for the whole of the survey and be within its capabilites for the whole of the survey

(e) No uncalibrated equipment or equipment components may be used during the survey

(f) The equipment shall always be installed and used according to the manufacturer's instructions

(g) Calibration techniques shall be approved by the company.

14.1.5 Static trials

Although methods vary greatly for individual systems, some forms of static and dynamic trials are possible for all systems. The static trial tests the functioning of the main system components in the field prior to deployment. It is important that these trials be conducted by the contractor and supervised by the company. By doing this both the company and the contractor begin to establish a relationship with the system and its peculiarities, and will know what to watch out for during the project.

For most radio positioning systems all that is required is a measured baseline of about 2km in length, preferably over a sea path but a land path is also suitable provided the land is flat with no obstructions, no grazing rays and not susceptible to interference from power lines, etc. A distance of 2km is too short for propagation errors to have a marked effect, and too long to risk damage to equipment or personnel through exposure to high-power transmissions.

Such a static trial will quantify equipment delays and test the general functionality of the system.

14.1.6 Dynamic trials

Dynamic trials are intended to determine the accuracy of the system when installed on the vessel in the final configuration. The system will be used in conjunction with the contractor's and QC's navigation software, so that the datum, delay, C–O and velocity correction parameters can also be checked.

Systems must be checked for repeatability and reliability in different parts of the prospect, and where positioning needs to be very accurate (3D projects) a secondary system must be used to calibrate the system.

14.1.7 Redundancy

Redundancy as agreed in the technical specifications must be maintained at all times. If there is a loss of redundancy, the survey may only continue at the discretion of the company.

See section 14.4 for more on redundancy.

14.1.8 Equipment records

A record book shall be kept for the duration of the survey containing the following details:

(a) A plan showing all base stations, antenna positions and equipment positions in relation to other survey sensors, including all coordinates.

(b) Equipment types and serial numbers, calibration data, receiver settings and any adjustments.

(c) Equipment downtime, reason for downtime and the action taken to remedy the problems.

(d) All changes to equipment, together with the new calibration information, etc.

On completion of the survey all records and log books shall be submitted to the company.

14.1.9 Argo

Specifications.

Manufacturer: Cubic Western Data, USA

Support services: Worldwide

Frequency: 1.6 — 2MHz

Range: 250 — 350km

Resolution: 0.01 lane

Accuracy: 0.05 lane

Mode: Ranging (can also be used in hyperbolic)

No. of users: 7 (using four ranges)

Operating areas: Worldwide

Specifications are the manufacturer's own.

Description

Argo (which stands for Automatic Ranging Grid Overlay) is nearly always used in its ranging mode, and works using a time-multiplexed system.

Argo must be calibrated using a secondary navigation system such as Miniranger or Micro-Fix. Such calibration must be used together with the method used to calibrate the partial lane counts on an Argo system. (The partial lane count is the fraction of a lane to be added to an observed lane count from a station to make that lane count equate to the correct range, and is a computed – observed (C–O) value due to instrumental delays at both the station and the vessel.)

Summary

The variation in velocity experienced, particularly in tropical waters where there are rapid seasonal and diurnal temperature and relative humidity variations, means that the system performance will **definitely be degraded** if regular calibration checks are not carried out.

14.1.10 Static calibration

Argo contains instrumental delays — they are in terms of lanes, not metres, and are known as 'partials' because they are corrected using partial lane corrections. Ideally the same short baseline calibration should be performed as with Syledis, but this is not practical because of the antenna size (30m), ground net size (radius up to 80m) and power requirements. Therefore the chain is set up ashore and the mobile unit is installed on the vessel.

Prior to sailing field-strength trials must be conducted on the vessel and the shore stations to determine if all parts of the system are transmitting at their optimum levels.

14.1.11 Dynamic calibration

If the mobilization port is within chain coverage and reasonably free of obstructions, alongside calibration and harbour trials can be carried out, but these will only give approximate values before the partials are known. The trials are of use as a confidence check during crew changes if the vessel returns to port.

Alongside calibration

The prime mobile is installed on the vessel, together with its assigned antenna and cable. If the wharf at the mobilization port is within the chain coverage, the following trial shall be performed.

The vessel is moored very tightly to the wharf, to restrict surging and change of heading to an absolute minimum. All crane and similar movements shall be stopped during the trial to reduce re-radiation and reflection effects to a minimum.

Harbour trials

Harbour trials shall be conducted when the vessel sails for the prospect, but are not recommended as a very accurate method. Pairs of beacons or other shore marks are selected to form transits along which the vessel steams. The two transits are selected so that they cut as closely as possible at right angles, preferably aligned north — south and east — west. Each of the marks forming the transits must be coordinated, so that the navigation computer can be programmed with the planned tracks. The vessel shall steam in both directions, towards and away from the marks, with the helmsman keeping the vessel on track by eye. The navigation system shall log the distance off track and plot the results.

The trial's success is dependent upon the helmsman's ability to stay on the transits, and for that reason is not always successful. But it will quantify the system's dynamic capability under reasonably controlled conditions.

Basline crossings

The baselines are multi-system, with a secondary station at one end and an Argo station at the other, but other than that the technique is the same as for Syledis. The baselines are crossed within 10km of the secondary station, so that there will be no propagation effect from that station. Then all the baseline error should be attributable to the Argo range and the propagation scale factor can be deduced.

Partial calibration

To calibrate Argo the contractor must use a secondary terrestrial radio-navigation system at least as accurate as Argo itself. Good candidates are Micro-Fix, Trisponder and Miniranger. The secondary system shall undergo a short baseline calibration and secondary remotes shall be installed about 10m up the Argo antenna towers, making sure that the Argo and secondary electrical centres are as close as possible and within 1m of each other. The calibration can be performed with the secondary system set on a satellite station, but the computation is a little more complicated and therefore more prone to errors. The secondary mobile is installed on the vessel, and the offsets between it and the Argo antenna and the NRP measured.

The vessel shall proceed to the vicinity of each station in turn and shall steam reciprocal lines about 2km long at a distance of 10 — 15km from the station. The intersection of the tracks shall be on the line between the stations and the prospect, and the lines shall be orientated in the direction of the primary sail line and at right angles to them.

The navigation computer shall be set to cycle at 10 second intervals, and on each cycle the Argo and secondary ranges shall be logged simultaneously. The difference between the two ranges will be equivalent to the partial lane correction in metres.

This computation shall be repeated for every cycle; most software available will do this automatically, and provide statistics for the partial corrections. An integrated system which can automatically deskew observations from multiple systems to the NRP can be used by giving the secondary system range a standard error of (say) 25 m and the Argo range a standard error of (say) 5 m. Then the partial correction in metres will be equal to the difference between the residuals for the two ranges.

The reasons for steaming lines in this configuration are as follows:

The intersection point is on the line between the station and the prospect so that it lies on the propagation path to be used for the work. There will be no abnormalities due to an induced land path, for instance.

The lines are run in the same direction as the primary sail lines because any phase error induced by the vessel running in that particular direction will become evident.

The resulting partial corrections from the four lines should be very close to each other (within 0.03 lane).

On-site calibration

If there is no suitable structure or island offshore to conduct baseline crossings, the propagation factor must be deduced from an on-site calibration.

The vessel shall steam two boxes inside the prospect. The box sides shall be about 5 km long and the sides shall be aligned with the primary sail lines. The vessel shall steam the boxes with the navigation system cycling at about 10 seconds, and logging both Argo and secondary system positions. Production of residual, variance and separation statistics between the two systems will point to any errors. The secondary system must be as accurate as Argo for this test to be successful; the secondary system most often used is DGPS, and great care must be taken that the DGPS is operated using optimum constellations, elevation masks, update rates and so on.

14.2 Guidelines for representatives in seismic operations

14.2.1 Aims

(a) To provide a structure within which the survey representative can work efficiently knowing his responsibilities and duties.

(b) To set out the procedures and minimum requirements required by the company.

(c) To accustom the survey representative to the briefing, debriefing and minimum reporting procedures required by the company.

14.2.2 Procedures

Planning and pre-survey preparation including calibrations.

Acquisition and quality control of data prior to acceptance.

Binning for coverage.

Infill work (if necessary)

Processing and checking.

Reporting and acceptance of survey data.

Storage and record keeping.

14.2.3 The company's responsibilities

(a) The company must approve and agree to any exception to the Contract Specifications prior to the start of the project.

(b) The company specifies the positioning control that shall be used in the project area and shall ascertain that the systems used are suitable and provide enough redundancy.

(c) The company specifies the parameters for the quality pPositioning plan and shall ascertain that the systems used are suitable and fulfil all requirements.

(d) The company is responsible for communicating all survey and positioning requirements to the contractor and suppliers prior to the commencement of the survey.

(e) The company shall ascertain that the angle of cut between any two lines of position shall be between 30 and 150 degrees and that at least three independent lines of position with good geometry are available for the whole survey area.

(f) The company must inform all parties of any navigational hazards/problems, e.g. wrecks.

(g) The company is responsible for ensuring that all personnel and equipment are mobilized and logistical problems sorted.

(h) The company shall ensure that the data ares acquired in the formats required by the companyy, as specified in the contract and technical specifications for the project.

14.2.4 Responsibilities of the company's seismic representative

(a) The seismic representative shall make daily progress reports via phone, fax or electronic mail.

(b) The seismic representative is responsible for the quality control of on-board positioning and must include the survey information in the daily report.

(c) The survey representative is responsible to the seismic representative.

d) The survey representative must inform the seismic representative when the specifications are not being met.

(e) The seismic representative is responsible for deciding when shooting can continue or when it must stop, but will accept advice from the survey representative in reaching his decisions.

(f) The survey representative shall keep the seismic representative informed if he considers that line criteria are not being met. This should be recorded and the company informed immediately.

(g) The seismic and survey representatives shall work together as a team and the seismic representative has overall responsibility for the day-to-day operations on board.

(h) The seismic and survey representatives should be briefed by the company and every effort taken to arrange their meeting prior to the commencement of the project. It is the responsibility of the company to ensure the seismic representative is fully informed and aware of the obligations between the company and the contractor.

(i) The survey representative shall keep the seismic representative informed of all potential problems in the survey field and of all decisions taken by the survey representative.

(j) The seismic representative is responsible for ensuring that suitable occupational hHealth and safety policies and procedures are adhered to by the contractor and sub-contractors.

14.2.5 Responsibilities of the company's survey representative

(a) The seismic and survey representatives should be briefed by the company and every effort taken to arrange their meeting prior to the commencement of the project. It is the responsibility of the company to ensure the survey representative is fully informed and aware of the obligations between the company and the contractor.

(b) The seismic and survey representatives shall work together as a team and the seismic representative has overall responsibility for the day-to-day operations on board.

(c) The survey representative is responsible for the quality control of the navigation data and shall use the QC software to acquire data from the raw data inputs and the contractor to compare with the contractor's data. The survey representative must report any significant deviation from the specifications to the seismic representative.

The survey representative must:

(d) Ensure that progress is reported daily to the company via the seismic representative.

(e) Be vigilant for anything unusual or abnormal as possible sources of error. These must immediately be investigated and documented and problems potential or actual must be reported and a resolution of the problem sought.

(f) Ensure that pre-project checks, e.g. calibrations and survey instrument checks are carried out fully by the contractor and to the standard required by the company.

(g) Ensure that all survey equipment, including computers, antennae, positioning systems, gyro, echo-sounders and shore stations, is maintained and performs to specification. Special care is to be taken with equipment that is repaired or replaced so that it is calibrated properly and performs up to the standard required.

(h) Ensure that the contractor's personnel carry out their duties competently and according to the contract requirements.

(i) Ensure that all procedures are carried out according to the letter and spirit of the contract, and to achieve the company's objectives.

(j) Ensure that all navigation data acquired is clean and is correctly input into the navigation computer by the contractor.

(k) Ensure that any filtering or smoothing of data is approved by the company and is kept to a minimum. It must not warp the data and therefore the computation of position.

(l) Support all data and changes in data by relevant documentation.

(m) Document and understand the computational methods used by the contractor to acquire and process the data.

(n) Use the quality control software to compute positions from the raw inputs to check the contractor's data continuously.

(o) Ensure that regular calibration and system checks on both primary and secondary positioning systems are carried out to maintain the systems in optimum working order.

(p) Ensure that all line criteria meet all specifications laid down in the contract.

(q) Check the contractor's logs regularly and document any annotation conventions used during the project.

(r) Ensure that any changes in data formats and causes are fully documented as well as any other incidents that affect the data.

(s) Take an active role in ensuring that all occupational health and safety requirements are adhered to, in particular climbing antennae towers, wearing of correct safety equipment and small boat safety, as well as the safety of the personnel manning any shore stations.

(t) Ensure that confidential information remains so.

(u) Be available for de-briefing and prepare comments and suggestions as well as a final survey representativ'es report.

The survey representative shall fulfil the following specific requirements related to 3D surveys.

(a) Compute the common mid points and bin the data for details of coverage.

(b) Monitor the streamer separation.

(c) Check the spread meets the contract requirements throughout the survey.

(d) use the QC system to ensure that the quality positioning plan criteria are met throughout the survey.

(e) Ensure that both absolute and relative positioning systems meet the operational criteria at all times.

14.2.6 Confidentiality

All data and information related to marine seismic surveys is confidential to the company.

14.2.7 Briefing /de-briefing checklist

Survey and seismic representative shall attend both briefing prior to the project and de-briefing as soon as possible after the survey. The briefing shall clarify all duties and responsibilities. The details to be discussed are listed in Table 14.1.

Table 14.1 Briefing/ debriefing checklist

Item	Brief	Debrief
Survey name and number		
Name of the Survey Vessel		
Names of the contractor and sub-contractors		
Port of mobilisation		
Date of mobilisation		
Survey area maps		
Start and end of line coordinates		
Projection, spheroid, and datum parameters to be used.		
Station coordinates and logistical details of shore stations		
Contractor's system capabilities		
Absolute positioning systems		
Relative systems to be used for streamer positioning		
Factors affecting the performance of the positioning systems.		
Calibration and system checks		
Navigation reference point		
Occupational health and safety requirements		
(C–0) values, zero delay constants		
Quality control hardware, software, interfacing and support		
Quality control acquisition, processing and binning		
Quality control Documentation.		
Data formats and media		
Methods of numbering shotpoints and lines		
NRP, offset and layback diagrams and measurements drawn to scale, showing all nodes.		
Navigation hazards		
Rig and pipeline coordinates		
Travel and accommodation details		
Contract and technical specifications		
Contractor's written procedures		
Daily reporting procedures and requirements		

Table 14.1 Briefing/debriefing checklist

Item	Brief	Debrief
The company's phone, fax and after hours contact numbers		
Vessel's Inmarsat phone and fax numbers		
Vessel's sea phone number and mobile phone number		
End of survey reporting, data and documentation handover		
Quality control reporting, documentation and data handover		

14.2.8 Quality control reporting

The seismic representative and the survey representative shall be fully briefed about the specifications and requirements during the pre-survey meeting. Procedures for daily reporting and reporting of problems shall be put in place.

14.2.9 Quality control checks

The survey representative shall be responsible for checking

(a) The calibration of the positioning systems

(b) Positioning systems performance

(c) Interfacing

(d) Computers

(e) Data formats and inputs

(f) Data quality and line checks

(g) Binning for coverage

(h) Recording and reporting.

14.2.10 Positioning systems

The survey representative shall:

(a) Ensure the correct positioning system is installed correctly and is working to the technical specifications.

(b) Ensure the system has been fully calibrated both statically and through dynamic trials.

(c) Ensure that where there is more than one positioning system used, it is preferable that the systems are fully integrated. If they are not the mismatches between the processed sub-networks shall be used as quality indicators.

N.B. Phased adjustment does not make full use of the redundancy available in the network.

(d) Ensure that operation of the positioning systems does not interfere with other users when used outside of their proper operating slot.

(e) Ensure that the angle of cut between any two lines of position shall be between 30 and 150 degrees and that at least three independent lines of position with good geometry are available for the whole survey area.

(f) Ensure that the best geometry is used for any line if there is a choice of stations and beacons.

14.2.11 Computers

The survey representative shall:

(a) Ensure that all computing, printing and data storage equipment is installed and working.

(b) Ensure that all computers, contractor's and quality control are operating as specified in the technical specifications.

14.2.12 Interfacing

The survey representative shall:

(a) Ensure that all computers are interfaced correctly and the data being acquired are clean and free from error, and that the update rate is fast enough.

14.2.13 Calibration

The survey representative shall:

(a) Ensure that all navigation systems, including all back-up systems and spares and including but not limited to gyro, echo-sounder, relative systems and absolute systems, are fully and rigorously calibrated according to their technical specifications.

(b) Navigation systems are calibrated before during and after the survey according to their technical specifications.

14.2.14 Starting data formats and data input

The survey representative shall:

(a) Ensure that all start of line (SOL) and end of line (EOL) coordinates are those agreed in the contract.

(b) Ensure that the correct projection and datums are being used.

(c) Ensure that all starting data for calibration, filters, station information and changes mobile numbers and vessel information are entered correctly.

(d) Ensure that all positioning systems are operating correctly.

(e) Ensure that all base information, i.e. positioning station coordinates, line data and positioning data, as shown in Table 14.2, are entered into the contractor's computer correctly and are correctly documented.

Table 14.2 Check list of basic data to be checked and documented

Datum and projection parameters	
Base station coordinates	
Base station descriptions	
Base station heights, names, pattern identifiers and slot numbers	
Base station delays, mobile delays and /or zero/delay constants	
Propagation Velocity	
Pattern (C–O)'s	
Baseline lane widths	
Frequencies	
Mobile delays	
Mobile serial numbers	
Beacon serial numbers	
Calibration data for streamer compasses	
Streamer compass corrections	
Filter settings	
Elevation masks	
Speed of sound in sea water	
Magnetic variation value	
Shotpoint Interval	
Cable configuration, lengths, stretch sections, compass locations, distance to tailbuoy	
Laybacks and offsets	
Tailbuoy numbering	

(f) Ensure that the offsets relative to the ship's head shown in Table 14.3 are agreed in writing between the survey representative and the contractor's party chief or coordinator, with particular attention being paid to sign conventions.

Table 14.3 Check list of offsets relative to ship's bearing

Primary antenna to acoustic transducers	
Primary antenna to towpoint(s) source and streamers	
Primary antenna to echo sounder transducer	
Primary antenna to secondary antenna (including GPS)	
Positioning antenna to vessel reference point	

(g) Ensure the layback distances shown in Table 14.4 are agreed in writing between the survey representative and the contractor's party chief, with particular attention being paid to sign conventions.

Table 14.4 Check list of layback distances for 2D and 3D surveys

Centre(s) of near receiver group(s) to centre of far receiver group(s)	
Centres of far receiver group(s) to tail buoy(s)	
Distance from centre of near receiver group(s) to each compass	
End of stretch section(s) to centre of near receiver group(s)	
Nominal stretch section length(s)	
Streamer towpoint(s) to centre of near receiver group(s)	
Streamer towpoint(s) to start of stretch section(s)	
Source towpoint to centre of gun array(s)	

For each streamer all offsets and laybacks detailed in the above tables shall be shown on a separate diagram indicating each measured value.

(h) Ensure that data logging commences at the SOL.

(i) Ensure that the first shotpoint coordinates are correct.

14.2.15 Data acceptance and rejection

Datasets

This is the most critical part of the survey representative's job. He shall apply the data acceptance and rejection criteria to the datasets and advise the seismic representative when tolerances are exceeded.

A dataset is in this context defined as the minimum set of time contiguous raw positioning data and derived coordinates which is yet large enough to yield a statistically meaningful sample of quality measurements. The dataset size shall be agreed by the company and the contractor. A value of 200 events will be used unless otherwise notified.

This mechanism is intended to capture and reject consistently poor positioning, while allowing the occasional outlier or spike in the raw data to be accepted on the basis that the spike or outlier can be (and should be) removed in on-board or post processing, avoiding the need to re-acquire the data in the latter case.

Acceptance/rejection procedure

Data, forming part of datasets satisfying all criteria defined in the section 14.2.16 shall be considered to be of adequate quality and shall be accepted.

Conversely, only those data forming part of datasets that do not satisfy one or more of the criteria defined in section 14.2.16 should be considered to be of inadequate quality. Such data shall be rejected.

14.2.16 Data quality criteria

A dataset is considered to satisfy the quality criteria of the survey if all of the following conditions are satisfied.

Unbiased data

At least 99% of the position fixes (= events) constituting the dataset have a **standard deviation of unit weight** not greater than its **tolerance**, whilst the effect of any spikes or outliers on the computed positions have been removed by processing, e.g. filtering, smoothing or interpolation techniques (the **tolerance** is a function of the **redundancy** or **degrees of freedom** in each position fix).

A short list of these tolerances is given in Table 14.5. If the degree of freedom is greater that 10, the relevant tolerance can be derived from any textbook on statistical calculus.

Table 14.5 Tolerance (99%) of the SD of unit weight as function of degree of freedom

Degree of freedom	99% tolerance
1	2.58
2	2.14
3	1.94
4	1.82
5	1.74
6	1.67
7	1.62

Table 14.5 Tolerance (99%) of the SD of unit weight as function of degree of freedom

8	1.58
9	1.55
10	1.52

The above values are derived from the (known) **random behaviour** of the positioning systems used and does not directly relate to the **position tolerance.**

Predicted coordinate precision

The **semi-major axis** of the **standard error ellipse** of each of the specified points, multiplied by the factor 3.03 is not greater than the **position tolerance** in any position fix which is part of the dataset: and:

Redundancy

The number of redundant **observations** in any position fix which is part of the dataset is not less than the agreed minimum redundancy, expressed as a percentage;

However, lack of redundancy, which is defined as the condition where the redundancy is below the agreed minimum, is acceptable if:

the company and contractor have agreed to accept a positioning configuration with an inherent lack of **redundancy,** or

the following two conditions are satisfied:

(a) any contiguous series of position fixes in which lack of **redundancy** occurs is not longer than the agreed maximum number of fixes, and

(b) the positioning satisfied all data quality criteria immediately before and immediately after such a period in which lack of redundancy in the dataset does not exceed the agreed maximum total number of fixes.

14.2.17 Summary of tolerances to be agreed — sample tables

Position tolerance

Table 14.6 Position tolerance

Point description	Tolerance
a)	
b)	
c)	
d)	
e)	

Minimum redundancy

Table 14.7 Minimum redundancy

(Sub-) Network description	Minimum (%)
a)	
b)	
c)	
d)	
e)	

Data acceptance

Table 14.8 Data acceptance

Dataset size (see datasets)	____fixes
Maximum number of contiguous fixes with lack of redundancy	____fixes
Maximum total number of fixes with lack of redundancy in any dataset	____fixes

The survey representative shall also:

(a) Compare antenna positions, offset positions, standard deviations, streamer positions, navigation reference point, nodes, sources, etc. with the contractor's information and pay particular attention to any discrepancies between the smoothed and raw calculations.

(b) Compare the delta standard deviations (DSDs) between the contractor's data and the QC dataset. The larger the DSD, the more noisy the pattern.

(c) Note any significant changes in the ship's speed and heading and check these are documented alongside any consequent smoothing and filtering.

(d) Compare the contractor's time plots of the standard deviation of fixes, shotpoint intervals and off-line distance with the quality control time plots to enable quick end of line assessment of whether the line is acceptable.

(e) Tally the number of shotpoints and record numbers recorded. Any discrepancies may indicate misfires or other problems and the seismic representative and contractor should be alerted.

(f) Ensure the gyro is checked daily.

(g) Ensure the feather angle and ship's heading are recorded continuously and accurately. Feather needs to be recorded to enable prediction to be carried out on systems capable of this function.

(h) Compare the statistics and time plots from the quality control system with the contractor's.

(i) If all line criteria are satisfactorily met, approve the line with the seismic representative.

14.2.18 Logging requirements

The survey representative shall:

(a) Ensure that each new project is logged on separate tape(s) or disc(s) and the contents of the tape/disc(s) are recorded.

(b) Ensure that media contents and access methods are fully documented.

(c) Ensure that sensible administrative measures are used.

(d) Ensure that the quality control computer has logged as required — station data for all patterns interfaced, pattern data for all patterns interfaced, gyro values, water depth values, point(s) coordinates, precise line names, all streamer positioning parameters.

Logging, 3D Surveys — specifics

The survey representative should be aware that methods used may differ between contractors. Flexibility is required to fulfil the same ends.

(a) The positions to be quality controlled are the NRP, the centre sources and all common mid-points.

(b) Every shot shall be logged so there is no possibility of systematic bias in the quality control data.

(c) All raw sensors shall be interfaced to the QC computer, including but not limited to closure, compasses, DGPS, radio-positioning systems, acoustics, laser, gyro and RGPS. Also the QC computer shall be able to directly compare NRP, time and shot point number.

14.2.19 Reporting

Daily reporting

Daily reporting shall be concise and free of conjecture so that the company can:

(a) Monitor progress.

(b) Evaluate the survey quantitatively and qualitatively.

(c) Avoid potential problems and plan ahead.

(d) Make accurate time and cost estimates.

End of Job Reporting

(a) Intermediate reports shall be submitted at every crew change. These reports shall contain a full description of all activities, problems, statistics, etc. during the previous swing.

(b) The final report shall be submitted within four weeks of the completion of the acquisition.

14.3 Positioning systems — acoustic (long, short baseline)

14.3.1 Aim

The aim of this specification is to set out the requirements for the equipment and procedures necessary for the calibration of long baseline acoustic systems and seabed array adjustments.

14.3.2 General

The contractor shall supply and operate a medium-frequency (MF) or high-frequency (HF) acoustic positioning system using the highest professional survey principles and standards. Transponders used shall be of the intelligent type.

14.3.3 Equipment

As a minimum the following positioning and survey equipment shall be provided by the contractor:

(a) Vessel transducers (MF or HF).

(i) The transducer(s) shall be rigidly mounted through a gate valve.

(ii) The transducer(s), in its (their) operational position, shall be clear of the hull, propellers, thrusters and other possible noise sources.

(iii) The manufacturer's recommended signal to noise ratio shall be conformed with. Upon the company's request noise level measurements shall be carried out using suitable equipment (e.g. a calibrated hydrophone).

(iv) A vessel mounted short baseline system shall consist of at least three transducers.

(b) Surface positioning system.

The surface positioning system shall be in compliance with the applicable company technical specifications.

(c) Gyro compass, installed and calibrated in accordance with the manufacturer's and the company's specifications.

(d) Survey computer.

The survey computer shall be interfaced to (a) the vessel transducers, (b) the surface positioning system, and (c) the gyro compass.

(e) Echo-sounder, installed and calibrated in accordance with the manufacturer's and the company's specifications.

(f) Temperature/salinity bridge, operated and calibrated in accordance with the manufacturer's and the company's specifications.

(g) In case of a vessel mounted short baseline system a Vertical Reference Unit shall be used, which shall be installed as close as possible to the centre of gravity of the vessel.

14.3.4 Procedures — general

(a) The manufacturer's installation and operation manuals shall be followed at all times.

(b) In the event that replacement transponder(s) have to be deployed during the survey, a recalibration of the array shall be undertaken (long baseline systems only).

(c) A minimum of four lines of position for each event (fix) shall be available at all times. If the number of lines of position is reduced to three, the acceptability of the event (fix) shall be at the discretion of the the company representative (long baseline systems only).

(d) All transponders shall be capable of being enabled/disabled and of being released acoustically from the vessel after use. They shall have an endurance of not less than 3 weeks.

(e) All transponders specified by the company shall have temperature, salinity and depth sensors fitted. These sensors shall be pre-calibrated and operating to within the accuracies stated in the manufacturer's specification.

(f) Transponders shall be installed using moorings and buoyancy collars approved by the company, or be deployed in rigid frames to prevent any movement, depending upon the specified positioning tolerance. If moorings are used the nylon strop between the seabed and the transponder shall be 2 m, and concrete sinkers weighing not less than 150kg in air shall be used unless otherwise specified by the company..

(g) The contractor shall be responsible for the recovery of all transponders, moorings and frames on completion of the survey.

(h) The contractor shall be responsible for the possible loss of transponder(s).

(i) All parameters entered into the acoustic control unit shall be logged by the survey computer in order to facilitate a fast recovery in the event of a partial or total power loss.

14.3.5 Long baseline system calibrations

Transponder box-in(s)

(a) The vertical separation between transducer and transponder shall be determined from echo-sounder readings and the fixed offsets for transducer depth and height of the transponder above the seabed.

(b) After deployment of the transponder at least 50 fixes shall be taken evenly distributed on a circle, with the transponder as its centre point. The radius of the circle shall be between two and three times the vertical separation of transducer and transponder.

(c) The vessel's position shall be computed and displayed on-line together with the raw acoustic ranges to ensure that the requirements under (b) are met.

The transponder position shall be calculated from the obtained data using the least squares method. The results of the adjustments shall consist of raw and adjusted data, residuals and standard deviation.

Array calibration

(d) Transponders shall be employed without hindering their acoustic line of sight.

(e) The distance between transponders shall not exceed:

500m for HF

1000m for MF

(f) Within each array, a minimum of three transponders shall be positioned by means of transponder box-ins as described in (a), (b) and (c).

(g) A minimum of 20 acceptable direct transponder baseline measurements shall be made between every two transponders (i.e. 10 measurements in each direction).

(h) The preferred method of calibration is by sequential least squares. In this method the vessel shall steam a cloverleaf pattern through the entire seabed array in such a way that every transponder is completely circled once. Ranges shall be logged continuously from all transponders within range, and absolute vessel positioning shall be carried out simultaneously. Approximate coordinates for each transponder shall be determined from the box-ins and direct measurements described in (b) and (c), and the final coordinates are determined as the vector of unknowns in the least squares algorithm. This method is the most efficient means of calibration.

(i) If sequential least squares is unavailable, the positions of the transponders shall be established by means of a least squares adjustment using the accepted transponder baseline measurements.

Array verification

Transponder arrays shall be verified prior to commencement of the survey by the following methods:

(j) Simultaneous and cyclic remote ranging to the array using an array transponder. Transponder coordinates derived from the array shall be compared with the coordinates derived from the array calibration.

(k) Simultaneous and cyclic direct ranging to the array using the transducer. Transducer coordinates derived from the array shall be compared with the coordinates derived from the surface positioning system.

14.3.6 Short baseline system calibration

The calibration procedures and requirements for ultra short baseline (USBL) calibrations as prescribed in the appropriate company survey specification shall apply.

14.3.7 Software

The software shall be capable of:

(a) Data storage and data retrieval.

(b) Computation of coordinates of boxed-in transponders by least squares including statistical analysis.

(c) Array adjustments by least squares including statistical analysis.

(d) Provision of residuals on acoustic ranges including statistical analysis.

14.3.8 Tolerance

The tolerance for long baseline acoustic systems is 2m.

14.3.9 Rejection criteria

Transponder box-ins

The transponder box-ins shall be rejected and repeated if:

(a) The standard deviation of the final transponder coordinates exceeds the required positioning tolerance as specified in the technical specifications.

(b) Five percent or more of the fixes are rejected because of excessive residuals.

(c) The residuals of the fix data show a pattern indicating the presence of a systematic error.

Array adjustment

The array adjustment shall be rejected and repeated if:

(a) The solution is mathematically unstable.

(b) The standard deviation of the positions obtained in 14.3.5(j) and (k) exceeds the positioning tolerance as specified in the technical specifications.

(c) The difference between the two positions obtained in 14.3.5(j) and (k) exceeds the positioning tolerance as specified in the technical specifications.

(d) The difference between the fixes obtained in 14.3.5(j) and (k) exceeds the positioning tolerance for the surface positioning systems as specified in the technical specifications.

14.3.10 Reporting

(a) The final report shall include a listing of each transponder detailing the following:

(i) Position and depth.

(ii) Interrogation and reply frequencies.

(iii) Installation and recovery dates.

(iv) Type and serial number.

(b) All calibration data shall be submitted with the final report.

14.3.11 Operating limits

No acoustic calibrations are to be taken in weather conditions which are seen to be impairing the quality of the data or records being obtained. The contractor's party chief shall inform the company survey representative if he considers the quality of the data or records to be impaired. However, the decision to accept the data from the acoustic calibrations shall be the responsibility of the The company survey representative.

14.4 Streamer and source positioning for 3D seismic surveys

14.4.1 System supply

The contractor shall supply, maintain and operate a streamer and source positioning system, capable of continuous operation and providing at least 30% redundancy at all network nodes. The contractor shall use the positioning systems in accordance with good survey principles and shall insure that no ambiguities occur in the position of the streamer(s) or source(s) . The type, deployment and proposed adjustment methods are subject to company approval..

The streamer position(s) shall be determined using a combination of radio positioning systems, laser systems, acoustics, digital compasses with DGPS receiver(s) or radio positioning system at the tailbuoy(s).

The following must be kept in mind at all times:

(a) Ensure that there are redundant observations for every facet of the entire spread.

(b) The standard deviation of all common Mid Points shall be better than 0.3 × nominal bin width.

(c) The inline offset from the NRP to sources and front receiver groups shall be as short as possible, to increase the accuracy of the sources and streamer heads.

The company will give preference to integrated systems, in other words those which are able to be computed as a single network and which spread their network nodes and observations evenly along the sources and streamers. It is conceded that there are few systems which can currently achieve this mode of operation, but the contractor shall nevertheless make every effort to comply with the requirements of the applicable company survey specifications, and in particular the Specification dealing with navigation quality control.

14.4.2 Calibration and operation of individual sensors

All systems shall be calibrated and operated according to the applicable company survey specification. The following notes augment those specifications.

Magnetic compasses in or on the seismic streamers

Operation and calibration shall follow the applicable company survey specification.

Radio positioning system

The appropriate specification shall apply both to the vessel and tailbuoy(s) installations and operation. In case where there are multiple tailbuoys, no line shall be started without at least one operational tailbuoy. If the last operational tailbuoy fails during a line, that line may be continued provided that the crossline position of the far group does not deviate by more than 5 m immediately after the tailbuoy fails. The tailbuoy shall then be repaired before commencement of the next line.

Gyro

The gyro shall be calibrated and operated in accordance with the applicable company survey specification. The gyro shall be checked (using sun azimuths or similar) on a daily basis wherever practicable.

Acoustic systems

The acoustic systems shall be calibrated and operated in accordance with the applicable company survey specification. All corrections, delays and propagation factors derived from the calibrations shall be entered into the navigation computer and shall be fully documented.

Laser

The laser shall be calibrated and operated in accordance with the applicable company survey sSpecification.

14.4.3 Real time quality control

The procedures for real time quality control are laid down in the applicable company survey specification. Sufficient observations shall always be available to provide the required redundancy.

If redundancy is lost then the decision as to whether or not work will stop will be made by the company representative onboard the vessel. In general, work will not stop if redundancy is lost for less than 1000m on any line, provided that full redundancy is maintained for the remainder of that line, and provided that time series plots of the remaining observations do not indicate any deterioration in streamer or source positioning.

14.4.4 Dataset size

The dataset size for determining quality shall be 200 shots. In other words, a line consisting of 1000 shots shall comprise five contiguous datasets.

14.4.5 Tolerance for streamer positions

Tolerances for streamer positions are as follows:

Common mid-points

Common mid points for 2D surveys shall have a tolerance of 15m.

Common mid points for 3D surveys shall have a tolerance of 0.3 × nominal bin width. This specification shall also hold for spreads in which the nominal bin width is tapered.

Separation

Tolerances for cable separation are as follows:

Front group separation 5m.

Mid group separation 7m.

Back group separation 10m.

Definitions for tolerance, standard deviation, etc are in the applicable company survey specification.

14.4.6 Tolerance for source positions

The tolerance for the centre of the active source is 3m.

14.4.7 On board recording

The contractor shall record all raw unfiltered positioning data in real time at each individual shot point in UKOOA P2/91 format. The latter is preferred since data will then be time-tagged. Contractor shall ensure that there are no anomalies between numbering of navigation and seismic records. If differences occur, they shall be properly annotated and documented.

All analogue survey records shall be annotated at least every 10 shots.

The contractor shall provide adequate facilities for interfacing an online quality control system to all navigation sensors, contractor's navigation system and the event closure. These facilities shall include but not be limited to a stable power supply, adequate secure computer space, cabling and (where applicable) data conversion units (for example IEEE488 to RS232-C).

14.4.8 Operations logs and reports

The company requires detailed logs and reports to be kept by the contractor's staff and forwarded to the company at each crew change and at the end of acquisition. These shall contain the following information:

(a) All calibration data including baseline crossings, equipment settings, filter settings, C–Os, delays, spheroid and projection parameters and propagation factors.

(b) Serial numbers of sensors, cables, antennae.

(c) Chronological record of all checks and calibrations performed on all equipment.

(d) Chronological line records.

(e) Description of operational procedures.

(f) All occurrences of equipment malfunction.

(g) All occurrences of data not meeting tolerances.

A comprehensive operations report covering the above requirements shall be forwarded to the company at each crew change and within 3 weeks of the end of acquisition.

Log sheets shall be maintained for each line, part line and re-shoot. These logs shall include the following data:

(a) Date.

(b) Start and end times.

(c) Start and end shot numbers.

(d) Line start and end coordinates.

(e) Line direction and length.

(f) Details of all observations used.

(g) Details of all observations outside tolerance.

(h) Mean line speed and standard deviation.

(i) Mean shot spacing and standard deviation.

(j) Mean distance off line and standard deviation.

(k) Mean feather and standard deviation for each streamer.

(l) A brief summary of overall system performance along the line.

14.4.9 Dual vessel operations

For dual vessel operations the contractor shall ensure that adequate means of synchronizing data acquisition are made.

All raw positioning sensor information shall be made available from the slave vessel to the master vessel using a suitable telemetric link in real time.

Appendix A UKOOA P1/90

A.1 General

The data required for conventional 2D seismic surveys is the position of shot points (energy source, common mid-point, etc.) as defined in the header.

In this document the term 'shotpoint' refers to the 'centre of source' and the term 'common mid point' refers to the 'mid-point between the centre of source and the near trace'.

For all other surveys there are two ways of exchanging data: 'pre-stacked' or 'post- stacked'. The post-stacked data are bin positions which are stored in the same way as 2D shotpoint positions.

The pre-stacked data should contain all the data that are required for binning; i.e. the position of the energy source and all receiver groups. This makes for consistency of data format/ content and allows for re-binning on a purely positional basis.

For 3D offshore surveys the coordinates of each defined receiver group are listed following each shotpoint. For onshore 3D surveys and for onshore 2D surveys requiring special processing it is necessary to establish three data files. A separate file for each of the two main elements, source positions and receiver positions, with a third file to define the relationship between source and receiver groups. These files are:

(a) Source — this is for a 2D shotpoint record.

(b) Receiver group positions — this is for a 2D shotpoint record.

(c) A relation record — this details which groups were being recorded at a specific shot.

In view of the large number of traces/receiver groups per shotpoint it is necessary to minimize storage. This is achieved by using only grid coordinates for trace positions, combining several traces in one record and by storing receiver group positions of onshore surveys only once.

It is assumed that this format is for the exchange of data from a single survey, and not for compiled databases including multiple surveys.

A.2 Tape specification

Half-inch magnetic tape	: IBM compatible
Number of tracks	: 9
Number of bytes per inch	: 6250 standard (1600 or 800 are permissible)
Mode	: Coded EBCDIC or ASCII
Record length	: 80 bytes
Block size	: 8000 bytes

Blocks physically separated by inter-record gap.

(1600 bytes for 1600/800 bpi)

Disk specification

Format	: MSDOS IBM PC compatible
Size/capacity/density	: 3.5 in./720k/double
Mode	: Coded ASCII
Record length	: 80 bytes with CR/LF after character 80.

In the interest of standardisation 3.5 in. × 720k disks have been chosen as standard. Other formats and media are acceptable by prior arrangement between the affected parties e.g. client, contractor, broker.

File description

A tape or disk may contain one or more files depending on the type of survey. Each file is started by a number of 'Header Records' followed by data records and closed by an end of file (EOF) statement in col 1 — 3 of the final record.

- A disk file is closed by one IBM file mark.
- A tape file must be closed by two IBM file marks.
- Multiple lines per file are allowed, as long as all data and header records are consistent.

Tape and disk labelling

Each tape or disk should be adequately labelled so that its format and content can be readily ascertained. This labelling shall include, as a minimum:

SURVEY AREA/NAME: CONTRACTOR			: SURVEY TYPE
TAPE DATA FORMAT: MODE		: DENSITY	: BLOCK SIZE: RECORD
e.g. IBM	: EBDIC	: 6250bpi	: 8000bytes : 80bytes
DISK DATA FORMAT: MODE		: DENSITY	: BLOCK SIZE: RECORD
e.g. MSDOS	: ASCII	: DOUBLE	: 720k : 80bytes

A.3 Header record specification

General

Each file should start with a number of header records which contain information about, and parameters controlling, all of the data records which follow.

The general format for header record is:

	Cols	Format
(a) Record Identifier "H"	1	A1
(b) Header Record Type	2 — 3	I2
(c) Header Record Type Modifier	4 — 5	I2
(d) Parameter Description	6 — 32	A27
(e) Parameter Data	33 — 80	see below

Header record types **0100 to 0800** and **1400 to 2000** inclusive are mandatory for all surveys even if a 'N/A' entry is required. Header record types **0900** and **1000** are additionally mandatory for all offshore surveys. Header record type **1100** is also mandatory for offshore 3D surveys but is not needed for other surveys. Header records of types **2100** to **2500** are mandatory as far as they are applicable to the projection used.

Text fields should be left justified, and numeric fields right justified unless otherwise stated.

Configuration details

For multi- vessel, multi-source, multi-streamer operations the format allows unique identification of each of these components.

Header records **H0101, H0102, H0103, H0104** are used to define the survey details, and the source/streamer/tailbuoy configuration associated with each vessel.

For consistency, the following order convention has been adopted:

From starboard, top, back through front, bottom, port

e.g. Vessel 1:	Sources 2:	Streamers 3 (stbd top & btm, port)
Vessel 2:	Sources 2:	Steamers 2 (mini, main)

	Name	Vessel ID	Srce ID	Strmr ID	TB ID	Other ID
H0102 Vessel Details	M/V Oil Finder	1				
H0103 Source Details	Stbd source	1	1			
H0103 Source Details	Port source	1	2			
H0104 Streamer Details	Stb upper 240 ch	1		1	1	
H0104 Streamer Details	Stb lower 240 ch	1		2	2	
H0104 Streamer Details	Port cable 120 ch	1		3	3	
H0102 Vessel Details	M/V Dryhole	2				
H0103 Source Details	Stb source	2	3			
H0103 Source Details	Port source	2	4			
H0104 Streamer Details	Back main 190 ch	2		4	4	
H0104 Streamer Details	Front mini 20 ch	2		5	5	
H0105 Other Details	Front nav. float	2				1

Vessel IDs should be used for all survey details, e.g. in **H09XX H0105.** Other details can be used when a towed body, such as float, is used for acoustic ranging. A full description should be put in **H2600.**

Line prefix

H0203 should be used only where the line name exceeds 12 characters.

Offset definitions

The offset code defines the type of offset data expected,

The code is: 1 for polar data
2 for rectangular data

Code 1: Polar : Offset A = radial distance from ship's reference point to the offset point.
Offset B = angle from ship's head (clockwise)

Code 2: Rectangular : Offset A = X axis offset across ship's axis, positive to starboard.

Offset B = Y axis offset along ship's axis, positive towards the bows.

Note that the offset orientation is always with the ship's head (gyro) and that the origin is the ship's reference point unless otherwise specified.

Datum and spheroid Information

H1600 and **H1601** require datum transformation parameters. These are defined by the Bursa-Wolfe transformation model:

$$\begin{bmatrix} X \\ Y \\ Z \end{bmatrix} = \begin{bmatrix} DX \\ DY \\ DZ \end{bmatrix} + (SCALE) \times \begin{bmatrix} 1 & -RZ & RY \\ RZ & 1 & -RX \\ -RY & RX & 1 \end{bmatrix} \times \begin{bmatrix} X \\ Y \\ Z \end{bmatrix}$$

where

X, Y, Z	are geocentric cartesian coordinates in metres.
DX, DY, DZ	are translation parameters in metres.
RX, RY, RZ	are clockwise rotations defined in arc secs, but converted to radians for use in the formula.
$SCALE$	= [1 + S(10e–6)] where S is in parts per million.

Example (For checking formula only)

From	Datum 1: WGS72	To Datum 2: WGS84
Semi Major Axis a	6378135.0 metres	6378137.0 metres
	$1/f$ 298.26	298.257223563
Latitude	39 13 26.5782 N	39 13 26.6976 N
Longitude	98 32 32.2870 W	98 32 31.7330 W
Spheroidal height	570.88m	573.249m
X	–734985.205	–734972.229
Y	–4893185.191	–4893188.272
i	4011976.605	4011982.012
DX	0.0	
DY	0.0	
DZ	+4.5 m	
RX	0.0	

RY	0.0	
RZ	+0.554 arc secs	= 0.000002686 radians
S	+0.2263 ppm	

Vertical datum

Header record **H1700** must specify the vertical datum.

e.g.LAT	**Lowest astronomic tide**
MSL	**Mean sea level**
SL	**Sea level**
ES	**Echo — sounder**

The units of measurement are specified in **H2001.** These should, whenever possible, be consistent with the position data.

Depths will be referred to the coordinated data point, unless otherwise stated in header record **H1700.**

e.g. **H1700**	LAT	: Centre of source
or **H1700**	SL	: Echo-sounder

Header **H2600** should be used to specify details of depth data reduction e.g. tide/velocity/ transducer correction.

Projection data

Projection data is specified in header records **H1800 — H2600.**

The following projection type codes have been defined:

001 — UTM Northern Hemisphere
002 — UTM Southern Hemisphere
003 — Transverse Mercator (north orientated)
004 — Transverse Mercator (south orientated)
005 — Lambert conic conformal, one standard parallel
006 — Lambert conic conformal, two standard parallels
007 — Mercator
008 — Cassini- Soldner
009 — Skew orthomorphic
010 — Stereographic

011 — New Zealand Map Grid

999 — Any other projection or non-standard variation of the above projections.

Requirements for projection definition include the following header records:

Transverse Mercator	2200 2301 2302 2401 2402
UTM	1900 2200
Oblique Mercator	2301 2302 2401 2402 2509
and	2506 or 2509 or 2507 or 2508
Lambert Conformal (1SP)	2100 2200 2301 2302 2401 2402
Lambert Conformal (2SP)	2100 2200 2301 2302 2401 2402
Stereographic	2301 2302 2401 2402

When a survey crosses the equator from south to north, and the whole survey is shot on a Southern Hemisphere UTM zone, then coordinates may exceed 9999999.9. The format cannot accept this, so a warning note must be written to **H2600** advising that 10000000 must be added to such coordinates.

Definition of units

H200 Grid unit code is 1 for metres, 2 for any other unit.

H2001 Height unit code is 1 for metres, 2 for any other unit.

H2002 Angular unit code is 1 for degrees, 2 for grads.

Other relevant information

Header record type **H2600** is a free-format statement of any other relevant information such as base station coordinates and geodetic control, description of additional data in receiver group records, survey adjustments done/not done, misclosures, etc. **H2600** may be repeated as often as required.

Formats of parameter data fields for each of the header record types are:

Types	*Item*	*Cols. format*
H0100	Description of survey area	33 — 80 A48
H0101	General survey details	33 — 80 A48
H0102	Vessel details — name: IDs	33 — 76 A24 5(I4)
H0103	Source details — name: IDs	33 — 76 A24 5(I4)
H0104	Streamer details — description: IDs	33 — 76 A24 5(I4)

H0105	Other Details — description: IDs	33 — 76 A24 5(I4)
H0200	Date of survey	33 — 80 A48
H0201	Date of issue of post- plot tape (d.m.y.)	33 — 80 A48
H0202	Tape version identifier	33 — 80 A48
H0203	Line prefix	33 — 80 A48
H0300	Details of client	33 — 80 A48
H0400	Details of geophysical contractor	33 — 80 A48
H0500	Details of positioning contractor	33 — 80 A48
H0600	Details of positioning processing contractor	33 — 80 A48
H0700	Descriptions of positioning and onboard computer system(s)	33 — 80 A48
H0800	Coordinate location, e.g. centre of source	33 — 80 A48
H0900	Offset from ship system position to coordinate location — vessel ID: code: A:B	33 — 56 I4,I4 2(F8.2)
H09XX	Other specified offsets e.g. antenna *XX* in range 1 — 99 — vessel ID: code: A:B	33 — 80 I4,I4 2(F8.2)
H1000	Clock time in respect of GMT (clock display in advance of GMT expressed as GMT + *N* hours)	33 — 80 A48
H1100	Number of receiver groups per shot	33 — 36 I4
H1400	Geodetic datum description as used for survey Datum name: spheroid name: *a*: 1/*f*	33 — 80 2(A12) F12.3, F12.7
H1401	Transformation parameters for H1400 to WGS84 *dx dy dz rx ry rz s*	33 — 78 3(F6.1) 3(F6.3), F10.7
H1500	Geodetic datum description as used for post Datum name: spheroid name: a: l/f	33 — 80 2(A12) F12.3,F12.7
H1501	Transformation parameters for H1500 to WGS84 *dx dy dz rx ry rz s*	33 — 78 3(F6.1) 3(F6.3), F10.7
H1510	Township system data flag (Type 2) and a description of the specific township used	33 — 80 A6, A42
H1600	Transformation parameters between H1400 (Datum 1)and H1500 (Datum 2) *dx dy dz rx ry rz s*	33 — 78 3(F6.1) 3(F6.3), F10.7
H1700	Vertical datum — Name: Origin	33 — 78 2(A24)
H1800	Projection code: description	33 — 80 A4,A44
H1810	Township relative coordinates	33 — 80 A48
H1900	Projection zone (including hemisphere for U.T.M.)	33 — 80 A48
H1910	For Township & Range, description of principal meridian	33 — 80 A48
H2000	Description of grid units — Code: Unit of Measurement: conversion factor to International Metres	33 — 72 I1, A24, F15.12
H2001	Description of height units — Code: Unit of Measurement: Conversion factor to International Metres	33 — 72 I1, A24 F15 — 12
H2002	Description of angular units — Code: Unit of Measurement	33 — 57 I1, A24

H2100	Latitude of standard parallel(s) (d.m.s. N/S)	33 — 56 2(I3,I2 F6.3, A1)
	(grads N/S)	2(F11.7, A1)
H2200	Longitude of central meridian (d.m.s. E/W	33 — 44 (I3,I2 F6.3, A1)
	(grads E/W)	F11.7, A1
H2301	Grid origin (latitude, longitude, d.m.s. N/E)	33 — 56 2(I3,I2 F6.3, A1)
	(grads N/E)	F11.7, A1
H2302	Grid co — ordinates at grid origin (E,N)	33 — 56 (F11.2 A1)
H2401	Scale factor	33 — 44 F12.10
H2402	Latitude/longitude at which scale factor is defined	33 — 56 2(I3, I2 F6.3, A1
	(grads N/E)	2F11.7, A1)
H2506	Latitude/longitude of two points defining Initial line of projection (d.m.s.)	33 — 80 4(I3, I2 F6.3, A1
	(grads)	4(F11.7, A1)
H2507	Circular bearing of initial line of projection (d.m.s.)	33 — 44 I3, I2 F7.4
	(grads)	F12.7
H2508	Quadrant bearing of initial line of projection (N/S/,d.m.s.,E/W)	33 — 44 A1,2I2 F6.3, A1
		A1, F10.7, A1
H2509	Angle from skew to rectified grid (d.m.s.)	33 — 44 I3, I2 F7.4
H2600	Any other relevant information	6 — 80 A74

N.B. See beginning of this section for detailed explanations of header information.

A.4 Data record specification

The data record will vary depending on the type of survey and the data content. The general content of offshore and onshore surveys is given separately. For conventional surveys a series of data records is required.

Where spare characters are available in the format, these can be used at the discretion of the client/contractor. The definition must then be included in **H2600.**

Two types of data record exist: Type 1 for coordinates quoted in terms of grid or graticule values, and Type 2 for those quoted as local offsets from townships or section markers. A file may contain either Type 1 or Type 2 data records but not a mixture of both. If Type 2 data are present, then the flag must be set in header record **H1510.**

The Type 2 record applies only to some North American onshore surveys.

Offshore surveys

Conventional 2D Surveys:

The dataset consists of one file with header records followed by a series of data records containing one shotpoint position each. Header record H0800 indicates whether the coordinated point represent the shotpoint, the common mid-point, or other defined location.

When one parameter changes the complete header record should be rewritten.

Other surveys:

The dataset contains one file. Following the header the position of the shotpoint is given in a data record and the positions of the receiver groups in receiver group records immediately following the data record.

Onshore surveys

Conventional 2D surveys:

The dataset consists of one file with data records. Each record contains data for one point (shotpoint, common mid-point, etc. as specified in header record **H0800**).

Other surveys:

The dataset consists of three file with an identical block of header records:

First file	: Data records with positions of receiver groups
Second file	: Data records with positions of shotpoints.
Third file	: Relation records specifying for each shot the relation between recording channel numbers and receiver groups.

In order to avoid ambiguities each physical position in the field (shotpoint or receiver group) must have a unique name.

Type 1: Grid or geographical coordinates

Item	*Description*	*Col. format*
1	Record identification S = Centre of source G = Receiver group Q = Bin centre A = Antenna position T = Tailbuoy position C = Common mid-point V = Vessel reference point	1A1

	E = Echo-sounder Z = Other, defined in H0800		
2	Line name (left justified, including reshoot code)		2 — 13 A12
3	Spare		14 — 16 A3
4	Vessel ID		17 A1
5	Source ID		18 A1
6	Tailbuoy/ Other Id		19 A1
7	Point number (right justified)		20 — 25 A6
8	Latitude (d.m.s. N/S)		26 — 35 2(I2), F5.2, A1
		(grads N/S)	F9.6, A1
9	Longitude (d.m.s.E/W)		36 — 46 I3,I2 F5.2, A1
		(grads E/W)	F10.6, A1
10	Map grid easting (metres)		47 — 55F9.1
		(non metric)	I9
11	Map grid northing (metres) (non metric)		56 — 64 F9.1 I9
12	Water depth (datum defined in H1700) or elevation	(non metric)	65 — 70F6.1 I9
13	Julian day of the year		71 — 73I3
14	Time (h.m.s., GMT or as stated in H1000		74 — 793I2
15	Spare		801X
16	Applicable to 3D offshore surveys see 1 — 80 ITEM 16 following		

ITEM 16. Receiver group records(3D offshore surveys)

Item	*Description*		*Col. format*
16a	Record identification 'R'		1A1
16b	Receiver group number		2 — 5 I4
16c	Map grid easting (metres)		6 — 14 F9.1
		(non metric)	I9
16d	Map grid northing (metres		15 — 23 F9.1
		(non metric)	I9
16e	Cable depth(metres) (additional information as specified in H2600)		24 — 27 F4.1
		(non metric)	I4
16f	Receiver group number		28 — 31 I4
16g	Map grid easting (metres)		32 — 40 F9.1
		(non metric)	I9
16h	Map grid northing (metric)		41 — 49 F9.1
16i	Cable depth (etc.)		50 — 53 F4.1
16j	Receiver group number		54 — 57 I4
16k	Map grid easting (metric)		58 — 66 F9.1
		(non metric)	I9
16l	Map grid northing (metric)		67 — 75 F9.1

		(non metric)	I4
16m	Cable depth (etc.)		76 — 79 I4
16n	Streamer ID		80I1

N.B. A cable 'depth' above the vertical datum (e.g. Transition Zone Survey) will be recorded as a negative value.

Type 2: Coordinate data as local offsets from township/section corners

Item	*Description*	*Col. format*
1	Record type identifier, 'L'	1 A1
2	Line name (left justified) including reshoot code	2 — 13 A12
3	Point number (right justified)	14 — 18 A5
4	Suffix to point number for fractional intervals (plus chainages, decimals of SP interval or suffixes) or point descriptors (e.g. for skidded point, no hole, etc.) as described in H2600 records	19 — 21 A3
5	Record identification **S = Centre of Source** **G = Receiver Group** **Q = Bin Centre** **A = Antenna Position** **T = Tailbuoy Position** **C = Common Mid Point** **V = Vessel Reference Point** **E = Echo Sounder** **Z = Other, defined in H0800**	22 A1
6	Offset of point from ref. point in northerly direction (N = +ve,S = –ve)	23 — 29 F7.1
7	Offset of point from ref. point in easterly direction (E = +ve,W = –ve)	30 — 36 F7.1
8	Elevation or water depth	37 — 42 F6.1
9	Reference point name	43 — 58 A16
10	Reference point latitude(d.m.s. N/S)	59 — 68 I2,I2

A.5 Example header and data file

```
H0100   SURVEY DESCRIPTION                                             Demonstration
H0101   SURVEY DETAILS                                           Two  boat 6 streamer
H0102   VESSEL DETAILS                               Demo1   101
H0102   VESSEL DETAILS                               Demo2   202
H0103   SOURCE DETAILS                                         1 301
H0103   SOURCE DETAILS                                         1 302
H0104   STREAMER DETAILS                                       1     201
H0104   STREAMER DETAILS                                       1     202
H0104   STREAMER DETAILS                                       1     203
H0104   STREAMER DETAILS                                       2     204
H0104   STREAMER DETAILS                                       2     205
H0104   STREAMER DETAILS                                       2     206
H0105   SLED/TB DETAILS                                        1         401
H0105   SLED/TB DETAILS                                        1         202
H0200   SURVEY DATE                                       1995   73   5  11     21.24
H0201   POSTPLOT DATE                                     1995  312  14  45     33.17
H0202   TAPE VERSION                 UKOOA P1/90 3D SEISMIC POST PLOT DATA
H0300   CLIENT                                                Demo 6-streamer spread
H0400   GEOPHYSICAL CONTRACTOR                                             A company
H0500   POSITIONING CONTRACTOR                                       Another company
H0600   PROCESSING CONTRACTOR                                                Unknown
H0700   ONBOARD SYSTEMS
H0900   OFFSET FROM SYSTEM  POS                                                  N/A
H1000   HOURS FROM GMT                                                             0
H1100   GROUPS PER SHOT                                                         1440
H1400   SURVEY SPHEROID                  WGS 84      WGS 84 6378137.000 298.2572236
H1401   DATUM SHIFT                    0.0    0.0    0.0 0.000 0.000 0.000 0.0000000
H1700   VERTICAL DATUM                                                           LAT
H1800   PROJECTION TYPE                                                            1
H1900   PROJECTION ZONE                                                          UTM
H2000   GRID UNITS              1                    metres 1.000000000000
H2001   HEIGHT UNITS            1                    metres 1.000000000000
H2002   ANGULAR UNITS           1           ddd mm ss.sss
H2200   CENTRAL MERIDIAN                                                 75 0 0.000E
H2301   GRID ORIGIN              0 0 0.000S 75 0 0.000E
H2302   GRID COORDS AT ORIGIN      500000.00E       0.00N
H2401   PROJECTION SCALE FACTOR 0.9996000000
H2402   COORDS OF SCALE FACTOR   0 0 0.000S 75 0 0.000E
V         2465    11     735192122.33N 72 245.16E 189638.42142894.9    0.0 88174156
S         2465     1     735192117.65N 72 246.44E 189673.52142750.2        88174156
R    1 189782.12142518.215.0    2 189781.92142505.715.0    3 189781.72142493.115.01
R    4 189781.52142480.615.0    5 189781.32142468.116.0    6 189781.02142455.615.01
R    7 189780.82142443.115.0    8 189780.62142430.616.0    9 189780.42142418.117.01
```

Appendix B UKOOA P2/91

B.1 Introduction

The UKOOA P2/91 data exchange format is designed to record positioning data for both 2D and 3D seismic surveys. Raw data is deemed to be the measurements taken by positioning sensors before the application of variable (C–O) corrections and/or variable scale correction, which may result from calibrations. However, for satellite systems provision is only made for the recording of co-ordinates as obtained from the satellite receiver or the associated dedicated data processing computer and for some additional associated relevant data. It is envisaged that a raw data capability will be added to P2/91 at a later stage.

The format allows individual time-tagging of observations. This is done in a way transparent to computer software not capable of reading and processing the time information.

The design objective of this format was to provide a flexible raw data format, allowing effective storage of positioning data from modern, ever-changing survey configurations, within the following framework:

- the format should enable effective data exchange;
- the format should allow computer processing of the data to take place with minimum operator intervention.

The first requirement calls for completeness and has been interpreted to require a text file format, which is sufficiently logical and structured to the human brain to allow some degree of visual interpretation and inspection.

The format uses a coded system of records so that certain record types may be omitted entirely if they are not relevant. Any physical data storage medium may be used by prior agreement between the parties involved in exchange of the data.

The P2/91 format is slowly being replaced by the P2/94 format. The two are identical except that P2/94 allows for recording of raw satellite data.

B.2 Logical file structure

Record length

The data is stored in 80 byte card image records, the columns of which are numbered 1 through 80.

Record types

The format defines four main types of record which are identified by the first character of the record:

H: survey header data

C: comments

E: event data (implicit time reference)

T: inter-event data (explicit time reference)

Every file/line must start with records H0000 to H00@9, in sequential order. Although no further sequence is imposed on the survey header records, it is strongly recommended that the definition sequence in this document be adhered to. Comment records are allowed to be inserted anywhere in a file, but not before record H00@9.

Record codes

Characters 2 — 5 contain a numeric code which describes the nature of the data stored in the record and allows easy grouping of related records. For example, the numbering of the E- and T-records runs parallel to the numbering of H-records in which the definition of the relevant data is stored. Hence, an E25@0 record contains streamer depth sensor data, while an H25@0 record contains the matching definition.

The vessel reference number is shown in the record code definitions as '@', where the data in the record refers to one vessel with its towed configuration in particular. It is provided merely to facilitate the sorting of the data according to vessel in multi-vessel surveys in the case the user wishes to process subjects of data per vessel. In all cases the '@' in the record code is redundant information.

Time records

T-records may be used to supplement or replace corresponding E-record, subject to client requirements. The sequence of E-records and T-records is strictly chronological: if the time recorded in a T-record is between event time 'i' and event time' j', it is inserted **after** the E-records relating to event time 'i', but before the general event record defining event time 'j'. It is stressed that, although absolute time is recorded on the T-records to allow unambiguous identification of the data, only the **relative** times are important.

Data fields

The following types of data fields are defined (x = total field length)

F$x.y$ Fixed format numeric fields; sign and decimal point included; y = number of digits after decimal point; usually specified only to indicate the number of significant digits required; in some cases, e.g. geographical co-ordinates, the field format is consistent to facilitate efficient computer conversion.

Nx Free-format numeric field; sign and decimal point included.

Ix Integer field.

Ax Text field.

One line per file

The data for each seismic line must be recorded as a separate file, starting with a complete set of header records. If any of the survey header data changes, a complete set of revised header records should be inserted but no new file should be started mid-line. This is required to allow easy transcription of data from high capacity storage media to lower-capacity media and to facilitate random access to individual lines for processing.

Data for a seismic line may in general not be split over different storage media such as tapes, diskettes etc.

Exceptions to the one line per file rule

Two exceptions exist:

(a) Very long lines

The data for very long lines may physically not fit on the chosen storage medium. The first option should be to consider a more suitable physical storage medium. However, if this impracticable, the data should be split over different media, starting on the new medium with a new file and hence a complete set of header records.

(b) Multi vessel surveys

Although it is strongly recommended that all data relating to one seismic event and to one seismic line be stored on the same physical storage medium, regardless of the number of vessels involved, it is realized that this principle may occasionally lead to practical difficulties.

Subject to client requirements it is therefore considered acceptable to split multi-vessel data according to acquisition vessel and store each subset on separate storage media as if it concerned different seismic lines, each subject to the above rules. However, the following conditions should then be satisfied

No data is stored more than once, except the following categories:

all survey header data common to all vessels;

general event data (E1000 record).

This data must be repeated on each of the vessel subsets of the line data.

The data for one seismic line relating to one vessel may not be split further over different storage media, except when the line is too long.

Complete, not over-complete, headers

The set of header records supplied for the line should only contain definitions for observations and elements of the survey spread that are intended to be used during the survey.

This rule is intended to prevent vastly over-complete sets of header data being supplied with e.g. all radio positioning systems in the North Sea defined.

The header records should therefore contain close to the minimum information required to define all recorded positioning data. However, the definitions of observations that are intended to be used in the survey but are missing on an exceptional basis do not need to be excluded from the block of headers for those lines for which the data are not available.

Redundant information

In a number of places the format requires redundant information to be recorded. The purpose of this is to allow integrity checks on the supplied data to take place. Redundant information should therefore not conflict with information supplied elsewhere in the format.

Nominal offsets

The complete nominal, or design, confirmation of the survey spread should be supplied in the header data. This specifically holds for points that are surveyed in, for example, the front end of the streamers.

B.3 Storage media and physical file specification

Any physical storage medium which is agreed by all parties involved in the data exchange is acceptable.

Preferred specifications for two common media are detailed below. Variations are acceptable by prior arrangement of the parties involved.

Tape	type	: 0.5 inch, 9-track, IBM standard;
	data density	: 6250 bpi
	record size	: 80 bytes
	block size	: 8000 bytes, blocks separated by an inter-record gap
	character code	: ASCII or EBCDIC

A tape file should be closed off by an IBM end-of-file mark, the last file on a tape by two consecutive IBM end-of-file marks.

Diskette	type	: 3.5 inch, DOS IBM-PC compatible
	capacity	: 1.44 Mb
	record size	: maximum 80 bytes, followed by a CR/LF
	character code	: ASCII

Each tape, diskette or other storage medium should be labelled clearly with the specifications of the stored data.

B.4 General rules

In addition to the rules given in sections B.2 and B.3, the following general rules shall apply:

(a) All records shall be 80 characters long, i.e. padded with spaces if necessary; all non-specified columns shall therefore contain blanks. (In the case of storage of data on DOS diskette this rule is waived: records shall be up to 80 characters long and shall be terminated by a CR/LF).

(b) Data fields or records for which no data is available may be omitted (records) or left blank (data fields).

(c) Nil-data returns from positioning sensors shall be recorded as blanks.

(d) All correction items shall be defined to add to the raw value.

(e) Files/lines should begin records H0000 to H00@9 in sequential order. The sequence of the remainder of the survey header records is not crucial but they should follow the logical groupings indicated in this document.

(f) Comment cards should be inserted as close as possible to the data items they refer to. They may not be inserted before record H00@9.

(g) An event occurs at the moment of the seismic shot. All data recorded for that event in E-records is assumed to apply to that moment in time.

(h) The time tags recorded for inter-event data shall refer to the time of the master vessel.

(i) Unless otherwise specified, all text items (specifier A) shall be left-adjusted and all numeric items (specifiers F, N and I) shall be right-adjusted.

(j) All feet referred to in this document are international feet, defined as follows: 1 international foot = 0.30480m.

B.5 Summary of record codes

H0... Survey definitions

H00.. General Definitions

H0000	Line name
H0001	Project name
H0002	Project description
H0003	Media and format specification
H0004	Client
H0005	Geophysical contractor
H0006	Positioning contractor
H0007	Positioning processing contractor
H00@8	Line parameters
H00@9	Additional waypoint definitions

C0001	Additional information — entire project related
C0002	Additional information — line related
C0003	Additional information — (inter-)event related

H01.. Geodetic definitions

H0100	Magnetic variation — general information
H0101	Magnetic variation — grid data
H011#	Datum and spheroid definitions
H0120	Seven parameter cartesian datum shifts
H0130	Other datum shift parameters
H0140	Projection type
H0150	(Universal) Transverse Mercator projection
H0160	Mercator projection
H0170	Lambert projection
H0180	Skew orthomorphic and oblique Mercator projection
H0181	Skew orthomorphic and oblique Mercator projection
H0190	Stereographic projection
H0199	Any other projection

H02.. Survey summary data

H0200	General summary information
H0210	Vessel summary information
H0220	Streamer summary information
H0230	Gun array summary information
H0240	Towed buoy summary information

H1... Vessel definitions

H10@0	Vessel reference point definition
H11@0	Steered point definition
H12@0	Onboard navigation system definition
H12@1	Definition of quality Indicators for field derived data
H13@0	Vessel time system definition
H14@#	Echo-sounder definition
H1500	Observed velocity of sound — definitions
H1501	Observed velocity of sound — profile
H16@0	USBL definition
H16@1	USBL definition (continued)
H16@2	Definition of quality indicator type for USBL
H17@0	Pitch, roll and heave sensor definitions
H17@1	Definition of quality indicator type for pitch, roll and heave

H2... Streamer definitions

H21@0	Streamer geometry definitions
H21@1	Streamer geometry definitions (continued)
H2102	Definition of quality indicator type for compasses
H2103	Definition of quality indicator type for depth sensors
H22@0	Compass locations
H2300	Compass correction derivation (static)
H23@0	Compass corrections (static)
H2301	Compass correction derivation (dynamic)
H23@1	Compass corrections (dynamic)
H24@0	Seismic receiver group definitions
H24@1	Auxiliary seismic channel definition
H25@0	Streamer depth Sensor definitions

H3... Gun array definitions

H31@0	Gun array geometry definitions
H31@1	Individual gun definition
H32@0	Description of gun array depth sensors
H32@1	Gun array depth sensor definitions
H32@2	Definition of quality Indicator type for depth sensors
H33@0	Definition of intended gun firing sequence
H34@0	Gun array pressure sensors definitions
H34@1	Description of gun array pressure sensors

H4... Other towed body definitions

H41@0	Towed buoy geometry definitions

H5... Survey network definitions

H5000	Node definition (fixed locations)
H51@0	Node definition (vessel, gun array, streamer, towed buoy)
H52##	Observation definition
H5306	Differential observation - follow up record
H5307	Composite range - follow up record
H54##	Observation definition (continued)
H5500	Definition of system specific quality indicator
H56@0	Instrument correction

H6... Satellite system definitions

H600#	Satellite system description
H610#	Definition of differential reference stations
H620#	Satellite receiver definition

H7... User-defined observation sets

H7000	Definition of user-defined observation sets
H7010	Data field definitions
H7020	User-defined observation parameters
H7021	Definition of quality indicator type for user-defined observations

E1... Vessel-related and general event data

E1000	General event data
E11@0	Field positioning derived data
E14@0	Echo-sounder data
E16@0	USBL acoustic data
E17@0	Pitch, roll and heave sensor data

E2... Streamer data

E22@0	Streamer compass data
E24@1	Auxiliary seismic channel data
E25@0	Streamer depth sensor data

E3... Gun array data

E32@0	Gun array depth sensor data
E33@0	Gun fired mask
E34@0	Gun pressure sensor data

E5... Network data

E52##	Network observations
E54##	Network observation parameters

E6... Satellite positioning data

E620#	GPS or DGPS data
E621#	GPS or DGPS data (continued)
E6303	TRANSIT satellite data
E640#	Satellite data (other systems)

E7... User-defined event data

E7010	User-defined observation set data

T1... Inter-event vessel-related and general event data

T14@0	Inter-event echo-sounder data
T16@0	Inter-event USBL data
T17@0	Inter-event pitch, roll and heave sensor data

T5... Inter-event network data

T52## Inter-event network observations
T54## Inter-event network observation parameters

T6... Inter-event satellite positioning data

T620# Inter-event GPS or DGPS data
T621# Inter-event GPS or DGPS data (continued)
T6303 Inter-event TRANSIT satellite data
T640# Inter-event satellite data (other systems)

T7... Inter-event user-defined positioning data

T7010 Inter-event user defined observation set data

B.6 Description of header records

B.6.1 Survey definitions

B.6.1.1 General Definitions

H0000 Line name

'Line name:'	[6,15]	A10	
Line name	[29,44]	A16	
Line sequence number	[46,49]	I4	
Line description	[50,80]	A31	free text

Note:

The line sequence number is a sequential number to be allocated to each line in the order it was shot, starting with 1. The line description should contain information about the type of line, e.g. straight, circle, cycloid, etc.

H0001 Project name

'Project Name:'	[6,18]	A13	
Project identifier	[29,36]	A8	
Project name	[38,62]	A25	free text
Start date of survey	[64,71]	I4,I2,I2	YYYYMMDD
End date of survey	[73,80]	I4,I2,I2	YYYYMMDD

Note:

Data may be generated and delivered before the end of the survey. In that case the 'End date of survey' field shall be left blank.

H0002 Project description

'Project description:'	[6,25]	A20	
Survey type, location	[29,80]	A52	free text

H0003 Media and format specification

'Media Specification:'	[6,25]	A20	
Date of issue	[29,36]	I4,I2,I2	YYYYMMDD
Media label	[38,47]	A10	
Prepared by	[49,64]	A16	free text
Format name	[66,76]	A11	e.g. UKOOA P2/91
Format revision code	[78,80]	F3.1	e.g. 1.0

H0004 Client

'Client:'	[6,12]	A7	
Description of client	[29,80]	A52	free text

H0005 Geophysical contractor

'Geophysical contractor:'	[6,28]	A23	
Description	[29,80]	A52	free text

H0006 Positioning contractor

'Positioning contractor:'	[6,28]	A23	
Description	[29,80]	A52	free text

H0007 Positioning processing contractor

'Processing contractor:'	[6,27]	A22	
Description	[29,80]	A52	free text

H00@8 Line parameters

@ = 1...9, vessel reference number

'Line parameters vessel:'	[6,28]	A23	
Vessel reference number	[30,30]	I1	
Flag for geog. or grid	[32,32]	I1	0 = geographical 1 = grid
Start of line latitude	[34,45]	I3,I2, F6.3,A1	dddmmss.sss N/S
Start of line longitude	[46,57]	I3,I2, F6.3,A1	dddmmss.sss E/W
or:			
Start of line northing	[34,44]	N11	
'N'	[45,45]	A1	
Start of line easting	[46,56]	N11	
'E'	[57,57]	A1	
First shotpoint number	[59,64]	I6	

Shotpoint number increment	[66,68]	I3	
Shotpoint interval	[70,75]	F6.2	
Length unit	[77,77]	I1	metres or feet 0 = metres 1 = feet
Number of additional waypoints defined in H00@9 records	[79,80]	I2	

Note:

The **start of line** is defined as the planned position of the vessel reference point at the first shot of the line.

The end of line point should, when appropriate, be defined in record H00@9.

The number of additional waypoints defined in H00@9 records shall not include the start of line, which is defined in this record.

In the case of a straight line only the end of line shall be defined as an additional waypoint and this number therefore equals 1 for straight lines.

Complex line shapes, such as circles and cycloids, should only have one waypoint defined, viz. the start of line. No H00@9 records should be supplied in that case. The properties of the complex line should be described in one or more C0002 records, following the H00@9 record

H00@9 Additional waypoint definitions

@ = 1...9, vessel reference number

Vessel reference number	[7, 7]	I1	
Waypoint number	[9,12]	I3	
Waypoint Latitude	[13,24]	I3,I2, F6.3,A1	dddmmss.sss N/S
Waypoint Longitude	[26,37]	I3,I2, F6.3,A1	dddmmss.sss N/S
or:			
Waypoint Northing	[13,23]	N11	
'N'	[24,24]	A1	
Waypoint Easting	[26,36]	N11	
'E'	[37,37]	A1	

May be repeated for one more waypoint definition in columns [39,67]. Vessel reference number is not repeated. Record may be repeated.

Note:

Waypoint coordinates should be supplied in the same type of coordinates as start of line (geog. or grid) and should define successive positions of the vessel reference points.

The end of line is defined as the planned position of the vessel reference point at the last shot of the line. The end of line should be the last of the waypoints defined.

C0001 Additional information - entire project related

Project related information	[6,80]	A75	free text

C0002 Additional information - line related

Line related information	[6,80]	A75	free text

C0003 Additional information - (inter-)event related

(Inter-)event related info.	[6,80]	A75	free text

Additional comments

Three comment records are available for general, free text comments, considered relevant to the survey.

C0001 — for information related to the entire project;
C0002 — for information related to the seismic line only;
C0003 — for information related to (inter-)event data.

Any number of these records may be inserted in the data. C0001 and C0002 records may appear anywhere among the other header records, but after record H00@9.

C0003 records may appear anywhere among the (inter-)event data, but after the general eEvent record E1000.

Common sense would dictate that whatever comment records are used, they are inserted as close as possible to the records to which the comments refer.

B.6.1.2 Geodetic definitions

H0100 Magnetic variation — general information

Date for which the Magnetic Variation values are valid	[7,14]	I4,I2,I2	YYYYMMDD
Number of points in grid	[16,19]	I4	
Defined in geog. or grid	[21,21]	0 = geographical 1 = rectangular	
Source of magnetic variation	[23,80]	A58	free text

H0101 Magnetic variation - grid data

Point number	[7,10]	I4	
If geographical co-ordinates:			
Latitude of point	[12,23]	I3,I2,F6.3,A	dddmmss.sss N/S

Longitude of point	[25,36]	I3,I2,F6.3,A	dddmmss.sss E/W
If rectangular coordinates:			
Northing	[12,22]	N11	
'N'	[23,23]	A1	
Easting	[25,35]	N11	
'E'	[36,36]	A1	
Magnetic Variation	[38,44]	F7.3	+/– degrees decimal
Secular change in Magnetic Variation in this point	[46,51]	F6.4	+/– degr.dec./year

Record may be repeated.

Note:

Records H0100 and H0101 together allow a grid of points to be defined to cater for varying magnetic variation over the survey area. The grid may be defined either in terms of geographical co-ordinates or in terms of rectangular co-ordinates (e.g UTM), as defined in records H0140...H0199.

H011# Datum and Spheroid Definitions

= 1...9, datum & spheroid number

Datum name	[7,24]	A18
Spheroid name	[25,43]	A19
Semi-major axis (*a*)	[44,55]	N12
Conversion factor to metres	[57,68]	N12
Inverse flattening (1/*f*)	[70,80]	N11

Note:

The conversion factor, multiplied by the semi-major axis, should yield the value of the axis in metres. Hence, if the semi-major axis is supplied in international feet, the conversion factor should equal 0.30480.

H0120 Seven parameter Cartesian datum shifts

From datum 1 to datum 2

Datum 1: spheroid number		[7, 7]	I1	
Datum 2: spheroid number		[9, 9]	I1	
Rotation convention		[11,11]	I1	0 = position vector rotation (Bursa-Wolfe) 1 = co-ordinate frame rotation
X shift	(*dX*)	[13,22]	F10.2 metres	
Y shift	(*dY*)	[24,33]	F10.2 metres	
Z shift	(*dZ*)	[35,44]	F10.2 metres	
X rotation	(*rx*)	[46,53]	F8.4 seconds of arc	
Y rotation	(*ry*)	[55,62]	F8.4 seconds of arc	
Z rotation	(*rz*)	[64,71]	F8.4 seconds of arc	

Scale correction (S) [73,80] F8.4 ppm

Note:

Up to nine different datum/spheroids may be defined. Datum/spheroid number 1 is referred to as the survey datum and shall apply to all coordinates recorded in this format where the datum/spheroid is implied, such as the coordinates of surface radio positioning systems.

Additional datum/spheroid definitions may be made only to cover different satellite systems. The appropriate datum/spheroid number should be included in the definition of any satellite system in record H600# (see section B.6.7), thus explicitly linking them to a datum different from the survey datum. The datum shift parameters actually used in the field should be recorded in these records.

Datum conversion formulae - rotation conventions

Two different conventions for rotation definitions are in use in the survey industry, which has led to considerable confusion. Nevertheless both are valid when applied consistently. For this reason the format allows datum shift parameters for either rotation convention to be recorded. It is advised to exercise great care in filling in this record and to verify the information supplied against the worked example included in this section.

The two rotation conventions can be referred to as:

(a) Position vector rotation (Bursa-Wolfe model, commonly used in Europe)

(b) Coordinate frame rotation (commonly used in the USA)

(a) Position vector rotation (Bursa-Wolfe model)

The set of conversion formulae associated with this convention is commonly referred to as the Bursa-Wolfe model. Rotations are defined as positive clockwise, as may be imagined to be seen by an observer in the origin of the co-ordinate frame, looking in the positive direction of the axis about which the rotation is taking place. However, the rotation is applied to the position vector. Hence a positive rotation about the Z-axis of rz will rotate the position vector further east. For example, after applying a datum shift describing **only** a **positive** rotation about the Z-axis from datum 1 to datum 2 the longitude of a point will therefore be **larger** on datum 2 than it was on datum 1.

The associated conversion formula is:

$$\begin{bmatrix} X \\ Y \\ Z \end{bmatrix} = \begin{bmatrix} \delta X \\ \delta Y \\ \delta Z \end{bmatrix} + (1 + S \times 10^{-6}) \times \begin{bmatrix} 1 & -\theta z & \theta y \\ \theta z & 1 & -\theta x \\ -\theta y & \theta x & 1 \end{bmatrix} \times \begin{bmatrix} X \\ Y \\ Z \end{bmatrix} \qquad \text{(B.1)}$$

Datum 2 Datum 1

where Datum 1 and Datum 2 must be defined in record H011#.

Note:

The rotation angles as supplied in record H0120 above must be converted to radians for use in the above formula.

Example

	Datum 1: WGS84	Datum 2: ED87
Semi-major axis (*a*)	6378137 metres	6378388 metres
Inverse flattening(1/*f*)	298.257	297.0
Latitude	57°00m 00s N	57° 00m 02.343s
Longitude	2°00m 00s E	2° 00m 05.493s
Spheroidal Height	100 metres	55.12 metres
X	3479923.02 m	3480006.35
Y	121521.59 m	121617.29
Z	5325983.97 m	5326096.93
δX	+ 82.98 m	
δY	+ 99.72 m	
δZ	+ 110.71 m	
θx	Note:---> + 0.1047" = + 0.5076*10E-6 rad	
θy	Note:---> - 0.0310" = - 0.1503*10E-6 rad	
θz	Note:---> - 0.0804" = - 0.3898*10E-6 rad	
S	+ 0.3143 = ppm	

(b) Co-ordinate frame rotation

The 3 × 3 matrix in the formula associated with this convention derives from a type of matrix known in mathematics as a rotation matrix. A rotation matrix describes a rotation of a right-handed coordinate frame about its origin. Frame rotations are defined as positive clockwise, as may be imagined to be seen by an observer in the origin of the coordinate frame, looking in the positive direction of the axis about which the rotation is taking place. Hence a positive rotation about the *Z*-axis of *rz* will cause the *X*-axis of datum 2 (and therefore the zero meridian) to lie east of the *X*-axis of datum 1 Therefore, after applying a datum shift describing only a **positive** rotation about the *Z*-axis from Datum 1 to Datum 2 the longitude of a point will therefore be **smaller** on datum 2 than it was on Datum 1.

The associated conversion formula is:

$$\begin{bmatrix} X \\ Y \\ Z \end{bmatrix} = \begin{bmatrix} \delta X \\ \delta Y \\ \delta Z \end{bmatrix} + (1 + S \times 10^{-6}) \times \begin{bmatrix} 1 & \theta z & -\theta y \\ -\theta z & 1 & \theta x \\ \theta y & -\theta x & 1 \end{bmatrix} \times \begin{bmatrix} X \\ Y \\ Z \end{bmatrix} \qquad \text{(B.2)}$$

Datum 2 Datum 1

where Datum 1 and Datum 2 must be defined in record H011#.

Note:

The rotation angles as supplied in record H0120 above must be converted to radians for use in the above formula.

Example:

	Datum 1: WGS84	Datum 2: ED87
Semi-major axis(a)	6378137 metres	6378388 metres
Inverse flattening (1/f)	298.257	297
Latitude	57° 00m 00s N	57° 00m 02.343s
Longitude	2° 00m 00s E	2° 00m 05.493s
Spheroidal Height	100 metres	55.12 metres
X	3479923.02 m	3480006.35m
Y	121521.59 m	121617.29m
Z	5325983.97 m	5326096.93m

δX		+ 82.98 m
δY		+ 99.72 m
δZ		+ 110.71 m
θx	Note:-->	−0.1047" = −0.5076x10E-6 rad
θy	Note:-->	+ 0.0310" = + 0.1503x10E-6 rad
θz	Note:-->	+ 0.0804" = + 0.3898x10E-6 rad
S		+ 0.3143 = + ppm

H0130 Other datum shift parameters

From Datum 1 to Datum 2

Datum 1: spheroid number	[7, 7]	I1
Datum 2: spheroid number	[8, 8]	I1
Sequence number	[9,10]	I2
'/'	[11,11]	A1
Total number of records	[12,13]	I2

Description conversion [15,80] A65 free text

Note

This record allows datum shifts to be defined using a model different from the seven-parameter Cartesian model in H0120. This may include such datum conversions described by polynomials. The information provided in these records shall contain a complete definition of the datum conversion and shall contain the following information as a minimum:

(a) a description of the datum conversion;

(b) the formulae used;

(c) the parameters required by these formulae.

Example

The example below describes the conversion from ED87 to ED50 as agreed between the mapping authorities of Norway, Denmark, Germany, The Netherlands and Great Britain in 1990. This agreement defines a two step conversion from WGS84 to ED50 in the North Sea, of which the example describes the second step. The following records are required to describe the complete datum conversion of WGS84 to ED50:

- 3 × H011#: defining WGS84, ED87 and ED50 as 3 datums;
- 1 × H0120: defining the 7-parameter cartesian co-ordinate conversion from WGS84 to ED87;
- 14 × H0130: defining the conversion of ED87 latitude and longitude to ED50 latitude and longitude.

```
H0130 12 1/14 CONVERSION OF ED87 LAT/LON TO ED50 BY 4TH DEGREE POLYNOMIAL; REF:
H0130 12 2/14 THE TRANSFORMATION BETWEEN ED50 AND WGS84 FOR EXPLORATION PURPOSES
H0130 12 3/14 IN THE NORTH SEA, B.G. HARSSON; STATENS KARTVERK NORWAY; 1990
H0130 12 4/14 CORR=10^-6*(A0+A1*U+A2*V+A3*U^2+A4*U*V+A5*V^2+A6*U^3+A7*U^2*V+
H0130 12 5/14 +A8*U*V^2+A9*V^3+A10*U^4+A11*U^3*V+A12*U^2*V^2+A13*U*V^3+A14*V^4)
H0130 12 6/14 U=ED87 LAT.(DEGREES) MINUS 55; V=ED87 LON.(DEGREES)
H0130 12 7/14 LAT.(ED50) = LAT.(ED87)+CORRECTION FROM LAT. COEFF. A0 TO A14
H0130 12 8/14 LON.(ED50) = LON.(ED87)+CORRECTION FROM LON. COEFF. A0 TO A14
H0130 12 9/14 LATITUDE POLYNOMIAL COEFFICIENTS A0 TO A14: 5.56098,1.55391
H0130 1210/14.40262,.509693,.819775,.247592,-.136682,-.186198,-.12335
H0130 1211/14 -.0568797,.00232217,.00769931,.00786953,.00612216,.00401382
H0130 1212/14 LONGITUDE POLYNOMIAL COEFFICIENTS A0 TO A14: -14.8944,-2.68191
```

H0130 1213/14 -2.4529,-.2944,-1.5226,-.910592,.368241,.851732,.566713,.185188

H0130 1214/14 -.0284312,-.0684853,-.0500828,-.0415937,-.00762236

H0140 Projection type

Projection type code	[7, 9]	I3	see note below record H0199
Coord conversion factor to m	[11,20]	N10	
Projection type and name	[22,80]	A59	free text

Note:

The coordinate conversion factor, multiplied by the coordinate values as supplied in the data, should yield the coordinates in metres.

H0150 (Universal) Transverse Mercator projection

Zone number	[7, 8]	I2	(UTM only)
Latitude of grid origin	[10,21]	I3,I2,F6.3,A1	dddmmss.sss N/S
Longitude of grid origin	[23,34]	I3,I2,F6.3,A1	dddmmss.sss E/W
Grid Northing at grid origin	[36,46]	N11	
'N'	[47,47]	A1	
Grid Easting at grid origin	[48,58]	N11	
'E'	[59,59]	A1	
Scale factor at long. of origin	[61,72]	N12	

H0160 Mercator projection

Latitude of grid origin	[7,18]	I3,I2,F6.3,A1	dddmmss.sss N/S
Longitude of grid origin	[20,31]	I3,I2,F6.3,A1	dddmmss.sss E/W
Grid Northing at grid origin	[33,43]	N11	
'N'	[44,44]	A1	
Grid Easting at grid origin	[45,55]	N11	
'E'	[56,56]	A1	
Scale factor at latitude of origin	[58,69]	N12	

H0170 Lambert projection

Latitude of (first) std. parallel	[7,18]	I3,I2,F6.3,A1	dddmmss.sss N/S
Latitude of second std. parallel	[20,31]	I3,I2,F6.3,A1	dddmmss.sss N/S
Longitude of grid origin	[33,44]	I3,I2,F6.3,A1	dddmmss.sss E/W
Grid Northing at grid origin	[45,55]	N11	
'N'	[56,56]	A1	
Grid Easting at grid origin	[57,67]	N11	
'E'	[68,68]	A1	
Scale factor at std. parallels	[69,80]	N12	

H0180 Skew orthomorphic and oblique Mercator projection

Latitude of start point	[7,18]	I3,I2,F6.3,A1	dddmmss.sss N/S
Longitude of start point	[19,30]	I3,I2,F6.3,A1	dddmmss.sss E/W
Latitude of end point	[31,42]	I3,I2,F6.3,A1	dddmmss.sss N/S
Longitude of end point	[43,54]	I3,I2,F6.3,A1	dddmmss.sss E/W
Bearing of initial line of projection in grid origin	[55,66]	N12	degrees decimal
Angle from skew to rect. grid	[67,78]	N12	degrees decimal, clockwise positive
Is scale factor at origin = 1?	[80,80]	I1	0 = No; 1 = Yes

Note:

The initial line of projection is often referred to as 'central line of projection' in the Oblique Mercator Projection.

If the scale factor at grid origin does **not** equal 1 an H0181 record must be supplied.

H0181 Skew orthomorphic and oblique Mercator projection (continued)

Scale factor at grid origin	[7,18]	N12

Note:

Only to be supplied if the scale factor does NOT equal 1.

H0190 Stereographic Projection

Latitude of grid origin	[7,87]	I3,I2,F6.3,A1	dddmmss.sss N/S
Longitude of grid origin	[19,30]	I3,I2,F6.3,A1	dddmmss.sss E/W
Grid northing at grid origin	[32,42]	N11	
'N'	[43,43]	A1	
Grid Easting at grid origin	[44,54]	N11	
'E'	[55,55]	A1	
Scale factor at origin	[56,67]	N12	
Std parallel (for polar only)	[69,80]	I3,I2,F6.3,A1	dddmmss.sss E/W

H0199 Any other projection

Sequence number of record	[7, 8]	I2	
'/'	[9, 9]	A1	
Total number of records	[10,11]	I2	
Map projection parameters	[13,80]	A68	free text

Additional comments

The following projection type codes have been defined. The relevant code as detailed in the table below must be entered into record H0140. The associated projection parameters should be recorded using one of the above records, H0150 to H0190 for the appropriate projection.

Code	*Projection description*	*Record*
001	UTM north	H0150
002	UTM south	H0150
003	Transverse Mercator (north oriented)	H0150
004	Transverse Mercator (south oriented)	H0150
005	Lambert conic conformal, one standard parallel	H0170
006	Lambert conic conformal, two standard parallels	H0170
007	Mercator	H0160
008	Cassini-Soldner	H0170
009	Skew orthomorphic and oblique Mercator	H0180/1
010	Stereographic	H0190
011	New Zealand map grid	H0160
999	Any other projection	H0199

Not every projection can be defined by these codes and the elements of the projection header records. The intention is that the majority of standard projections can be defined in a computer-interpretable form.

Projections not covered by one of the codes 001 to 011 should be indicated by code 999. The associated projection parameters should be provided using several of H0199 records. These records must unambiguously define the map projection.

Grid origin is the origin or centre of the projection, not the origin of the grid coordinate system, which may be offset from the grid origin.

Scale factors must be given in real numbers (e.g. 0.9996 as opposed to –400 ppm).

For surveys in a UTM zone crossing the equator from the Southern to the Northern Hemisphere 10000000 is commonly added to the Northings on the Northern Hemisphere to avoid discontinuity in these coordinates. In that case a warning must be given (in an inserted C0001 record) to explain that convention.

For the definition of US State Plane Co-ordinate Systems (SPCS) reference is made to Transverse Mercator or Lambert projection definitions.

B.6.1.3 Survey summary data

H0200 General summary information

Number of survey vessels	[7, 7]	I1
Number of relay vessels or buoys	[9,10]	I2
Number of external network nodes	[12,13]	I2
Number of datums/spheroids defined	[15,15]	I1
Offset mode	[17,17]	I1 0 = polar

			1 = rectangular
Offset measurement units: for offset distances	[19,19]	I1	0 = metres
			1 = feet
Offset measurement units: for offset angles decimal	[21,21]	I1	0 = degrees
			1 = grads

Note:

Up to nine survey vessels may be defined (see *additional comments* below). Vessel 1 must be defined for all surveys and is the master vessel in multi-vessel surveys.

Relay vessels are purely considered as carriers of network nodes, assisting in the positioning of the seismic spread. A relay vessel or buoy carries one or more radio-positioning beacons of which the signals are used in the positioning of the seismic spread, while the relay vessel or buoy itself is continually positioned.

The number of external network nodes refers to the number of network nodes outside the survey vessel(s) and its/their towed configurations but include the nodes defined on relay vessels.

H021@ Vessel summary Information

@ = 1...9, vessel reference number
@ = 0 if relay vessel

Vessel name/description	[6,40]	A35	free text
Vessel reference number	[43,44]	I2	
Number of streamers	[50,51]	I2	
Number of gun arrays	[53,54]	I2	
Number of buoys	[56,57]	I2	
Number of echo sounders	[59,59]	I1	
Pitch/roll sensors	[61,61]	I1	0 = no; 1 = yes
Number of USBL systems	[63,63]	I1	
Number of satellite receivers	[65,66]	I2	
Number of network nodes	[68,70]	I3	

Note:

The number of buoys refers to the buoys towed directly from the vessel, hence it does not include the number of tailbuoys or buoys towed from gun arrays here; they should be defined in record H0220 or H0230 respectively.

The number of sensors and nodes refers only to those on the vessel.

H022@ Streamer summary lInformation

@ = 1...9, vessel reference number

Streamer description	[6,40]	A35	free text
Streamer reference number	[42,44]	I3	
Towed by ref. number	[46,48]	I3	

Number of buoys	[56,57]	I2
Number of network nodes exclusive of magnetic compasses	[68,70]	I3
Number of mag. compasses	[72,73]	I2
Number of depth sensors	[75,76]	I2
Number of receiver groups	[78,80]	I3

Note:

The number of buoys refers to the buoys towed by the streamer, normally just the tailbuoy.

The number of network nodes is exclusive of the magnetic compasses. These are counted separately in columns [72,73].

H023@ Gun array summary information

@ = 1...9, vessel reference number

Gun array description	[6,40]	A35	free text
Gun array ref. number	[42,44]	I3	
Towed by ref. number	[46,48]	I3	
Number of buoys	[56,57]	I2	
Number of satellite receivers	[65,66]	I2	
Number of network nodes	[68,70]	I3	
Number of depth sensors	[75,76]	I3	

Note:

The number of network nodes is exclusive of the satellite receivers. These are counted separately in columns [65,66].

H024@ Towed buoy summary information

@ = 1.9, vessel reference number

Towed buoy description	[6,40]	A35	free text
Towed buoy ref. number	[42,44]	I3	
Towed by ref. number	[46,48]	I3	
Number of other buoys towed by this buoy	[56,57]	I2	
Number of satellite receivers	[65,66]	I2	
Number of network nodes	[68,70]	I3	

Note:

The number of network nodes is exclusive of the satellite receivers. These are counted separately in columns [65,66].

Tailbuoys, front buoys, etc. should be defined as separate buoys.

Additional comments

The summary information in records H0200 to H0240 concentrate redundant information in the front end of the file with a dual purpose:

- to facilitate a quick overview of the survey by visual inspection; for that purpose the summary information is arranged in a columnar way: similar data items appear in the same columns for all records above.

- to enable automated (or visual) format integrity checking.

Each vessel, streamer, gun array and buoy must be given a reference number. The relationship between e.g. vessel and streamer is defined by providing the reference number of the towing vessel in record H0220.

This method has been chosen to provide flexibility in the sense that the towing object does not need to be the vessel.

Reference numbers shall be allocated in accordance to the following convention:

Survey vessels	: 1... 9
Relay vessels/buoys	: 10... 99
Streamers	: 200...299
Gun arrays	: 300...399
Other towed buoys	: 400..499

B.6.1.4 **Offset conventions**

Definition of co-ordinate axes

Throughout the document right-handed cartesian co-ordinate frames are maintained to express offset.

The axes of the co-ordinate frames are defined as follows:

Y-axis — parallel to the vessel's longitudinal axis, positive towards the bow.

The direction of the positive Y-axis is also referred to in this document as 'ship's head'.

X-axis — horizontal axis, perpendicular to the Y-axis, positive towards starboard.

Z-axis — perpendicular to the two horizontal axes, X and Y, the Z-axis completes a right-handed X, Y, Z coordinate frame. Hence, positive Z is upwards, synonymous with height.

Reference points

Each ship has its own coordinate frame, with its origin defined as the **ship's reference point** in record H10@0.

All towed objects, such as streamers, gun arrays and buoys, have their own **local reference point** and points on these objects are described in terms of **local offsets** relative to the local reference point.

Tow points

The header records of the P2/91 format describe the nominal, or design geometry of the spread, in which all towed objects are towed parallel to the longitudinal axis of the towing vessel, the *Y*-axis of its coordinate frame.

Each towed object 'streams' from a towpoint, which may be offset from the vessel by means of a paravane. This towpoint is defined in P2/91 as the **towpoint-in-sea**, as opposed to the point on the vessel the towed object is attached to, which is called the **towpoint-on-towing-body**.

This distinction is only relevant in the case where a paravane is used to offset the towed object from the vessel. When the object is towed direct from the vessel or from a boom rigidly attached to the vessel, the 'towpoint-in-sea' and the 'towpoint-on-towing-body' are coincident.

Some towed objects are in turn towing another towed object. Examples are a tailbuoy towed by a streamer and a front buoy towed by a gun array. Assuming that the buoy is towed straight behind the streamer or gun array, the towpoint-in-sea coincides with the towpoint-on-towing-body, which is the point on the streamer or gun array the buoy is being towed from. Note that the 'towing vessel' in this case is **not** the ship, but the streamer or gun array.

Local offsets and reference points

Each towed object has its own local co-ordinate frame of which the axes are parallel to the ship's *X*, *Y* and *Z* axes (that is: in the nominal, or deign, geometry). However, the local offsets on each towed object are measured from its **local reference point.**

For all towed objects except the streamers the local reference point is the towpoint-in-sea, defined for that object. The towpoint-in-sea may of course coincide with the towpoint-on-towing-body, as explained above. The only exception to this rule is the local reference point of a streamer, which is defined as the centre of the near receiver group of that streamer, the receiver group closest to the towpoint. Note that therefore *Y*-offsets along a streamer are negative and decreasing towards the tailbuoy behind the centre of the near receiver group.

Offset mode: polar or rectangular

Horizontal offsets may be given as either polar or rectangular co-ordinates.

The offset mode must be consistent for all offsets defined.

Polar mode:

Offset A = radial distance from ship's or local reference point to the point defined;

Offset B = angle, measured in the ship's or local reference point, clockwise from ship's head to the point defined.

Rectangular mode:

Offset A = *X*-axis offset from ship's or local reference point to the point defined, measured positive to starboard.

Offset B = *Y*-axis offset from ship's or local reference point to the defined point, measured positive towards the bows.

Z-axis offset or height

The third offset co-ordinate, along the Z-axis, is always positive upwards. Depths (of e.g. acoustic transducers) are therefore recorded as **negative heights**, with the minus sign included in the field provided.

Units of measurement

Offset distances may be expressed in metres decimal or international feet.

Offset angles may be expressed in degrees decimal or in grads.

The same measurement units must be used consistently for all offsets defined.

B.6.2 Vessel definitions

H10@0 Vessel reference point definition

@ = 1...9, vessel reference number

Height above sea level	[7,10]	F4.1	metres
Description of reference point	[12,80]	A69	free text

H11@0 Steered point definition

@ = 1...9, vessel reference number			
Description of steered point	[7,80]	A74	free text

H12@0 Onboard navigation system description

@ = 1...9, vessel reference number			
Details of onboard navigation & processing systems	[7,80]	A74	free text

Record may be repeated to describe more than one onboard system, such as an additional system used for quality control purposes.

H12@1 Definition of quality indicators for field positioning derived data

@ = 1...9, vessel reference number			
Record sequence number	[7, 8]	I2	
Definition of quality indicator types for field positioning derived data	[10,80]	A71	free text

Record may be repeated.

Note:

In record E12@0, field positioning derived data, three fields are provided to record quality indicators describing the quality of that data. These quality indicators should describe the quality of the processed positioning data. Examples are: the standard deviations of northing and easting and the standard deviation of unit weight.

A full descriptive record of these three indicators must be provided in this record.

Up to 99 E12@0 records may be supplied, each identified by its record sequence number, which needs to be recorded here to provide a link for the definitions of the quality indicators and the actual data in the E12@0 record.

H13@0 Vessel Time System Definition

@ = 1...9, vessel reference number			
Time correction to ship's time to convert to GMT	[7,12]	F6.2	+/- hours
Time correction to vessel's time system to convert to mastertime system	[14,21]	N8	seconds

Note:

The first time correction is a 'time zone' correction and is therefore defined to fractions of hours only. The correction is added to ship's time and therefore this convention is opposite to that used in the UKOOA P1/90 format.

The second correction enables the supplier of the data to record synchronization differences between the clocks used for the measurement systems on board the different survey vessels in a multi-vessel survey, should this information be required. It can be defined in fractions of seconds.

H14@# Echo-sounder definition

@ = 1...9, vessel reference number
# = 1...9, echo-sounder reference number

Offset A to transducer	[7,13]	F7.1	
Offset B to transducer	[15,21]	F7.1	
Offset Z from reference point to transducer	[23,28]	F6.1	
Propagation velocity used	[30,36]	N7	m/s or ft/s
Calibrated propagation velocity	[38,44]	N7	m/s or ft/s
Velocity unit	[46,46]	I1	0 = m/s 1 = ft/s
Water depth reference level	[47,47]	I1	0 = transducer 1 = sea level
Heave compensated depths?	[48,48]	I1	0 = depths not heave compensated 1 = depths heave compensated
Echo sounder description	[50,80]	A31	free text

Note:

Two propagation velocities are to be given for the echo-sounder; that at which the sounder was set during the survey or part-survey covered by this file and the velocity determined during calibration. Both velocities should be specified even if they are the same. Raw depths recorded by the echo-sounder may relate to the transducer or may have been corrected to sea level: the water depth reference level flag should be set appropriately.

The heave compensation flag should be set to 1 if the echo-sounder is interfaced to a heave-compensator, resulting in an output of heave compensated depths.

A description of the heave-compensator should then be supplied in record H17@0.

H1500 Observed velocity of sound — definitions

Profile number	[7, 8]	I2	
Date	[10,17]	I4,I2,I2	YYYYMMDD
Time (Master Vessel)	[19,22]	I2,I2	HHMM
Latitude	[24,35]	I3,I3,F6.3,A1	dddmmss.sss N/S
Longitude	[36,47]	I3,I3,F6.3,A1	dddmmss.sss E/W
Depth units	[48,48]	I1	0 = metres 1 = feet
Velocity units	[49,49]	I1	0 = metres/sec 1 = feet/sec
Temperature units	[50,50]	I1	0 = degrees Celsius 1 = degrees Fahrenheit

Salinity/Conductivity	[51,51]	I1	0 = promille (10^{-3}) (salinity) 1 = mmho/cm (conductivity) 2 = Siemens/metre (conductivity)
Instrument description	[53,80]	A28	free text

H1501 Observed velocity of sound — profile

Profile number	[7, 8]	I2
Depth	[10,15]	F6.1
Velocity	[16,21]	F6.1
Temperature	[22,26]	F5.1
Salinity or conductivity	[27,31]	F5.2

May be repeated for two more observations at [33,54] and [56,77] for the same profile; the profile number is not repeated.

Record may be repeated.

Note:

Up to 99 velocity of sound depth profiles may be defined by repeating one H1500 and as many H1501 records as required for every profile defined.

H16@0 USBL System definition

@ = 1...9, vessel reference number

USBL system ref. number	[7, 7]	I1	
Quality indicator type	[9, 9]	I1	
Sign convention for *Z*-axis data upward	[11,11]	I1	0 = positive (height) 1 =positive downward (depth)
Recorded data corrected for:			
Turn around delays?	[12,12]	I1	O = no, 1 = yes
Velocity of propagation?	[13,13]	I1	0 = assumed, 1 = calibrated
Horizontal alignment?	[14,14]	I1	0 = no, 1=ship's axis, 2 = raw gyro
Pitch alignment?	[15,15]	I1	0 = no, 1 = raw VRU, 2 = corrected VRU
Roll alignment?	[16,16]	I1	0 = no, 1 = raw VRU, 2 = corrected VRU

Reduction to ship's reference point?	[17,17]	I1	0 = no, 1 = yes

Note: the quality indicator type defines the type of quality indicator used in the (inter-)event fields.

The following types are available:

0 = no quality information recorded

1 = standard deviation

2 = signal/noise ratio

3 = system specific

4 = subjective scale

In the case where code 1 is chosen, a descriptive definition of the way the standard deviation is derived must be supplied in record H16@2.

In the case where code 3 is chosen, a descriptive definition must be supplied in record H16@2 of the following aspects of the system specific quality indicator:

- the range of values of the variable;
- the interpretation of its values.

USBL systems may give relative coordinate data with the sign convention of the Z-coordinates opposite the convention maintained throughout this format. If the USBL system produces Z data positive downwards (= depth) the flag should be set to 1. Note though that the definition of the transducer location in record H16@1 should adhere to the P2/91 convention that Z-offsets are positive upwards.

Since USBL systems do not generally provide raw data, the extent of the processing by the system should be defined in columns [11,16].

H16@1 USBL system definition (continued)

@ = 1...9, vessel reference number

USBL system reference number	[7, 7]	I1	
Transducer node identifier	[9,12]	I4	
Offsets from ship's reference point to USBL transducer:			
Offset A	[14,20]	F7.1	
Offset B	[22,28]	F7.1	
Offset Z	[30,35]	F6.1	
Correction to hor. alignment	[37,41]	N5	degrees decimal
Correction to pitch alignment	[43,47]	N5	degrees decimal
Correction to roll alignment	[49,53]	N5	degrees decimal
Assumed vel. of propagation	[55,61]	N7	
Calibrated vel. of propagation	[63,69]	N7	
Velocity measurement units	[71,71]	I1	0 = metres/second 1 = feet/second
Turn around delay	[73,80]	N8	milliseconds

Note:

The three offsets define the logical location of the USBL transducer. Thus if the device corrects to the ship's reference point, the offsets should be zero.

The node identifier should be a unique positive number (>0).

If the USBL system has **not** been set up to reduce for heading, pitch and roll corrections (C–O) these corrections should be supplied in this record. The horizontal alignment correction is that angle required to reduce the USBL system orientation to the ship's head.

The pitch and roll corrections are those required to correct the vertical reference unit (VRU)

Pitch corrections should be **positive** for the **bow down.**

Roll corrections should be **positive** for the ship **heeling to port.**

H16@2 Definition of quality indicator type for USBL

@ = 1...9, vessel reference number

USBL system ref. number	[7, 7]	I1	
Definition of quality indicator	[9,80]	A72	free text

Record may be repeated.

H17@0 Pitch, roll and heave sensor definitions

@ = 1...9, vessel reference number

Sensor reference number	[7, 7]	I1	
Rotation convention pitch	[9, 9]	I1	0 = positive up 1 = positive down
Rotation convention roll	[10,10]	I1	0 = positive stbd. 1 = positive port
Angular variable measured	[11,11]	I1	0 = pitch/roll angle 1 = sine of angle
Angular measurement units	[12,12]	I1	3 = deg. decimal 4 = grads 9 = other
Measurement units heave	[13,13]	I1	0 = metres 1 = feet 9 = other
If angular measurement units = 9:			
Conversion to degrees decimal	[15,22]	N8	
If heave measurement units = 9:			
Conversion factor to metres	[24,31]	N8	
Quality indicator pitch and roll	[33,33]	I1	
Quality indicator heave	[34,34]	I1	
(C–O) pitch observation	[36,42]	N7	

(C–O) roll observation	[44,50]	N7	
(C–O) heave observation	[52,58]	N7	
Description of pitch, roll, heave system	[60,80]	A21	free text

Note:

This record should be used in case data are recorded from a pitch/roll/heave system. The standard rotation conventions for both pitch and roll ought to be code 0 as that convention is consistent with the definition of positive rotations in right-handed Cartesian coordinate frames, which the vessel's *X, Y, Z* system is.

The conversion factor for pitch and roll should be multiplied with the raw readings for pitch and roll to yield degrees decimal values.

The conversion factor for heave should be multiplied with the raw readings for heave to yield values in metres.

The quality indicator type defines the type of quality indicator used in the (inter-)event fields.

The following types are available:

0 = no quality information recorded
1 = standard deviation
2 = signal/noise ratio
3 = system specific
4 = subjective scale

In the case where code 1 is chosen, a descriptive definition of the standard deviation is derived must be supplied in record H17@1.

In the case where code 3 is chosen, a descriptive definition must be supplied in record H17@1 of the following aspects of the system specific quality indicator:

- the range of values of the variable;
- the interpretation of its values.

H17@1 Definition of quality indicator type for pitch, roll and heave

@ = 1..9, vessel reference number

Sensor reference number	[7, 7]	I1	
Definition of quality indicator	[9,80]	A72	free text

Record may be repeated.

B.6.3 Streamer definitions

H21@0 Streamer geometry definitions

@ = 1...9, vessel reference number

Streamer reference number	[7, 9]	I3

Offsets from ship's reference point to:

towpoint-on-towing-vessel:		
Offset A	[11,17]	F7.1
Offset B	[19,25]	F7.1
Offset *Z*	[27,32]	F6.1
towpoint-in-sea:		
Offset A	[35,41]	F7.1
Offset B	[43,49]	F7.1
Offset *Z*	[51,56]	F6.1

Local offsets from the centre of the near receiver group to the towpoint-in-sea:

Local *Y*-offset	[58,64]	F7.1
Local *Z*-offset	[66,71]	F6.1

Note:

The offsets to the towpoints in columns 11 to 56 are measured relative to the ship's reference point.

The local offsets in columns 58 to 71 are measured relative to the local reference point: the centre of the near seismic receiver group.

The towpoint-in-sea is expected to be defined at sea level.

The local Z-offset is measured vertically from the local reference point (the near receiver group) to the towpoint-in-sea.

As the towpoint-in-sea, when defined at sea level, is higher then the local reference point, its local Z-offset is a positive figure and equals the nominal depth of the streamer.

H21@1 Streamer geometry definitions — continued

@ = 1...9, vessel reference number

Streamer reference number	[7, 9]	I3
Nominal front stretch length	[11,15]	F5.1
Nominal rear stretch length	[17,21]	F5.1
Number of active sections	[23,25]	I3
Length of each active section	[27,31]	F5.1
Next six parameters for **inserted** compass, acoustic and depth sections only:		
Number of compass sections	[33,35]	I3
Length of each compass section	[37,41]	F5.1
Number of acoustic sections	[43,45]	I3
Length of each acoustic section	[47,51]	F5.1

Number of depth sections	[53,55]	I3
Length of each depth section	[57,61]	F5.1
Quality indicator for compasses	[63,63]	I1
Quality indicator for depth sensors	[65,65]	I1

Note:

The lengths of the streamer sections given in this record should be supplied in the same units as the offsets.

Inserted sections should only be recorded in this record if they are separate sections. However, if they are integrated with another unit they should **not** be recorded in this record to prevent inserted sections from being counted twice. Add a C0001 comment record if that is the case.

The quality indicator type defines the type of quality indicator used in the (inter-)event data records for all compasses and should be one of the following codes:

0 = no quality information recorded

1 = standard deviation

2 = signal/noise ratio

3 = system specific

4 = subjective scale

In the case code 1 is chosen a descriptive definition of the way the standard deviation is derived must be supplied in record H21@2 and/or record H21@3, as appropriate.

In the case code 3 is chosen a descriptive definition must be supplied in record H21@2 and/or record H21@3 of the following aspects of the system-specific quality indicator:

the range of values of the variable;

the interpretation of its values.

H21@2 Definition of quality indicator type for streamer compasses
@ = 1...9, vessel reference number

Definition quality indicator type	[7,80]	A74	free text

Record may be repeated.

H21@3 Definition of quality indicator type for streamer depth sensors
@ = 1...9, vessel reference number

Definition quality indicator type	[7,80]	A74	free text

Record may be repeated.

H22@0 Compass locations
@ = 1...9, vessel reference number

Streamer reference number	[7, 9]	I3	
Node identifier	[11,14]	I4	
Compass serial number	[16,23]	A8	
Local offset to centre of compass	[25,32]	F8.1	
Clipped-on or inserted?	[34,34]	I1	0 = clipped-on 1 = inserted

May be repeated for one more compass on the same streamer at [36,59]; the streamer reference number is not repeated.

Record may be repeated.

Note:

The node number must be a unique positive number (>0) in the context of the entire survey configuration.

The nodes at which the compasses are located should **not** be defined again in record H51@0.

Local offsets are *Y*-axis offsets measured from the centre of the near seismic receiver group and are therefore negative towards the tailbuoy. The relevant fields should include the sign.

H2300 Compass correction derivation (static)

Description of the origin of the correction, incl. method, location, date.	[7,80]	A74	free text

Note:

Static compass corrections are derived on a magnetically undisturbed onshore site by calibrating the compass by means of e.g. a theodolite.

H23@0 Compass corrections (static)

@ = 1...9, Vessel reference number

Node identifier	[7,14]	A8	
Fixed correction to reading	[15,20]	F6.1	degrees decimal
Line direction 1	[21,23]	I3	degrees
Correction to reading	[24,27]	F4.1	degrees decimal
Line direction 2	[28,30]	I3	degrees
Correction to reading	[31,34]	F4.1	degrees decimal
.			
.			
.			
Line direction 8	[70,72]	I3	degrees
Correction to reading	[73,76]	F4.1	degrees decimal

Record may be repeated.

Note:

The correction to a compass reading may be defined as a single fixed correction and/or as a correction applicable for a particular line direction. Corrections for up to eight approximate line directions may be defined.

$$CorrectedCompass = RawCompass + FixedCorrection + CorrectionforLineCorrection \quad (B.3)$$

H2301 Compass correction derivation (dynamic)

Add to static corrections flag	[7, 7]	I1	0 = no; 1 = yes
Description of the algorithm used for the derivation of the corrections	[9,80]	A72	free text

Note:

Dynamic compass corrections are derived while the compasses are deployed on the streamer(s). Most of the methods presently used derive these corrections from the raw individual compass readings according to some model. As it is not usually possible to (easily) reconstruct these corrections from the recorded data, the format allows recording of these corrections in record H23@1.

Dynamic compass corrections may have been derived from the compass readings that had already been corrected with the static corrections. In that case the dynamic corrections **add** to the static corrections for the relevant line direction to give the total (C–O) to be added to the recorded compass readings.

Alternatively the dynamic corrections may have been determined from the **uncorrected** compass readings, in which case they would entirely **replace** the static corrections.

This option is expressed in the flag at column 7.

H23@1 Compass corrections (dynamic)

@ = 1...9, vessel reference number

Streamer reference number	[7, 9]	I3	
Node identifier	[11,18]	A8	
Compass correction	[20,24]	F5.1	degrees decimal

May be repeated for 3 more compasses on the same streamer at [26,39], [41,54]... [56,69]; the streamer reference number is thereby not repeated.

Record may be repeated.

H24@0 Seismic receiver group definitions

@ = 1...9, vessel reference number

Streamer reference number	[7, 9]	I3
Reference number first seismic receiver group in regular section	[11,14]	I4
Local offset of centre of first receiver group	[16,23]	F8.1
Reference number last seismic receiver group in regular section	[25,28]	I4
Local offset of centre of last receiver group	[30,37]	F8.1
Number of seismic receiver groups in section	[39,41]	I3
Distance between centres of receiver groups	[43,48]	F6.1

Record may be repeated.

Note:

This record allows a group or section of regularly spaced and regularly numbered seismic receiver groups to be defined. Breaks in the regularity of spacing of the seismic receiver groups may occur when compass sections or acoustic sections are inserted in the streamer.

Local offsets are *Y*-axis offsets measured from the centre of the near seismic receiver group and are therefore negative towards the tailbuoy. The relevant fields should include the sign.

The seismic receiver group reference numbers may include zero and may be incremented in either direction. The distance between centres of receiver groups must be given in the same measurement units as the distance offsets.

H24@1 Auxiliary seismic channel definition

@ = 1...9, vessel reference number

Streamer reference number	[7, 9]	I3
Auxiliary channel ref. number	[11,14]	I3
Auxiliary channel type	[16,16]	I1 0 = timebreak 1 = waterbreak 2...9 = user def. specify on C0001 record
Local offset to centre of auxiliary channel	[18,25]	F8.1
Description	[27,80]	A54 free text

Note:

The purpose of this record is to allow the recording of travel data from the seismic source to locations within the receiver array. These data can be used to confirm source — receiver geometry.

Timebreak channels record a zero offset figure for the reduction of waterbreak data; if no timebreaks are defined, waterbreak data is referenced to zero.

Include in the description the resolution to which the timebreak and waterbreak data are sampled.

H25@0 Streamer depth sensor definitions

@ = 1...9, vessel reference number

Streamer reference number	[7, 9]	I3
Depth sensor reference serial number	[11,18]	A8
Local offset to centre of depth sensor	[20,27]	F8.1
Depth correction (C-O)	[29,33]	F5.1
Clipped-on or inserted?	[35,38]	F4.1 0 = clipped-on 1 = inserted

Record may be repeated for one more depth sensor on the same streamer at [37,61]; the streamer reference number is thereby not repeated.

Record may be repeated.

Note:

When the depth sensor is integrated with a compass unit and inserted in the streamer, the length of the inserted section should be recorded in record H22@1 as 'Length of each inserted compass section'.

Local offsets are *Y*-axis offsets measured from the centre of the near seismic receiver group and are therefore negative towards the tailbuoy. The relevant fields should include the sign.

B.6.4 Gun array definitions

H31@0 Gun array geometry definitions

@ = 1...9, vessel reference number

Gun array reference number	[7, 9]	I3	
Offsets from towing vessel's reference point to:			
towpoint-on-towing-body:			
Offset A	[11,17]	F7.1	
Offset B	[19,25]	F7.1	
Offset *Z*	[27,32]	F6.1	
towpoint-in-sea:			
Offset A	[34,40]	F7.1	
Offset B	[42,48]	F7.1	
Offset *Z*	[50,55]	F6.1	
Local offsets from the towpoints-in-sea to the horizontal centre of the gun array:			
Local offset A	[57,63]	F7.1	
Local offset B	[64,70]	F7.1	
Nominal firing pressure	[72,77]	N6	
Pressure units code	[78,78]	I1	0 = kgf/sq.cm 1 = lbs/sq.in 2 = bar
Volumes units code	[79,79]	I1	0 = cu.cm 1 = cu.in
Depth units code	[80,80]	I1	0 = metres 1 = feet

H31@1 Individual gun definition

@ = 1...9, vessel reference number

Gun array reference number	[7, 9]	I3
Gun reference number	[11,13]	I3
Local offset A	[15,21]	F7.1
Local offset B	[23,29]	F7.1
Local offset *Z*	[31,36]	F6.1
Gun volume	[38,43]	I6

May be repeated for one more gun in the same array at [45,77]; the gun array reference number is not repeated. Record may be repeated.

Note:

The gun reference number should be unique within the array.

H32@0 Description of gun array depth sensors

@ = 1...9, vessel reference number

Gun array reference	[7, 9]	I3	
Quality indicator type	[11,11]	I1	
Description of depth sensors	[13,80]	A62	free text

Note:

The quality indicator type defines the type of quality indicator used in the (inter-)event data records for all gun depth sensors on this vessel and should be one of the following codes:

0 = no quality information available

1 = standard deviation

2 = signal/noise ratio

3 = system specific

4 = subjective scale

In the case where code 1 is chosen, a descriptive definition of the way the standard deviation is derived must be supplied in record H32@2.

In the case where code 3 is chosen, a descriptive definition must be supplied in record H32@0 of the following aspects of the system specific quality indicator:

the range of values of the variable;

the interpretation of its values.

H32@1 Gun array depth sensor definitions

@ = 1...9, vessel reference number

Gun array reference	[7, 9]	I3
Sensor number	[11,12]	I2
Sensor serial number	[14,21]	A8
Local offset A	[23,29]	F7.1
Local offset B	[31,37]	F7.1
Depth correction (C–O)	[39,44]	F6.1

May be repeated for one more depth sensor on the same gun array at [46,79]; the gun array number is not repeated.

Record may be repeated.

H32@2 Definition of quality indicator type for gun array depth sensors

@ = 1...9, vessel reference number

Gun array reference number	[7, 9]	I3	
Definition of quality ind. type	[11,80]	A70	free text

Record may be repeated.

H33@0 Definition of intended gun firing sequence

@ = 1...9, vessel reference number

Gun array reference number	[7, 9]	I3	
Starting gun number	[11,13]	I3	
Active gun mask	[15,80]	66*I1	0 = inactive 1 = active

Record may be repeated in case more than 66 guns need to be defined or when the discontinuity in the gun numbers spans more than 66.

Note:

This record defines which guns within a gun array are intended to fire (active guns) from those guns defined in the H31@1 records.

The starting gun number is the gun reference number of the first gun of a contiguous series of up to 66 guns for which the mask is provided.

Guns not explicitly set active in this record are not enabled.

The intended sequence of gun array firing is **not** defined in the format; any relevant information should be supplied in comment records

H34@0 Gun array pressure sensor definitions

@ = 1...9, vessel reference number

Gun array reference number	[7, 9]	I3
Gun number	[11,13]	I3
Sensor serial number	[15,22]	A8
Sensor correction (C–O)	[24,28]	F5.1

May be repeated for 2 more pressure sensors on the same gun array at [30,47] and [49,66]; the gun array reference number is not repeated.

Record may be repeated.

Note:

The gun number serves as a sensor identifier.

No offsets are defined, as sensor data is independent of position.

H34@1 Description of gun array pressure sensors

@ = 1...9, vessel reference number

Gun array reference number	[7, 9]	I3	
Description of gun array pressure sensors	[11,80]	A70	free text

B.6.5 Towed buoy definitions

H41@0 Towed buoy geometry definitions

@ = 1...9, vessel reference number

Towed buoy ref. number	[7, 9]	I3	
Towed by: ref. number	[11,13]	I3	
Offsets from towing body's reference point to:			
towpoint-on-towing-body:			
Offset A	[15,21]	F7.1	
Offset B	[23,29]	F7.1	
Offset *Z*	[31,36]	F6.1	
towpoint-in-sea:			
Offset A	[39,45]	F7.1	
Offset B	[47,53]	F7.1	
Offset *Z*	[55,60]	F6.1	
Description of the towed buoy	[62,80]	A19	free text

Note:

The required offsets should be defined in the coordinate frame of the object which tows the buoy (the towing body). This may be a ship, a streamer, a gun array or even another buoy.

The vessel reference number (@) in the record code defines the ship that directly or indirectly tows the buoy.

B.6.6 Survey network definitions

B.6.6.1 Introduction

The network approach adopted in the UKOOA P2/91 format allows a significant segment of the data to be described and emphasizes the geometric nature of the positioning data, rather than the physical measurement principles underlying the data. The network consists of a group of points: **network nodes** with observations defined between these nodes. Most observations define a geometric relationship between two or more nodes, e.g. a range between two nodes or a range difference between three, whereas some measure a geometric relationship between a part of the network and the outside world, e.g. the vessel's gyro compass, which describes the orientation of the ship's head with respect to the earth's rotational axis.

Network architecture allows great flexibility in the definition of observations such as sing around ranges and acoustic ranges between streamers, possibly towed by different vessels. It reflects an integrated approach to positioning.

A number of observations have not been included in the network approach, for reasons of clarity and because nothing would be gained by including them. Examples are data from USBL systems, which are stand-alone systems found on vessels only because of their bulk, gun depths, water depth (echo-sounder) and streamer compass data.

The network architecture necessitates a generalized approach to the definition of observations. A raw observation can generally be reduced

to fit the processing model by applying two corrections: an addition correction (C–O) and a scale correction (C/O), expressed in the following equation:

$$\text{Obs(reduced)} = \text{C/O} * \{\text{Obs(raw)} + (\text{C–O})\} \quad (\text{B.4})$$

An explanatory note follows the definition of the network header records, rather than the customary additional comments, because of the volume of the explanatory text and consequently the need to arrange it in a more structured fashion.

B.6.6.2 Definition of survey network header records

H5000 Node definition (fixed locations)

Node identifier	[7,10]	I4	
Name / description	[12,27]	A16	free text
Flag for geog. or grid coords.	[29,29]	I1	0 = geographical 1 = grid
Latitude	[31,42]	I3,I2,F6.3,A1	dddmmss.sss N/S
Longitude	[44,55]	I3,I2,F6.3,A1	dddmmss.sss E/W
Northing	[31,41]	N11	
'N'	[42,42]		A1
Easting	[44,54]	N11	
'E'	[55,55]	A1	
Height	[57,63]	N7	metres or feet
Height measurement unit	[65,65]	I1	0 = metres 1 = feet
Height datum	[67,67]	I1	0 = MSL 1 = LAT 2 = LLWS 3 = sea level 4 = spheroidal 5 = other: define in separate C0001 record

H51@0 Node definition (vessel, gun array, streamer, towed buoy)

@ = 1...9, vessel reference number
@ = 0 if relay vessel or buoy

Node identifier	[7,10]	I4	
Name / description	[12,27]	A16	free text

Located on: ref. number	[29,31]	I3	
(Local) offset A	[33,39]	F7.1	
(Local) offset B	[41,47]	F7.1	
(Local) offset Z	[49,55]	F6.1	

H52## Observation definition

= observation type

Observation identifier	[7,10]	I4	
Observation description	[12,27]	A16	free text
At node identifier	[29,32]	I4	
To node 1 identifier	[34,37]	I4	
To node 2 identifier	[39,42]	I4	
Measurement unit code	[44,45]	I2	
Positioning system identifier	[47,49]	I3	
Positioning system description	[51,80]	A30	free text

H5306 Differential observation — follow up record

Differential observation identifier	[7,10]	I4	
Observation 1 identifier	[12,15]	I4	
Observation 2 identifier	[17,20]	I4	
Differential obs. description	[22,80]	A59	free text

Note:

This record should, when used, immediately follow the corresponding H5206 record.

H5307 Composite range — follow up record

Observation identifier	[7,10]	I4	
To Node identifier	[12,15]	I4	
+ (addition) or – (subtraction)?	[17,17]	I1	0 = negative range section 1 = positive range section

May be repeated at [19,24], [26,31], [33,38], etc.; the Observation identifier is not repeated.

Note:

The last node identifier will generally be the same as the 'At' node identifier in the corresponding H5207 record, closing the loop.

This record should, when used, immediately follow the corresponding H5207 record.

H54## Observation definition (continued)

= observation type

Observation identifier	[7,10]	I4	
Propagation speed	[12,23]	N12	m/s or ft/s
Lane width on baseline or freq.	[25,36]	N12	m or ft or Hz

Defined length unit	[38,38]	I1	0 = metres 1 = feet
Lanewidth or frequency?	[40,40]	I1	0=lanewidth on baseline 1=comparison frequency
Scale factor	[42,53]	N12	
Fixed system (C-O)	[55,64]	N10	
Variable (C-O)	[66,73]	N8	
A priori standard deviation	[75,78]	N4	
Quality indicator type	[80,80]	I1	0 = no qual. info. recorded 1 = standard deviation 2 = signal/noise ratio 3 = system specific 4 = subjective scale

H5500 Definition of system specific quality indicator

Positioning system identifier	[7, 9]	I3	
Definition of quality indicator	[11,80]	A70	free text

H56@0 Instrument correction

@ = 1...9, vessel reference number
@ = 0 fixed or relay station

Node identifier	[7,10]	I4	
Positioning system identifier	[12,14]	I3	
Instrument correction	[16,26]	N11	
Instrument description [28,80]		A53	free text

B.6.6.3 Clarification of network nefinition necords

Node definitions (records H5000 and H51@0)

General

The Node identifier must be a unique positive number (>0); no duplicates are allowed.

Coordinates and offsets should define the antenna electrical centre or transducer plate centre.

Fixed nodes (record H5000)

Record H5000 should be used to define fixed nodes, such as shore stations of radio positioning systems. Also moored buoys for which the co-ordinates are assumed to be fixed, should be defined using this record.

The coordinates of fixed nodes must be defined on the Survey Datum.

Heights need only be supplied when relevant to the position calculation.

Heights for nodes under the height datum, e.g. long baseline (LBL) acoustic beacons on the sea floor, are negative and the sign should be included in the field.

Heights should be above sea level (code 3) in the case where the node is mounted on a fixed moored buoy.

Nodes on the seismic spread and relay vessels (record H51@0)

Record H51@0 should be used to define any node on the seismic spread. It should also be used for nodes on relay vessels or relay buoys.

Relay vessels and relay buoys are defined as carrying a radio-positioning beacon of which the signals are used to position the seismic spread, whilst its own position is continually determined, e.g by shore-based or satellite radio-positioning means.

The vessel reference number in the header code, represented by the '@' character, defines the vessel that tows, directly or indirectly, the object on which the node is located.

In the case where the node is located on that survey vessel itself, the vessel reference number is repeated in column 31.

When the node is located on a relay vessel or relay buoy: @ = 0 (zero) and the relay vessel reference number is entered in columns 30 and 31. Also in this case @ = 0 in record H56@0.

The 'located on' field requires the unique reference number of the vessel, gun array, etc. the node is located on.

The local offsets define the nominal or design location of the node.

Observation definitions (records H52## - H56@0)

General comments

The **observation identifier** must be a unique positive number (>0): no duplicates are allowed. This includes differential observations and composite ranges.

The **positioning system identifier**, also a unique positive number, allows grouping of observations that share certain characteristics or may be affected by the same type of errors. It may for instance be used to distinguish acoustic ranges from electromagnetic ranges. Electromagnetic and acoustic ranges can be further subdivided by positioning system. A bearing and distance, derived e.g. by differencing the GPS position of a tailbuoy with the GPS position of the towing vessel may be linked by means of the pPositioning system identifier.

A minimum of two records, H52## and H54##, are required to define one network observation. A differential observation or a composite range requires an additional H5306 or H5307 record respectively.

All nodes referred to in any observation definition must be defined in H5100 or H51@0 records, with the exception of USBL transducer nodes and streamer compass nodes, which are defined in records H16@1 and H22@0 respectively.

More than one observation may refer to one particular node. In such a case the node should only be defined once.

Observation types

Observations may be defined as one of the oObservation types and observed in one of the measurement units defined below.

Observation type codes:

1 = range

2 = hyperbolic, formula 1 (usually phase difference measurement)

3 = hyperbolic, formula 2 (usually time difference measurement)

4 = pseudo-range, common clock bias

5 = pseudo-range, clock bias per pseudo-range

6 = differential observation

7 = composite observation (e.g. sing around range)

8 = angle

9 = direction

10 = bearing, magnetic

11 = bearing, true

12 = differential true bearing (rate gyros)

Measurement unit codes:

00 = (international) metres

01 = international feet (1 int. foot = 0.30480 int. metre)

02 = lanes

03 = degrees

04 = grads

05 = radians

06 = arc minutes

07 = arc seconds

08 = seconds (time)

09 = milliseconds (time)

All feet referred to in this document are international feet, as defined above.

Code 01 — Range observation

A range defines a geometric relationship between two nodes. If the client or supplier wishes to distinguish between ranges that are measured according to different physical principles (electromagnetic, optical, acoustic), this should be done by attributing different Positioning System Identifiers to those different groups.

Codes 02, 03 — Hyperbolic observation

Only hyperbolic observations taken from multi-user hyperbolic positioning systems should be recorded as observation type 0 or 03. Simple differences between two ranges should be defined using code 06.

A hyperbolic observation of code 02 or 03 defines a geometric relationship between three nodes, usually a node on the vessel (indicated below by V) and two fixed nodes: often two shore transmitting stations (indicated below by S_1 and S_2).

The hyperbolic observation essentially describes a range difference involving the range S_1V (from station 1 to vessel) and S_2V (from station 2 to vessel). However, which range is subtracted from which depends on the way the positioning system operates and this gives rise to two different formulae to model the two different types of hyperbolic observation. The two formulae have been indicated in record H52## by equation (5) (code 02) and equation () (Code 03).

Code 02 — equation (B.5)

Equation (B.5) should be applied when the reading of a hyperbolic pattern 1 — 2 **increases** when moving over the baseline from station 1 to station 2. Most phase difference measurement systems satisfy this requirement.

Equation (B.5) for an observation L_{12}, made at node V of hyperbolic pattern 1 — 2, is:

$$L_{12} = \frac{(S_1S_2 + S_1V - S_2V)}{2\lambda} + (C-O)_{fixed} + (C-O)_{var} \qquad \text{(B.5)}$$

Code 03 — equation (B.6)

Equation (B.6) should be applied when the reading DECREASES when moving over the baseline from station 1 to station 2. Most time difference measurement systems satisfy the latter requirement.

Equation (B.6) for an observation L_{12}, made at node V of hyperbolic pattern 1 — 2, is:

$$L_{12} = \frac{(S_1S_2 - (S_1V + S_2V))}{2\lambda} + (C - O)_{fixed} + (C - O)_{var} \qquad \text{(B.6)}$$

where (in both formulae):

L_{12}	= hyperbolic observation
S_1S_2	= distance between Stations 1 and 2
S_1V	= distance between Station 1 and vessel
S_2V	= distance between Station 2 and vessel
λ	= lanewidth at the baseline
$(C\text{-}O)_{fixed}$	= fixed (C–O), see below
$(C\text{–}O)_{variable}$	= variable (C–O), see below

Codes 04, 05 — Pseudo-range observation

A pseudo-range is derived from the measured one-way signal travel time between two nodes. A different time-standard used by the transmitter and the receiver causes the observation to be contaminated by a clock offset, which is the difference between the two time systems. These clock offsets need to be solved for in the position computation process.

Pseudo-ranges thus measured by one receiver should be grouped by allocating the same positioning system identifier to each of them. If a second receiver is used to measure pseudo-ranges from the same transmitting stations, the second group of pseudo-ranges should be allocated a different positioning system identifier. This is regardless of the location of the two receivers: they may be on the same vessel.

Two types of pseudo-ranging systems may be distinguished, leading to two types of pseudo-range observations:

Code 04 — Time-synchronized transmitters

The first type of pseudo-ranging system operates with time-synchronised transmitters, which requires one clock offset to be solved for in the position calculation process That clock offset is common to all pseudo-ranges of that system observed by the same receiver. An example of a system operating according to this principle is GPS.

Code 05 — Free-running transmitters

The second type of pseudo-ranging system has transmitters that are driven by free-running, but highly accurate clocks. Such systems require one clock-offset per observed pseudo-range to be solved for in the position calculation process. Pseudo-ranging systems of this type are also referred to as rho-rho systems.

Code 06 — Differential observation

A differential observation is defined as the arithmetic difference between two simultaneously measured observations, termed the parent observations below.

The simultaneity should be interpreted in a practical sense: some difference in timing may be acceptable if that has no appreciable impact on the practical interpretation of the differential observation.

A differential observation, together with its two parent observations, is defined by means of the following records:

H52## — Observation definition: parent observation 1

H54## — Observation definition (continued): parent obs. 1

H52## — Observation definition: parent observation 2

H54## — Observation definition (continued): parent obs. 2

H5306 — Differential observation definition

The H5306 record should define the differential observation in the sense:

$$\text{Diff.Obs} = \text{Obs1} - \text{Obs2} \tag{B.7}$$

The Variable (C–O) and, if relevant, the Fixed System (C–O) of the differential observation are then defined as follows:

$$(C-O)_{\text{Diff.Obs}} = (C-O)_{\text{Obs1}} - (C-O)_{\text{Obs2}} \tag{B.8}$$

with $(C-O)_{\text{Obs1}}$ and $(C-O)_{\text{Obs2}}$ recorded on the H54## records of each of the two parent observations.

However, if the differential observation is calibrated itself, rather than its two parent observations, a separate H5406 record should be added, which should be blank but for the first field, which should contain the differential observation identifier and the seventh field, which contains the residual (C–O). The variable (C–O) fields in the H54## records defining the parent observations should then be left blank.

Code 07 — Composite range

This observation type is intended to allow definition of sing-around ranges. Sing-around ranges are achieved by relaying a ranging signal over a series of beacons, starting at node A, then to node B, node C etc. and may be available both in electromagnetic and underwater acoustic positioning systems.

The receiver/interrogator normally measures the one-way total travel-around time, hence the sum of the lengths of the legs of the polygon the signal travelled around.

However, the record does not only allow polygon legs to added to the total sing-around range, but also subtracted, hence a generalization of the name to 'composite range'.

A composite range is defined by means of the following records:

H5207 – Observation definition

H5307 – Composite range – follow up record
(repeated if necessary)

H5407 – Observation definition (continued)

Code 08 – Angle

An angle defines a geometric relationship between three nodes: the node at which the angular measurement device is located and two target nodes, node 1 and node 2. By convention the angle is measured clockwise from node 1 to node 2.

In marine applications angular measurement devices are often aligned to ship's head, not to a specific node on the vessel. Ship's head is in this context identified with an imaginary node with identifier 0 (zero). When node 0 is defined the angle must be measured at the vessel.

Most vessel mounted laser systems presently used for fixing targets in the vicinity of the vessel's stern are set up in this way and the angles measured by such systems should be recorded as explained in this section.

Code 09 – Direction

An angle can be seen as the difference between two directions. A direction requires two nodes in order to be defined, the station and the target. However, one direction alone is meaningless: at least one more direction to another target needs to be defined so that the angle between the two target nodes, measured from the station, can be derived.

Defining angular relationships in terms of directions rather than angles (code 08) is only meaningful if:

the angular measurement device is unorientated, and

the device measures to multiple targets.

All such directions will share the same orientation unknown, much in the same way as a number of pseudo-ranges may share the same clock-offset, as in GPS.

Codes 10, 11 – Bearing

Measured at any station, the bearing from that station to a target is the clockwise angle between the North direction (true or magnetic) and the direction to the target. In general, a bearing defines a geometric relationship between two nodes and the earth's magnetic field.

In the case of a compass mounted on a ship the target node is implied, namely the ship's head.

In the case of compass mounted on a streamer the target is also implied and is in this case the tangent to the curve the streamer assumes at the time of observation.

Bearings may also measured — or rather derived — by satellite positioning systems operated in local differential or relative mode as is presently the case with many GPS tailbuoy tracking systems. The bearing output produced by such systems indeed requires two nodes to be explicitly defined: the station and the target. This application is covered by Code 11.

Code 12 — True bearing with unknown index (rate gyros)

Although not yet used (extensively) in the offshore survey industry, rate gyros (e.g ring laser gyros) are able to measure bearing differences, that are continually integrated by the unit. The output is essentially a bearing with an unknown index or addition constant. Code 12 is intended to describe that type of observation.

Observation definitions: parameters

Parameters assumed fixed (records H54##, H5500, H56@0)

(a) Propagation speed

The propagation speed is the speed built into the receiver where conversion to metric units occurs in the equipment, or the assumed/measured propagation speed where such is not the case.

(b) Lane-width at baseline / signal transmission frequency

The lane-width at baseline is only relevant for hyperbolic observations. In this field the supplier has the option to record either the lane-width at the baseline or the comparison frequency.

Many positioning systems designed for hyperbolic mode are often used in range-range mode, providing as output ranges expressed in units of half its wavelength (lanes).

In spite of that the flag lane-width or frequency in record H 54## should in that case be set to 1 and the comparison frequency should be entered in columns [25,36]. See under *scale factor* below how to define the measurement unit of such systems.

(c) Scale factor (C/O):

The scale factor or C/O should correct the raw observation measurement unit to the standard measurement unit, defined in the measurement unit code field of record H52##.

Normally no scale correction needs to be made, in which case the value of the C/O needs to be recorded as 1 (unity). Three notable exceptions are mentioned below.

(i) A range — range system providing output in lanes requires the following approach:

- Set measurement unit code in record H52## to metres or feet (Code 0 or Code 1).
- Set lanewidth or frequency flag to 1 (frequency) and enter the comparison frequency in columns [25,36], both in record H54##.
- The lane-width of the system, which in this case is the propagation speed divided by twice the comparison frequency should then be entered into record H54## as the scale factor.

(ii) Another application of the scale factor may occur when ranges are reduced in the measurement device for two-way travel, while the signal has only travelled one-way, as can be the case with sing-around ranges. In such cases the scale factor needs to be set to 2.

(iii) The converse happens when a ranging system measuring two-way travel time does not compensate for that, but treats the measurement as a one-way travel time. In that case the scale factor needs to be set to 0.5.

(d) Fixed (C-O)

The fFixed (C-O) correction is determined by the mode of operation of the relevant positioning system or sensor. More often than not hyperbolic systems have fixed (C-O) corrections associated with them. fixed (C-O) corrections do not vary over time, or with location. The fixed (C-O) must be recorded in the same measurement unit as the observation it refers to.

(e) Variable (C-O)

The variable (C-O) correction is related to systematic minor deviations of the measurements from the assumptions underlying the measurement process. Examples are unmodelled signal propagation variations and instrumental variations. Variable (C-O) corrections are determined by calibration. They may be instrument-specific and/or time/location dependent. The variable (C-O) must be recorded in the same unit as the observation it refers to.

(f) Variable (C-O) by Instrument Correction (H56@0)

The norm is to supply the calibration correction to an observation in the form of one variable (C-O). However, for some systems, notably some ranging systems, the variable (C-O) is split up into component parts and expressed as instrument or sensor corrections, often derived from bench calibrations of the sensors. These corrections are commonly supplied in the form of receiver, beacon or transponder delays.

Instrument corrections can be supplied in record H56@0 when relevant. They should add to the range measured to/from the relevant node and therefore equal MINUS the instrument delays. When instrument corrections are supplied the variable (C-O) fields in the H54## records of the affected observations should be left blank. The total variable (C-O) for such a range between nodes A with instrument *i* and node B with instrument *j* is:

$$(C\text{–}O)_{var} = Instr.Corr_i + Instr.Corr_j \tag{B.9}$$

Reduction of observations

(a) General reduction formula

The general observation reduction equation is:

$$Obs_{reduced} = C/O * \{Obs_{raw} + (C\text{–}O)_{fixed} + (C\text{–}O)_{var}\} \tag{B.10}$$

The (C-O) and C/O corrections are in principle obtained from record H54##, and, where relevant from records H56@0.

(b) Changes to (C-O) and C/O within a line

This format allows changes to (C–O)s and C/Os which occur within a seismic line, to be recorded without having to insert a new block of header records. This option is implemented by means of the E54## and T54## records for (inter-)event observationpParameters.

Variable (C–O) And/or C/O (scale) corrections supplied in (inter-)event records take precedence over the values supplied in record H54## And replace the lLatter.

However, a propagation speed value supplied in an (inter-)event record will replace the propagation speed supplied in record H54##, but the scale correction in record H54## will then still be valid and should be applied in the reduction formula.

A change in e.g. a (C–O) during a line needs to be recorded only once by inserting one E54## or T54## record. The new value will be deemed to valid until:

it is changed again by means of an E54## or T54## record for a later point in time,

or

the end of the line is reached.

If the new (C–O) value is still valid at the beginning of the next line, the value will need to be consolidated in the relevant H54## record for the new line.

It is important to realise that if data processing is started not at the beginning of the line but at a point further down the line reading of the header data will not be sufficient to set the values of the observation parameters: the data file will have to be scanned from its start for all E54## and T54## records.

Quality parameters

Provision has been made in the format for the recording of two types of observation quality indicator:

1. the **a priori** or **expected** quality, as for instance assumed in the design of the network;
2. the **actual** quality, as occurring during the survey.

The recording of quality indicators is subject to client requirements and data availability.

(a) A priori or expected accuracy

The A Priori Standard Deviation must, when available, be recorded in the same units as the observation and expresses the expected precision of the observation. The decimal point must be included.

(b) Actual quality

The quality indicator may be recorded in the (inter-)event records of the network observations and some other observations.

A four character field is provided, and is intended to hold a numeric value with decimal point where appropriate.

The following options are available.

Code	Name	Definition
0		No quality information recorded.
1	Standard deviation	Standard statistical variable; measure of the noise level of the observation. Unit: same as the observation unit of measurement.
2	Signal/noise ratio	Standard physical variable. Unit: db.
3	System specific	Positioning system specific quality indicator (see below), often a measure of signal strength. When used, it should apply to all observations grouped into the same positioning system and an H5500 record must be provided defining this parameter.

4	Subjective scale:	0 = poor quality; unusable;
		1 = poor quality but usable;
		2 = fair quality;
		3 = good quality.

(c) System-specific quality indicator (record H5500)

Many positioning systems provide a parameter with each observation which is a measure of the quality of the signal, often signal strength. For that reason an H5500 record should be inserted when code 3 is defined for an observation quality indicator. This record should provide a descriptive definition of the following aspects of the system-specific quality indicator:

the range of the variable;

the interpretation of its values.

The same code 3 should then apply to all observations of the relevant positioning system.

B.6.7 Satellite system definitions

H600# Satellite system description

= 1...9, satellite system reference number

Name	[7,14]	A8	free text
Datum and spheroid number	[16,16]	I1	
Diff. system operator	[18,35]	A18	free text
Diff. system name	[37,46]	A10	free text
Software description and version number and additional information	[48,80]	A33	free text

Note:

Satellite system numbers must be numbered according to the following convention:

1 = GPS autonomous positioning

2 = Differential GPS (DGPS)

3 = TRANSIT

4...9 = User definable

The datum and spheroid referred to must be defined in record H011#.

H610# Definition of differential reference stations

= 2, 4...9, satellite system reference number

Reference station number	[7, 7]	I1	
Reference station name	[9,20]	A12	free text
Latitude	[22,33]	I3,I2,F6.3,A1	dddmmss.sss N/S

Longitude	[35,46]	I3,I2,F6.3,A1	dddmmss.sss E/W
Spheroidal height	[48,54]	F7.2	metres
Geoid-Spheroid separation	[56,62]	F7.2	metres
Geoidal model	[64,80]	A17	free text

Note:

The coordinates of the reference station must be defined on the same datum/spheroid as the satellite system.The geoid-spheroid separation in the station and the geoidal model from which the separation was derived is only relevant if the coordinates of the reference station have been converted from a local datum to the satellite system datum. If they were determined by means of observations from that same satellite system, the spheroidal height would have been determined direct and no geoid — spheroid separation and geoidal model need be recorded in this record.

H620# Satellite receiver definition

\# = 1...9, satellite system reference number

At node identifier	[7,10]	I4	
Receiver number	[12,12]	I1	
Located on: ref. number	[14,16]	I3	
Offset A	[18,24]	F7.1	
Offset B	[26,32]	F7.1	
Offset Z	[34,39]	F5.1	
Receiver name, description and additional information	[41,80]	A41	free text

Note:

The node identifier must be a unique positive number. The second field is the reference number of the vessel, gun array, streamer or buoy the receiver is mounted on. Offsets define the nominal location of the antenna electrical centre.

B.6.8 User defined observation sets

H7000 Definition of user defined observation sets

Observation set reference number	[7, 9]	I3	
Number of data fields associated with this set	[11,12]	I2	
Description of observation set	[14,80]	A67	free text

Record may be repeated.

Note:

This record type allows the definition of observations with supporting data, which are not covered by the format otherwise. The observations with their supporting data, such as the coordinates at which the observations refer to. Date and time, etc. should be defined as one observation set.

Include sensor type, make, serial number, calibration details and other relevant information. Expand by repeating the record if necessary, leaving the second field (number of data fields associated with this set) blank.

H7010 Data field definitions

Observation set ref. number	[7, 9]	I3	
Data field number	[11,12]	I2	
Data field width	[14,15]	I2	
Data field description	[17,80]	A64	free text

Record may be repeated.

Note:

Include items such as sensor channel number, units of measurement and any other information required for interpretation and processing. Repeat the record if some more descriptive space is required, leaving the third field (data field width) blank.

H7020 User defined observation parameters

Observation set ref. number	[7, 9]	I3
Data field number	[11,12]	I2
Quality indicator type	[14,14]	I1
(C–O) correction	[16,..]	Nx

Note:

The quality indicator type defines the type of quality indicator used in the (inter-)event data records for the relevant data field; it should be one of the following codes:

0 = no quality information recorded

1 = standard deviation

2 = signal/noise ratio

3 = system specific

4 = subjective scale

In the case where code 1 is chosen, a descriptive definition of the way the standard deviation is derived must be supplied in record H7021.

In the case where code 3 is chosen, a descriptive definition must be supplied in record H7021 of the following aspects of the system specific quality indicator:

the range of values of the variable;

the interpretation of its values.

The width of the (C–O) field must be the same as the width of the data field it refers to, as defined in record H7010.

H7021 Definition of quality indicator type for user defined observations

Observation set ref. number	[7, 9]	I3	
Data field number	[11,12]	I2	
Definition quality indicator type	[14,80]	A67	free text

Record may be repeated if necessary.

Example of user defined observations

Header data

H7000 001 01 Gravity data: standard sensor S/N 31

H7000 001 Last in-port gravity tie: Aberdeen 31 October 1991

H7010 001 01 06 Gravity count in milligals

H7010 001 02 06 Spring tension in milligals

H7010 001 03 04 Average beam

H7010 001 04 04 Total cross coupling

H7010 001 05 04 Total correction

H7010 001 06 04 Vertical cross coupling

H7010 001 07 04 Along cross coupling

H7010 001 08 04 Across cross coupling

H7010 001 09 04 Vertical correction

H7010 001 10 04 Average across acceleration

H7010 001 11 04 Average along acceleration

H7010 001 12 04 Second order cross coupling

H7010 001 13 04 Offset calibration

Event data

E70100010112341234560212341234560312341234

E701000104123412340512341234061234123407 12341234

E7010001081234123409123412341012341234111 2341234

E701000112123412341312341234

Inter-event data

T7010001011234010203112345602123401020311 23456

T70100010312340102031 1234

B.7 Event data records (implicit time tag)

B.7.1 General and vessel related event data

E1000 General event data

Line name	[7,22]	A16	
Shot/event number	[24,31]	I8	
Seismic record number	[33,48]A16		
Year, month, day	[50,57]	I4,I2,I2	YYYYMMDD
Time	[59,66]	I2,I2,F4.1	HH,MM,SS.S
Gun array fired	[68,70]	I3	

Note:

Only one general event data record is required regardless of the number of vessels. however, this can only be achieved if all data relating to one event plus the inter-event data observed after that but before the next event are stored on one medium. in the case that the data is divided by vessel over various storage media the general event data record must be repeated for each storage medium containing data related to that event.

The gun array fired field contains the gun array number defined on the H31@0 record. It is provided to allow redundancy and to cover those cases where individual gun data is not available.

The seismic record identifier would typically contain both file and reel identifier in the 16 character field provided.

All positioning data in subsequent records and explicitly time-tagged is assumed to relate to the event time, defined in this record.

E12@0 Field positioning derived data

@ = 1...9, vessel reference number

Record sequence number	[6, 7]	I2	
Node identifier	[8,11]	I4	
Flag for geographical or grid co-ordinates	[12,12]	I1	0 = geographical 1 = grid
If geographicals:			
Latitude	[13,24]	I3,I2,F6.3,A1	dddmmss.sss N/S
Longitude	[25,36]	I3,I2,F6.3,A1	dddmmss.sss E/W
If grid:			
Northing	[13,24]	N11	
'N'	[24,24]	A1	
Easting	[25,35]	N11	
'E'	[36,36]	A1	
Course made good or (nominal) ship's heading	[37,42]	F6.2	degrees decimal
Flag for course made good or ship's heading	[43,43]	I1	0 = course made good

			1 = ship's heading
Quality indicator 1	[44,47]	N4	
Quality indicator 2	[48,51]	N4	
Quality indicator 3	[52,55]	N4	
Calculation details	[56,80]	A25	free text

Record may be repeated.

Note:

The field positioning derived data record allows the position of any node, as computed on-board, to be recorded. Alternatively, the coordinates of the same node, but computed for the secondary, tertiary, etc. positioning system may be recorded, details of which may be entered in the last field of the record.

This is achieved by allowing up to 99 of these records to be included, to be numbered uniquely by means of the record sequence number. A record sequence should refer to the same type of data throughout the line, e.g. primary system antenna position, secondary system antenna position, tailbuoy position, etc.

The quality indicators should describe the quality of the processed positioning data. Examples are: the standard deviations of northing and easting and the standard deviation of unit weight. A full descriptive definition of the quality indicator(s) recorded should be supplied in record H12@1.

E14@0 Echo-sounder data

@ = 1...9, vessel reference number

Echo sounder ref. number	[6, 6]	I1	
Echo sounder reading	[7,12]	F6.1	metres

May be repeated for four more echo sounders mounted on the same vessel at [21,27], [36,42], [51,57] and [66,72].

E16@0 USBL acoustic data

@ = 1...9, vessel reference number

USBL system ref. number	[6, 6]	I1	
Target Node identifier	[7,10]	I4	
X coordinate of target	[11,17]	N7	metres
Y coordinate of target	[18,24]	N7	metres
Z coordinate of target	[25,31]	N7	metres
Quality indicator	[32,35]	N4	

May be repeated for one more USBL system mounted on the same vessel at [44,73].

Record may be repeated.

Note:

The Z coordinate should conform to the sign convention defined in record H16@0.

E17@0 Pitch, roll and heave sensor data

@ = 1...9, vessel reference number

Sensor reference number	[6, 6]	I1	
Pitch angle	[7,16]	N10	
Roll angle	[17,26]	N10	
Heave	[27,36]	N10	
Quality indicator pitch	[37,40]	N4	
Quality indicator roll	[41,44]	N4	
Quality indicator heave	[45,48]	N4	

B.7.2 Streamer event data

E22@0 Streamer compass data

@ = 1...9, vessel reference number

Streamer reference number	[6, 8]	I3	
Node identifier	[9,12]	I4	
Compass reading	[13,17]	F5.1	degrees decimal
Quality indicator	[18,21]	N4	

May be repeated for 4 more compasses on the same streamer at [22,34], [35,47], [48,60], [61,73]; the streamer reference number is not repeated.

Record may be repeated.

E24@1 Auxiliary seismic channel data

@ = 1...9, vessel reference number

Auxiliary channel ref. number	[6, 9]	I3	
Time observed	[10,17]	N8	milliseconds
May be repeated for five more channels at	[18,29]... [66,77].		

NOTE:

This record is intended to contain first arrival travel time information to enable checking of source versus receiver group geometry. A timebreak channel gives a zero offset to reduce any waterbreak data. If no timebreak channel is defined, waterbreak data starts at zero.

E25@0 Streamer depth sensor data

@ = 1...9, vessel reference number

Streamer reference number	[6, 8]	I3	
Node identifier	[9,12]	I4	
Depth reading	[13,17]	N5	metres
Quality indicator	[18,21]	N4	

May be repeated for four more depth sensors on the same streamer at [22,34], [35,47],

[48,60], [61,73]; the streamer reference number is not repeated.

Record may be repeated.

B.7.3 Gun array event data

E32@0 Gun array depth sensor data

@ = 1...9, vessel reference number

Gun array reference number	[6, 8]	I3
Sensor reference number	[9,10]	I2
Depth reading	[11,15]	N5
Quality indicator	[16,19]	N4

May be repeated for 5 more depth sensors on the same gun array, at [20,30], [31,41]... [64,74]; the gun array reference number is not repeated.

Record may be repeated.

E33@0 Gun fired mask

@ = 1...9, vessel reference number

Gun array reference number	[6, 8]	I3	
Starting gun number	[9,11]	I3	
Guns fired mask	[15,80]	66*I1	0 = not fired 1 = fired

Record may be repeated as necessary to define all guns that have fired on any event.

Note:

Guns not explicitly set as fired in this record are deemed not to have fired.

This record should be supplied in addition to the gun array fired code in the E1000 record to allow cross-checking against the array definition in record H31@1 and the defined gun firing sequence in record H33@0. If individual gun data is not available, array sequence checking is only available via the array numbers supplied in the E1000 records.

E34@0 Gun pressure sensor data

@ = 1...9, vessel reference number

Gun array reference number	[6, 8]	I3
Gun number	[9,11]	I3
Pressure reading	[12,17]	N6

May be repeated for 7 more sensors in the same gun array at [18,26], [27,35]... [72,80]; the gun array reference number is not repeated.

Network event data

E52## Network observations

## = Observation type		
Observation identifier	[6, 9]	I4
Observation	[10,19]	N10
Quality indicator	[20,23]	N4

May be repeated at [31,48] and at [56,73] for two more observations:

of the same observation type, and provided that:

the observations were also observed at the same event time.

Record may be repeated.

E54## Network observation parameters

## = Observation type		
Observation identifier	[6, 9]	I4
Variable (C–O)	[10,17]	N8
C/O or propagation speed	[18,29]	N12
Flag for C/O or speed	[30,30]	I1 0 = C/O (=scale factor) 1 = propagation speed

May be repeated for one more set of observation parameters at [38,62]:

relating to the same observation type, and provided that:

the change in observation parameters relate to the same event time.

Record may be repeated.

Note:

(a) This record should only be inserted at events at which the above parameters change. The parameters are deemed to be valid from that event onward until the end of the line or until the event related to the next E54## or T54## record.

(b) Variable (C-O) replaces variable (C–O) in record H54##.

(c) Scale (C/O) replaces scale factor in record H54##.

(d) Propagation speed, if supplied, replaces propagation speed in record H54##; the scale factor in record H54## is then still valid.

B.7.4 Satellite positioning event data

E620# GPS or DGPS Data
= 1, GPS
= 2, DGPS

At node identifier	[6, 9]	I4	
Receiver reference number	[10,10]	I1	
Latitude	[11,22]	I3,I2,F6.3,A1	dddmmss.sss N/S
Longitude	[23,34]	I3,I2,F6.3,A1	dddmmss.sss E/W
Height	[35,40]	F6.1	metres
Height datum	[41,41]	I1	0=height above ellipsoid 1=height above geoid(~MSL)
Satellites used numbers)	[42,67]	13 x I2	(GPS SV
Reference stations use	[68,71]	4 x I1	
Position calculation mode	[72,73]	I2	see below

Note:

If GPS is used in autonomous mode the field 'reference stations used' should be left blank.

Although it is not envisaged, nor recommended that DGPS data is usable without a time tag, the implicit time of observation is here, as per convention, event time.

The identifiers used to indicate the satellites used in the position fix must the official GPS SV numbers.

The 'reference stations used' field refers to record H610#. Only those reference station identifiers should be recorded here of which the differential corrections have been used in the calculation of the position recorded in this record.

Position calculation mode codes

1 = 3D solution; 4+ SVs

2 = 3D solution; 3+ SVs; height fixed

3 = 3D solution; 3+ SVs; height aided

4 = 3D solution; 3+ SVs; clock aided

5 = 3D solution; 2+ SVs; height aided and clock aided

6 = 3D solution; 2+ SVs; height fixed and clock aided

7 = 2D solution; 3+ SVs

8 = 2D solution; 2+ SVs; height fixed

9 = 2D solution; 2+ SVs; height aided

10 = 2D solution; 2+ SVs; clock aided

11 = 2D solution; 1+ SVs; height aided and clock aided

12 = 2D solution; 1+ SVs; height fixed and clock aided

E621# GPS or DGPS Data (continued)

= 1, GPS
= 2, DGPS

At node identifier	[6, 9]	I4	
Receiver reference number	[10,10]	I1	
Standard deviations of:			
latitude:	[11,15]	N5	metres
longitude:	[16,20]	N5	metres
height:	[21,25]	N5	metres
DOP type	[26,26]	I1	0 = GDOP 1 = PDOP 2 = HDOP 3 = TDOP 4 = VDOP 5..9 = user defined on comment records
DOP figure	[27,30]	N4	unitless

The DOP type and DOP figure fields may be repeated for other DOPs or quality indicators in columns [31,35]... [51,55] as required.

Note:

The standard deviations of latitude, longitude and height should be estimates of the quality of the fix, as produced by the onboard software.

The standard deviation of the height should be zero in the case where height is not solved for in the position calculation, such as in 'height fixed' mode.

E6303 TRANSIT Satellite Data

At node identifier	[6, 9]	I4	
Latitude	[10,21]	I3,I2,F6.3,A1	dddmmss.sss N/S
Longitude	[22,33]	I3,I2,F6.3,A1	dddmmss.sss E/W
Position includes dead reckon?	[34,34]	I1	0 = no 1 = yes
Standard deviations of last accepted satellite fix:			
latitude	[35,39]	N5	metres
longitude	[40,44]	N5	metres

Record may be repeated.

Note:

The position supplied should refer to the satellite receiver antenna electrical centre.

If a dead reckoning system involving TRANSIT is being used, then the estimated current position should be recorded and the dead reckoning flag should be set.

No attempt is made to record raw data for the TRANSIT system and the standard deviations of the coordinates are recorded as an estimate of the quality of the fix, as supplied by the (on board) processing software.

E640# Satellite data (other systems)

= 4...9, satellite system reference number

At node identifier	[6, 9]	I4	
Latitude	[9,20]	I3,I2,F6.3,A1	dddmmss.sss N/S
Longitude	[21,32]	I3,I2,F6.3,A1	dddmmss.sss E/W
Height	[33,38]	F6.1	metres
Height data	[39,39]	I1	0 = height above ellipsoid 1 = height above geoid (~MSL)
Standard deviations of:			
latitude	[40,44]	N5	metres
longitude	[45,49]	N5	metres
height	[50,54]	N5	metres

Note:

The standard deviations of latitude, longitude and height should be estimates of the quality of the fix, as produced by the onboard software.

User-defined event data

E7010 User-defined observation set data

Observation set ref. number	[6, 8]	I3
Data field number	[9,10]	I2
Quality indicator	[11,14]	N4
Observation	[15,..]	Nx

Note:

Observation data field width as specified in the H7010 record, hence no column can be specified here.

The last three fields (triplet) in this record may be repeated until the record is full. However, partially filled triplets are not permitted.

B.7.5 Inter-event data records (explicit time tag)

B.7.5.1 Inter-event vessel related data

T14@0 Inter-event echo sounder data

@ = 1...9, vessel reference number

Echo sounder ref. number	[6, 6]	I1	
Echo sounder reading	[7,12]	F6.1	metres
Time of observation	[13,19]	I2,I2,I3	HH,MM,SSs

May be repeated for four more echo sounders mounted on the same vessel at [21,34], [36,49], [51,64] and [66,79].

Record may be repeated.

T16@0 Inter-event USBL acoustic data

@ = 1...9, vessel reference number

USBL system ref. number	[6, 6]	I1	
To Node number	[7,10]	I4	
X range to node	[11,17]	N7	metres
Y range to node	[18,24]	N7	metres
Z range to node	[25,31]	N7	metres
Quality indicator	[32,35]	N4	
Time of observation	[36,42]	I2,I2,I3	HH,MM,SSs

May be repeated for one more USBL system mounted on the same vessel at [44,80].

Record may be repeated.

T17@0 Inter-event pitch, roll and heave sensor data

@ = 1...9, vessel reference number

Sensor reference number	[6, 6]	I1	
Pitch angle	[7,16]	N10	
Roll angle	[17,26]	N10	
Heave	[27,36]	N10	
Quality indicator pitch	[37,40]	N4	
Quality indicator roll	[41,44]	N4	
Quality indicator heave	[45,48]	N4	
Time of observation	[49,55]	I2,I2,I3	HH,MM,SSs

B.7.5.2 Inter-event network data

T52## Inter-event network data

= observation type

Observation Identifier	[6, 9]	I4	
Observation	[10,19]	N10	
Quality indicator	[20,23]	N4	
Time of observation	[24,30]	I2,I2,I3	HH,MM,SSs

May be repeated at [31,55] and at [56,80] for two more observations of the same observation type.

T54## Inter-event network observation parameters

= observation type

Observation Identifier	[6, 9]	I4	
Variable (C–O)	[10,17]	N8	
C/O or propagation speed	[18,29]	N12	
Flag for C/O or speed	[30,30]	I1	0 = C/O (=scale factor) 1 = propagation speed
Time of observation	[31,37]	I2,I2,I3	HH,MM,SSs

May be repeated for one more set of observation parameters at [38,69] relating to the same observation type.

Record may be repeated.

Notes:

(a) This record is optional and should only be inserted for points in time at which one or all of the above parameters change. The new parameters shall be valid from the time recorded in this record until the time recorded in the next T54## or E54## record inserted.

(b) Variable (C–O) replaces variable (C–O) in record H54##.

(c) Scale (C/O) replaces scale fFactor in record H54##.

(d) Propagation speed, if supplied, replaces propagation speed in record H54##; the scale factor in record H54## is then still valid.

B.7.5.3 Inter-event satellite data

T620# Inter-event GPS or DGPS data

= 1, GPS
= 2, DGPS

"At" Node identifier	[6, 9]	I3	
Receiver reference number	[10,10]	I1	
Latitude	[11,22]	I3,I2,F6.3,A1	dddmmss.sss N/S
Longitude	[23,34]	I3,I2,F6.3,A1	dddmmss.sss E/W
Height	[35,40]	F6.1	metres
Height datum	[41,41]	I1	0=height above ellipsoid 1=height above geoid (~MSL)
Satellites used numbers)	[42,67]	13*I2	(GPS SV
Reference stations used	[68,71]	4*I1	
Position calculation mode	[72,73]	I2	see below
Time of observation	[74,80]	I2,I2,I3	HH,MM,SSs

Note:

If GPS is used in autonomous mode the field 'reference stations used' should be left blank.

The time recorded should be in the same time system as other time tagged observations.

The identifiers used to indicate the satellites used in the position fix must have the official GPS SV numbers.

The 'reference stations used' field refers to record H610#. Only those reference station identifiers should be recorded here of which the differential corrections have been used in the calculation of the position recorded in this record.

Position calculation mode codes

1 = 3D solution; 4+ SVs

2 = 3D solution; 3+ SVs; height fixed

3 = 3D solution; 3+ SVs; height aided

4 = 3D solution; 3+ SVs; clock aided

5 = 3D solution; 2+ SVs; height aided and clock aided
6 = 3D solution; 2+ SVs; height fixed and clock aided
7 = 2D solution; 3+ SVs
8 = 2D solution; 2+ SVs; height fixed
9 = 2D solution; 2+ SVs; height aided
10 = 2D solution; 2+ SVs; clock aided
11 = 2D solution; 1+ SVs; height aided and clock aided
12 = 2D solution; 1+ SVs; height fixed and clock aided

T621# Inter-event GPS or DGPS data (continued)

# = 1, GPS			
# = 2, DGPS			
At node identifier	[6, 9]	I4	
Receiver reference number	[10,10]	I1	
Standard deviations of:			
latitude:	[11,15]	N5	metres
longitude:	[16,20]	N5	metres
height:	[21,25]	N5	metres
DOP type	[26,26]	I1	0 = GDOP 1 = PDOP 2 = HDOP 3 = TDOP 4 = VDOP 5..9 = user defined on comment cards
DOP figure	[27,30]	N4	unitless
Time of Observation	[74,80]	I2,I2,I3	HH,MM,SSs

The DOP type and DOP figure fields may be repeated for other DOPs or quality indicators in columns [31,35]... [51,55] as required.

Note:

The standard deviations of latitude, longitude and height should be estimates of the quality of the fix, as produced by the onboard software.

The standard deviation of the height should be zero in the case where height is not solved for in the position calculation, such as in 'height fixed' mode.

T6303 Inter-event TRANSIT satellite data

At node identifier	[6, 9]	I4	
Latitude	[10,21]	I3,I2,F6.3,A1	dddmmss.sss N/S
Longitude	[22,33]	I3,I2,F6.3,A1	dddmmss.sss E/W
Position includes dead reckon?	[34,34]	I1	0 = no

			1 = yes
Standard deviations of last accepted satellite fix:			
latitude	[35,39]	N5	metres
longitude	[40,44]	N5	metres
Time of observation	[45,51]	I2,I2,I3	HH,MM,SSs

Record may be repeated.

Note:

The position supplied should refer to the satellite receiver antenna electrical centre.

If a dead reckoning system involving TRANSIT is being used, then the estimated current position should be recorded and the 'dead reckoning flag' should be set. The Time of Observation then refers to this dead reckoned position, not the bare satellite fix.

No attempt is made to log raw data for the TRANSIT system and the standard deviations of the co-ordinates are recorded as an estimate of the quality of the fix, as supplied by the (onboard) processing software.

T640# Inter-event satellite data (other systems)

= 4...9, satellite system reference number

At node identifier	[6, 9]	I4	
Latitude	[10,21]	I3,I2,F6.3,A1	dddmmss.sss N/S
Longitude	[22,33]	I3,I2,F6.3,A1	dddmmss.sss E/W
Height	[34,39]	F6.1	metres
Height datum	[40,40]	I1	0 = height above ellipsoid 1 = height above geoid (~MSL)
Standard deviations of:			
latitude:	[41,45]	N5	metres
longitude:	[46,50]	N5	metres
height:	[51,55]	N5	metres
Time of observation	[56,62]	I2,I2,I3	HH,MM,SSs

Note:

The standard deviations of latitude, longitude and height should be estimates of the quality of the fix, as produced by the onboard software.

Inter-event user defined data

T7010 Inter-event user defined observation set data

Observation set ref. number	[6, 8]	I3	
Data field number	[9,10]	I2	
Quality indicator	[11,14]	N4	
Time of observation	[15,21]	I2,I2,I3	HH,MM,SSs
Observation	[22,..]	Nx	

Note:

Observation data field width as specified in the H7010 record, hence no column width can be specified here.

The last four fields (quadruplet) in this record may be repeated until the record is full. However, partially filled quadruplets are not permitted.

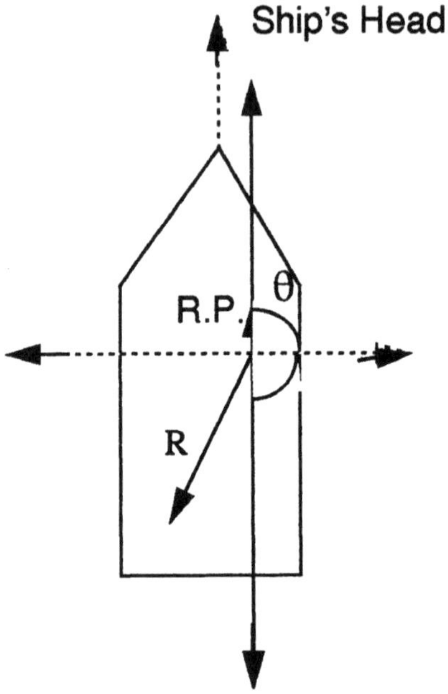

Polar Offset Mode

Offset A = R

Offset B = θ

R.P. = Ship Reference Point.

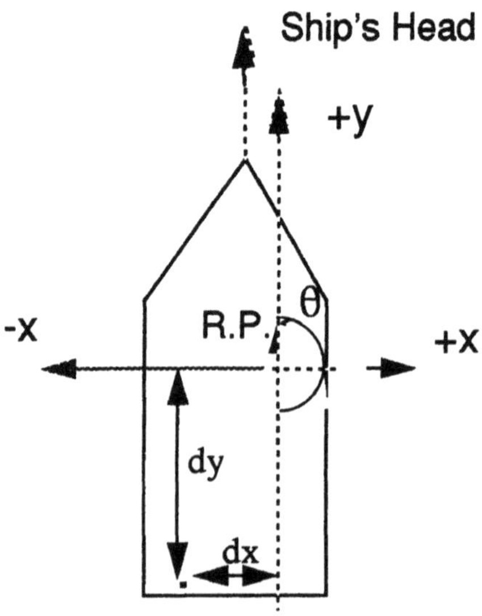

Rectangular Offset Mode

Offset A = -dx

Offset B = -dy

B.8 Example P2/91 header and data

```
H0000Line name:                                    0
H0001Project name:            Demonstration                                1995 91
H0002Project description:     Two-boat 6-streamer
H0003Media specification:     1995 91  QSEA Innovations                    UKOOA P2/91 v1.0
H0004Client:                  Demo 6-streamer spread
H0005Geophysical contractor: A company
H0006Positioning contractor: Another company
H0006Processing contractor:  Unknown
H0018Line parameters vessel: 1 0  0 0 0.00S  0 0 0.00E 0     1   0.00   0 1
H0019 1 1    0 0 0.00S   0 0 0.00E
C0001  Opus4X-generated header V1.0     0  1995  331   0  29    26.19
H0100 1995  91    1 0
H0101     11918 0.00N  72 0 0.00E  -1.100
H0111              WGS 84              WGS 84 6378137.000        1.000     298.257
H0112              WGS 84              WGS 84 6378137.000        1.000     298.257
H0120 1 2 0        0.00         0.00        0.00   0.0000   0.0000   0.0000   0.0000
H0140   1  1.000000 UTM
H0150     0 0 0.00S  75 0 0.00E        0.00N  500000.00E      0.999600
H0200 2 0  0 2 1 0 0
H0211                              Demo1  00       3  2  1 1 0 0  6  10
H0212                              Demo2  00       3  0  0 1 0 0 12 132
H0221                                   201   1        1            6 14  0 240
H0221                                   202   1        1            6 14  0 240
H0221                                   203   1        1            6 14  0 240
H0222                                   204   2        1            6 14  0 240
H0222                                   205   2        1            6 14  0 240
H0222                                   206   2        1            6 14  0 240
H0231                                   301   1        0        1   3     0
H0231                                   302   1        0        0   4     0
H0241                                   401   1        0        2   2
H0241                                   202   1        0        0   0
H1010 0.0  NRP
H1110 NRP
H1210
H1211 01
H1310     0        0
H1411     0.0     0.0     0.0     0.0     0.0 000
H1010 0.0  NRP
H1110 NRP
H1210
H1211 01
H1310     0        0
H1411  -450.0    90.0     0.0     0.0     0.0 000
H1500 01 1970    1  0 0             0 0 0.00S  0 0 0.00E0000
H1501 1 0.0    0.0   0.0  0.00
H2110 201        0       0       0   150.0  -369.0    0.0      6.0     8.0
```

```
H2111 201 100.0 100.0 240   12.5   14    0.0     0    0.0 0 0
H2112
H2113
H2210 201   81      1101       54.1 0    82      1102        0.1 0
H2210 201   83      1103      -75.0 0    84      1104     -299.4 0
H2210 201   85      1105     -598.9 0    86      1106     -898.5 0
H2210 201   87      1107    -1198.0 0    88      1108    -1497.5 0
H2210 201   89      1109    -1797.0 0    90      1110    -2096.6 0
H2210 201   91      1111    -2396.1 0    92      1112    -2695.6 0
H2210 201   93      1113    -2920.5 0    94      1114    -2991.6 0
H2410 201    1    -375.0   240   -3368.8 240    12.5
H2110 202          0         0         0    0.0  -369.0      0.0       6.0      8.0
H2111 202 100.0 100.0 240   12.5   14    0.0     0    0.0 0 0
H2112
H2113
H2210 202   95      1201       54.1 0    96      1202        0.1 0
H2210 202   97      1203      -75.0 0    98      1204     -299.4 0
H2210 202   99      1205     -598.9 0   100      1206     -898.5 0
H2210 202  101      1207    -1198.0 0   102      1208    -1497.5 0
H2210 202  103      1209    -1797.0 0   104      1210    -2096.6 0
H2210 202  105      1211    -2396.1 0   106      1212    -2695.6 0
H2210 202  107      1213    -2920.5 0   108      1214    -2991.6 0
H2410 202    1    -375.0   240   -3368.8 240    12.5
H2110 203          0         0         0 -150.0  -369.0      0.0       6.0      8.0
H2111 203 100.0 100.0 240   12.5   14    0.0     0    0.0 0 0
H2112
H2113
H2210 203  109      1301       54.1 0   110      1302        0.1 0
H2210 203  111      1303      -75.0 0   112      1304     -299.4 0
H2210 203  113      1305     -598.9 0   114      1306     -898.5 0
H2210 203  115      1307    -1198.0 0   116      1308    -1497.5 0
H2210 203  117      1309    -1797.0 0   118      1310    -2096.6 0
H2210 203  119      1311    -2396.1 0   120      1312    -2695.6 0
H2210 203  121      1313    -2920.5 0   122      1314    -2991.6 0
H2410 203    1    -375.0   240   -3368.8 240    12.5
H2120 204          0         0         0  150.0  -420.0      0.0       6.0      8.0
H2121 204 200.0 100.0 240   12.5   14    0.0     0    0.0 0 0
H2122
H2123
H2220 204  123      1401       54.1 0   124      1402        0.1 0
H2220 204  125      1403      -75.0 0   126      1404     -299.4 0
H2220 204  127      1405     -598.9 0   128      1406     -898.5 0
H2220 204  129      1407    -1198.0 0   130      1408    -1497.5 0
H2220 204  131      1409    -1797.0 0   132      1410    -2096.6 0
H2220 204  133      1411    -2396.1 0   134      1412    -2695.6 0
H2220 204  135      1413    -2920.5 0   136      1414    -2991.6 0
H2420 204    1    -426.0   240   -3419.8 240    12.5
```

```
H2120 205        0        0        0     0.0  -420.0     0.0      6.0      8.0
H2121 205 200.0 100.0 240   12.5   14    0.0    0    0.0 0 0
H2122
H2123
H2220 205  137      1501      54.1 0  138      1502       0.1 0
H2220 205  139      1503     -75.0 0  140      1504    -299.4 0
H2220 205  141      1505    -598.9 0  142      1506    -898.5 0
H2220 205  143      1507   -1198.0 0  144      1508   -1497.5 0
H2220 205  145      1509   -1797.0 0  146      1510   -2096.6 0
H2220 205  147      1511   -2396.1 0  148      1512   -2695.6 0
H2220 205  149      1513   -2920.5 0  150      1514   -2991.6 0
H2420 205    1   -426.0  240   -3419.8 240    12.5
H2120 206        0        0        0  -150.0  -420.0     0.0      6.0      8.0
H2121 206 200.0 100.0 240   12.5   14    0.0    0    0.0 0 0
H2122
H2123
H2220 206  151      1601      54.1 0  152      1602       0.1 0
H2220 206  153      1603     -75.0 0  154      1604    -299.4 0
H2220 206  155      1605    -598.9 0  156      1606    -898.5 0
H2220 206  157      1607   -1198.0 0  158      1608   -1497.5 0
H2220 206  159      1609   -1797.0 0  160      1610   -2096.6 0
H2220 206  161      1611   -2396.1 0  162      1612   -2695.6 0
H2220 206  163      1613   -2920.5 0  164      1614   -2991.6 0
H2420 206    1   -426.0  240   -3419.8 240    12.5
H3110 301        0        0        0    37.5  -145.0    -6.0      0.0      0.0
H3110 302        0        0        0   -37.5  -145.0    -6.0      0.0      0.0
H4110 401   1       0        0        0     -25.0  -332.0    -8.0
H4110 202   1       0        0        0       0.0 -3471.7    -8.0
H5110     1             V1G1 101      0.0      0.0     26.2
H5110     2               V1 101      0.0      0.0      0.0
H5110     3             V1G2 101     -0.2     -0.1     25.9
H5110     4            TB2G1 202      0.0  -3471.7     -6.0
H5110     5              TB2 202      0.0  -3471.7     -8.0
H5110     6               S2 202      0.0   -375.0     -8.0
H5110     7             A2G1 301     37.5   -143.2      0.1
H5110     8              So1 301     37.5   -145.0     -6.0
H5120     9             V2G1 102      0.7      0.0     19.9
H5120    10               V2 102      0.0      0.0      0.0
H5110    11            TB1G1 201    150.0  -3471.7     -6.0
H5110    12              TB1 201    150.0  -3471.7     -8.0
H5110    13               S1 201    150.0   -375.0     -8.0
H5110    14            TB3G1 203   -150.0  -3471.7      2.0
H5110    15              TB3 203   -150.0  -3471.7      0.0
H5110    16               S3 203   -150.0   -375.0     -8.0
H5120    17            TB4G1 204    150.0  -3522.7      2.0
H5120    18              TB4 204    150.0  -3522.7      0.0
H5120    19               S4 204    150.0   -426.0     -8.0
```

H5120	20	TB5G1	205	0.0	-3522.7	-6.0
H5120	21	TB5	205	0.0	-3522.7	-8.0
H5120	22	S5	205	0.0	-426.0	-8.0
H5120	23	TB6G1	206	-150.0	-3522.7	-6.0
H5120	24	TB6	206	-150.0	-3522.7	-8.0
H5120	25	S6	206	-150.0	-426.0	-8.0
H5110	26	A5G1	302	-37.5	-143.2	0.1
H5110	27	So2	302	-37.5	-145.0	-6.0
H5110	28	V1O1	101	-1.7	-21.1	7.0
H5110	29	A2O1	301	37.5	-143.0	-0.5
H5110	30	A5O1	302	-37.5	-143.0	-0.5
H5120	31	V2O1	102	0.0	0.2	15.8
H5110	32	A5T1	302	-37.5	-146.0	-8.0
H5110	33	A2T1	301	37.5	-146.0	-8.0
H5110	34	S1T5	201	150.0	-367.3	-8.0
H5110	35	S2T5	202	0.0	-367.3	-8.0
H5110	36	S3T5	203	-150.0	-367.3	-8.0
H5120	37	S4T5	204	150.0	-418.3	-8.0
H5110	38	S1T1	201	150.0	-3463.1	-8.0
H5110	39	S1T2	201	150.0	-3357.9	-8.0
H5110	40	S2T1	202	0.0	-3463.5	-8.0
H5110	41	S2T2	202	0.0	-3357.9	-8.0
H5120	42	S5T2	205	0.0	-3408.9	-8.0
H5120	43	S4T1	204	150.0	-3513.7	0.0
H5120	44	S4T2	204	150.0	-3408.9	-8.0
H5120	45	S4T3	204	150.0	-3333.7	-8.0
H5120	46	S6T1	206	-150.0	-3513.2	-8.0
H5120	47	S6T2	206	-150.0	-3408.9	-8.0
H5120	48	S6T3	206	-150.0	-3333.7	-8.0
H5110	49	S3T3	203	-150.0	-3282.7	-8.0
H5110	50	S2T3	202	0.0	-3282.7	-8.0
H5120	51	S5T5	205	0.0	-418.3	-8.0
H5120	52	S6T4	206	-150.0	-488.3	-8.0
H5120	53	S6T5	206	-150.0	-418.3	-8.0
H5110	54	S2T4	202	0.0	-437.3	-8.0
H5120	55	S4T4	204	150.0	-488.3	-8.0
H5120	56	S5T1	205	0.0	-3513.7	-8.0
H5110	57	S3T2	203	-150.0	-3357.9	-8.0
H5110	58	S3T1	203	-150.0	-3463.5	0.0
H5110	59	S3T4	203	-150.0	-437.3	-8.0
H5120	60	S5T3	205	0.0	-3333.7	-8.0
H5120	61	S5T4	205	0.0	-488.3	-8.0
H5110	62	S1T3	201	150.0	-3282.7	-8.0
H5120	63	V2T1	102	-1.9	-1.4	-6.9
H5110	64	V1T1	101	-1.9	6.2	-7.5
H5110	65	S1T4	201	150.0	-437.3	-8.0
H5110	66	N1G1	401	-25.0	-332.0	2.0

H5110	67	N1	401	-25.0	-332.0	0.0
H5110	68	N1O1	401	-25.0	-332.0	2.0
H5110	69	N1T1	401	-25.0	-332.0	-6.0
H5110	70	QMA_1	101	0.0	0.0	0.0
H5110	81	1101	201	150.0	-320.9	-8.0
H5110	82	1102	201	150.0	-374.9	-8.0
H5110	83	1103	201	150.0	-450.0	-8.0
H5110	84	1104	201	150.0	-674.4	-8.0
H5110	85	1105	201	150.0	-973.9	-8.0
H5110	86	1106	201	150.0	-1273.5	-8.0
H5110	87	1107	201	150.0	-1573.0	-8.0
H5110	88	1108	201	150.0	-1872.5	-8.0
H5110	89	1109	201	150.0	-2172.0	-8.0
H5110	90	1110	201	150.0	-2471.6	-8.0
H5110	91	1111	201	150.0	-2771.1	-8.0
H5110	92	1112	201	150.0	-3070.6	-8.0
H5110	93	1113	201	150.0	-3295.5	-8.0
H5110	94	1114	201	150.0	-3366.6	-8.0
H5110	95	1201	202	0.0	-320.9	-8.0
H5110	96	1202	202	0.0	-374.9	-8.0
H5110	97	1203	202	0.0	-450.0	-8.0
H5110	98	1204	202	0.0	-674.4	-8.0
H5110	99	1205	202	0.0	-973.9	-8.0
H5110	100	1206	202	0.0	-1273.5	-8.0
H5110	101	1207	202	0.0	-1573.0	-8.0
H5110	102	1208	202	0.0	-1872.5	-8.0
H5110	103	1209	202	0.0	-2172.0	-8.0
H5110	104	1210	202	0.0	-2471.6	-8.0
H5110	105	1211	202	0.0	-2771.1	-8.0
H5110	106	1212	202	0.0	-3070.6	-8.0
H5110	107	1213	202	0.0	-3295.5	-8.0
H5110	108	1214	202	0.0	-3366.6	-8.0
H5110	109	1301	203	-150.0	-320.9	-8.0
H5110	110	1302	203	-150.0	-374.9	-8.0
H5110	111	1303	203	-150.0	-450.0	-8.0
H5110	112	1304	203	-150.0	-674.4	-8.0
H5110	113	1305	203	-150.0	-973.9	-8.0
H5110	114	1306	203	-150.0	-1273.5	-8.0
H5110	115	1307	203	-150.0	-1573.0	-8.0
H5110	116	1308	203	-150.0	-1872.5	-8.0
H5110	117	1309	203	-150.0	-2172.0	-8.0
H5110	118	1310	203	-150.0	-2471.6	-8.0
H5110	119	1311	203	-150.0	-2771.1	-8.0
H5110	120	1312	203	-150.0	-3070.6	-8.0
H5110	121	1313	203	-150.0	-3295.5	-8.0
H5110	122	1314	203	-150.0	-3366.6	-8.0
H5120	123	1401	204	150.0	-371.9	-8.0

```
H5120  124            1402 204   150.0  -425.9    -8.0
H5120  125            1403 204   150.0  -501.0    -8.0
H5120  126            1404 204   150.0  -725.4    -8.0
H5120  127            1405 204   150.0 -1024.9    -8.0
H5120  128            1406 204   150.0 -1324.5    -8.0
H5120  129            1407 204   150.0 -1624.0    -8.0
H5120  130            1408 204   150.0 -1923.5    -8.0
H5120  131            1409 204   150.0 -2223.0    -8.0
H5120  132            1410 204   150.0 -2522.6    -8.0
H5120  133            1411 204   150.0 -2822.1    -8.0
H5120  134            1412 204   150.0 -3121.6    -8.0
H5120  135            1413 204   150.0 -3346.5    -8.0
H5120  136            1414 204   150.0 -3417.6    -8.0
H5120  137            1501 205     0.0  -371.9    -8.0
H5120  138            1502 205     0.0  -425.9    -8.0
H5120  139            1503 205     0.0  -501.0    -8.0
H5120  140            1504 205     0.0  -725.4    -8.0
H5120  141            1505 205     0.0 -1024.9    -8.0
H5120  142            1506 205     0.0 -1324.5    -8.0
H5120  143            1507 205     0.0 -1624.0    -8.0
H5120  144            1508 205     0.0 -1923.5    -8.0
H5120  145            1509 205     0.0 -2223.0    -8.0
H5120  146            1510 205     0.0 -2522.6    -8.0
H5120  147            1511 205     0.0 -2822.1    -8.0
H5120  148            1512 205     0.0 -3121.6    -8.0
H5120  149            1513 205     0.0 -3346.5    -8.0
H5120  150            1514 205     0.0 -3417.6    -8.0
H5120  151            1601 206  -150.0  -371.9    -8.0
H5120  152            1602 206  -150.0  -425.9    -8.0
H5120  153            1603 206  -150.0  -501.0    -8.0
H5120  154            1604 206  -150.0  -725.4    -8.0
H5120  155            1605 206  -150.0 -1024.9    -8.0
H5120  156            1606 206  -150.0 -1324.5    -8.0
H5120  157            1607 206  -150.0 -1624.0    -8.0
H5120  158            1608 206  -150.0 -1923.5    -8.0
H5120  159            1609 206  -150.0 -2223.0    -8.0
H5120  160            1610 206  -150.0 -2522.6    -8.0
H5120  161            1611 206  -150.0 -2822.1    -8.0
H5120  162            1612 206  -150.0 -3121.6    -8.0
H5120  163            1613 206  -150.0 -3346.5    -8.0
H5120  164            1614 206  -150.0 -3417.6    -8.0
H5201    1 TB1G1R          1   11       0   11                      Direct range
H5401    1     0.000000    0.000000 0 0     1.000000    0.000000       0 2.00 0
H5211    2 TB1G1B          1   11       3   1                            Azimuth
H5411    2     0.000000    0.000000 0 0     1.000000    0.000000       0 2.00 0
H5201    3 TB2G1R          1   4        0   1                       Direct range
H5401    3     0.000000    0.000000 0 0     1.000000    0.000000       0 2.00 0
```

```
H5211      4 TB2G1B          1    4      3   1                  Azimuth
H5411      4      0.000000  0.000000 0 0   1.000000   0.000000   0 2.00 0
H5201      5 TB3G1R          1   14      0   1             Direct range
H5401      5      0.000000  0.000000 0 0   1.000000   0.000000   0 2.00 0
H5211      6 TB3G1B          1   14      3   1                  Azimuth
H5411      6      0.000000  0.000000 0 0   1.000000   0.000000   0 2.00 0
H5201      7 TB4G1R          1   17      0   1             Direct range
H5401      7      0.000000  0.000000 0 0   1.000000   0.000000   0 2.00 0
H5211      8 TB4G1B          1   17      3   1                  Azimuth
H5411      8      0.000000  0.000000 0 0   1.000000   0.000000   0 2.00 0
H5201      9 TB5G1R          1   20      0   1             Direct range
H5401      9      0.000000  0.000000 0 0   1.000000   0.000000   0 2.00 0
H5211     10 TB5G1B          1   20      3   1                  Azimuth
H5411     10      0.000000  0.000000 0 0   1.000000   0.000000   0 2.00 0
H5201     11 TB6G1R          1   23      0   1             Direct range
H5401     11      0.000000  0.000000 0 0   1.000000   0.000000   0 2.00 0
H5211     12 TB6G1B          1   23      3   1                  Azimuth
H5411     12      0.000000  0.000000 0 0   1.000000   0.000000   0 2.00 0
H5201     13 MALR            1    9      0   1             Direct range
H5401     13      0.000000  0.000000 0 0   1.000000   0.000000   0 2.00 0
H5211     14 MALB            1    9      3   1                  Azimuth
H5411     14      0.000000  0.000000 0 0   1.000000   0.000000   0 2.00 0
H5201     15 A2G1R           1    7      0   1             Direct range
H5401     15      0.000000  0.000000 0 0   1.000000   0.000000   0 2.00 0
H5211     16 A2G1B           1    7      3   1                  Azimuth
H5411     16      0.000000  0.000000 0 0   1.000000   0.000000   0 2.00 0
H5201     17 A5G1R           1   26      0   1             Direct range
H5401     17      0.000000  0.000000 0 0   1.000000   0.000000   0 2.00 0
H5211     18 A5G1B           1   26      3   1                  Azimuth
H5411     18      0.000000  0.000000 0 0   1.000000   0.000000   0 2.00 0
H5201     19 NAVR            1   66      0   1             Direct range
H5401     19      0.000000  0.000000 0 0   1.000000   0.000000   0 2.00 0
H5211     20 NAVB            1   66      3   1                  Azimuth
H5411     20      0.000000  0.000000 0 0   1.000000   0.000000   0 2.00 0
H5201     21 A501R          28   30      0   1             Direct range
H5401     21      0.000000  0.000000 0 0   1.000000   0.000000   0 2.00 0
H5209     22 A501B          28   30      3   2                Direction
H5409     22      0.000000  0.000000 0 0   1.000000   0.000000   0 2.00 0
H5201     23 N101R          28   68      0   2             Direct range
H5401     23      0.000000  0.000000 0 0   1.000000   0.000000   0 2.00 0
H5209     24 N101B          28   68      3   2                Direction
H5409     24      0.000000  0.000000 0 0   1.000000   0.000000   0 2.00 0
H5201     25 A201R          28   29      0   2             Direct range
H5401     25      0.000000  0.000000 0 0   1.000000   0.000000   0 2.00 0
H5209     26 A201B          28   29      3   2                Direction
H5409     26      0.000000  0.000000 0 0   1.000000   0.000000   0 2.00 0
H5201     27 V201R          28   31      0   2             Direct range
```

```
H5401    27       0.000000     0.000000 0 0    1.000000    0.000000         0 2.00 0
H5209    28 V2O1B               28    31       3    2                        Direction
H5409    28       0.000000     0.000000 0 0    1.000000    0.000000         0 2.00 0
H5201    29 A5T1-A2T1           33    32       0    2                     Direct range
H5401    29       0.000000     0.000000 0 0    1.542000    0.000000         0 2.00 0
H5201    30 A5T1-S1T5           32    34       0    3                     Direct range
H5401    30       0.000000     0.000000 0 0    1.542000    0.000000         0 3.00 0
H5201    31 A5T1-S2T5           32    35       0    3                     Direct range
H5401    31       0.000000     0.000000 0 0    1.542000    0.000000         0 3.00 0
H5201    32 A5T1-S3T5           32    36       0    3                     Direct range
H5401    32       0.000000     0.000000 0 0    1.542000    0.000000         0 3.00 0
H5201    33 S1T1-S1T2           38    39       0    3                     Direct range
H5401    33       0.000000     0.000000 0 0    1.542000    0.000000         0 3.00 0
H5201    34 S1T1-S2T1           38    40       0    3                     Direct range
H5401    34       0.000000     0.000000 0 0    1.542000    0.000000         0 3.00 0
H5201    35 S1T1-S2T2           38    41       0    3                     Direct range
H5401    35       0.000000     0.000000 0 0    1.542000    0.000000         0 3.00 0
H5201    36 A2T1-S1T5           33    34       0    3                     Direct range
H5401    36       0.000000     0.000000 0 0    1.542000    0.000000         0 3.00 0
H5201    37 A2T1-S2T5           33    35       0    3                     Direct range
H5401    37       0.000000     0.000000 0 0    1.542000    0.000000         0 3.00 0
H5201    38 A2T1-S3T5           33    36       0    3                     Direct range
H5401    38       0.000000     0.000000 0 0    1.542000    0.000000         0 3.00 0
H5201    39 S3T3-S2T2           49    41       0    3                     Direct range
H5401    39       0.000000     0.000000 0 0    1.542000    0.000000         0 3.00 0
H5201    40 S3T3-S2T3           49    50       0    3                     Direct range
H5401    40       0.000000     0.000000 0 0    1.542000    0.000000         0 3.00 0
H5201    41 S1T2-S2T1           39    40       0    3                     Direct range
H5401    41       0.000000     0.000000 0 0    1.542000    0.000000         0 3.00 0
H5201    42 S1T2-S2T2           39    41       0    3                     Direct range
H5401    42       0.000000     0.000000 0 0    1.542000    0.000000         0 3.00 0
H5201    43 S1T2-S2T3           39    50       0    3                     Direct range
H5401    43       0.000000     0.000000 0 0    1.542000    0.000000         0 3.00 0
H5201    44 S3T5-S2T4           36    54       0    3                     Direct range
H5401    44       0.000000     0.000000 0 0    1.542000    0.000000         0 3.00 0
H5201    45 S3T5-S2T5           36    35       0    3                     Direct range
H5401    45       0.000000     0.000000 0 0    1.542000    0.000000         0 3.00 0
H5201    46 S3T2-S2T1           57    40       0    3                     Direct range
H5401    46       0.000000     0.000000 0 0    1.542000    0.000000         0 3.00 0
H5201    47 S3T2-S2T2           57    41       0    3                     Direct range
H5401    47       0.000000     0.000000 0 0    1.542000    0.000000         0 3.00 0
H5201    48 S3T2-S2T3           57    50       0    3                     Direct range
H5401    48       0.000000     0.000000 0 0    1.542000    0.000000         0 3.00 0
H5201    49 S3T2-S3T1           57    58       0    3                     Direct range
H5401    49       0.000000     0.000000 0 0    1.542000    0.000000         0 3.00 0
H5201    50 S3T4-S2T4           59    54       0    3                     Direct range
H5401    50       0.000000     0.000000 0 0    1.542000    0.000000         0 3.00 0
```

```
H5201   51 S3T4-S2T5      59   35      0   3                 Direct range
H5401   51      0.000000  0.000000 0 0  1.542000  0.000000     0 3.00 0
H5201   52 S1T5-S2T4      34   54      0   3                 Direct range
H5401   52      0.000000  0.000000 0 0  1.542000  0.000000     0 3.00 0
H5201   53 S1T5-S2T5      35   34      0   3                 Direct range
H5401   53      0.000000  0.000000 0 0  1.542000  0.000000     0 3.00 0
H5201   54 N1T1-S1T4      69   65      0   3                 Direct range
H5401   54      0.000000  0.000000 0 0  1.542000  0.000000     0 3.00 0
H5201   55 N1T1-S1T5      69   34      0   3                 Direct range
H5401   55      0.000000  0.000000 0 0  1.542000  0.000000     0 3.00 0
H5201   56 N1T1-S2T4      69   54      0   3                 Direct range
H5401   56      0.000000  0.000000 0 0  1.542000  0.000000     0 3.00 0
H5201   57 N1T1-S2T5      69   35      0   3                 Direct range
H5401   57      0.000000  0.000000 0 0  1.542000  0.000000     0 3.00 0
H5201   58 N1T1-S3T4      69   59      0   3                 Direct range
H5401   58      0.000000  0.000000 0 0  1.542000  0.000000     0 3.00 0
H5201   59 N1T1-S3T5      69   36      0   3                 Direct range
H5401   59      0.000000  0.000000 0 0  1.542000  0.000000     0 3.00 0
H5201   60 S1T3-S2T2      62   41      0   3                 Direct range
H5401   60      0.000000  0.000000 0 0  1.542000  0.000000     0 3.00 0
H5201   61 S1T3-S2T3      62   50      0   3                 Direct range
H5401   61      0.000000  0.000000 0 0  1.542000  0.000000     0 3.00 0
H5201   62 V2T1-V1T1      63   64      0   3                 Direct range
H5401   62      0.000000  0.000000 0 0  1.542000  0.000000     0 3.00 0
H5201   63 V2T1-S3T5      63   36      0   3                 Direct range
H5401   63      0.000000  0.000000 0 0  1.542000  0.000000     0 3.00 0
H5201   64 S1T4-S2T4      65   54      0   3                 Direct range
H5401   64      0.000000  0.000000 0 0  1.542000  0.000000     0 3.00 0
H5201   65 S1T4-S2T5      65   35      0   3                 Direct range
H5401   65      0.000000  0.000000 0 0  1.542000  0.000000     0 3.00 0
H5201   66 S3T1-S2T1      58   40      0   3                 Direct range
H5401   66      0.000000  0.000000 0 0  1.542000  0.000000     0 3.00 0
H5201   67 S3T1-S2T2      58   41      0   3                 Direct range
H5401   67      0.000000  0.000000 0 0  1.542000  0.000000     0 3.00 0
H5201   68 V1T1-A2T1      64   33      0   3                 Direct range
H5401   68      0.000000  0.000000 0 0  1.542000  0.000000     0 3.00 0
H5201   69 V1T1-A5T1      64   32      0   3                 Direct range
H5401   69      0.000000  0.000000 0 0  1.542000  0.000000     0 3.00 0
H5201   70 V1T1-S1T4      64   65      0   3                 Direct range
H5401   70      0.000000  0.000000 0 0  1.542000  0.000000     0 3.00 0
H5201   71 V1T1-S1T5      64   34      0   3                 Direct range
H5401   71      0.000000  0.000000 0 0  1.542000  0.000000     0 3.00 0
H5201   72 V1T1-S3T4      64   59      0   3                 Direct range
H5401   72      0.000000  0.000000 0 0  1.542000  0.000000     0 3.00 0
H5201   73 V1T1-S3T5      64   36      0   3                 Direct range
H5401   73      0.000000  0.000000 0 0  1.542000  0.000000     0 3.00 0
H5201   74 S2T1-S2T2      40   41      0   3                 Direct range
```

```
H5401   74       0.000000    0.000000 0 0    1.542000   0.000000          0 3.00 0
H5201   75 A5T1-S4T5           32    37       0   3                       Direct range
H5401   75       0.000000    0.000000 0 0    1.542000   0.000000          0 3.00 0
H5201   76 S5T2-S4T1           42    43       0   4                       Direct range
H5401   76       0.000000    0.000000 0 0    1.542000   0.000000          0 3.00 0
H5201   77 S5T2-S4T2           42    44       0   4                       Direct range
H5401   77       0.000000    0.000000 0 0    1.542000   0.000000          0 3.00 0
H5201   78 S5T2-S4T3           42    45       0   4                       Direct range
H5401   78       0.000000    0.000000 0 0    1.542000   0.000000          0 3.00 0
H5201   79 S5T2-S6T1           42    46       0   4                       Direct range
H5401   79       0.000000    0.000000 0 0    1.542000   0.000000          0 3.00 0
H5201   80 S5T2-S6T2           42    47       0   4                       Direct range
H5401   80       0.000000    0.000000 0 0    1.542000   0.000000          0 3.00 0
H5201   81 S5T2-S6T3           42    48       0   4                       Direct range
H5401   81       0.000000    0.000000 0 0    1.542000   0.000000          0 3.00 0
H5201   82 S3T3-S4T2           49    44       0   4                       Direct range
H5401   82       0.000000    0.000000 0 0    1.542000   0.000000          0 3.00 0
H5201   83 S3T3-S4T3           49    45       0   4                       Direct range
H5401   83       0.000000    0.000000 0 0    1.542000   0.000000          0 3.00 0
H5201   84 S5T5-S4T4           51    55       0   4                       Direct range
H5401   84       0.000000    0.000000 0 0    1.542000   0.000000          0 3.00 0
H5201   85 S5T5-S4T5           51    37       0   4                       Direct range
H5401   85       0.000000    0.000000 0 0    1.542000   0.000000          0 3.00 0
H5201   86 S5T5-S6T4           51    52       0   4                       Direct range
H5401   86       0.000000    0.000000 0 0    1.542000   0.000000          0 3.00 0
H5201   87 S5T5-S6T5           51    53       0   4                       Direct range
H5401   87       0.000000    0.000000 0 0    1.542000   0.000000          0 3.00 0
H5201   88 S3T5-S4T4           36    55       0   4                       Direct range
H5401   88       0.000000    0.000000 0 0    1.542000   0.000000          0 3.00 0
H5201   89 S3T5-S4T5           36    37       0   4                       Direct range
H5401   89       0.000000    0.000000 0 0    1.542000   0.000000          0 3.00 0
H5201   90 S5T1-S4T1           56    43       0   4                       Direct range
H5401   90       0.000000    0.000000 0 0    1.542000   0.000000          0 3.00 0
H5201   91 S5T1-S4T2           56    44       0   4                       Direct range
H5401   91       0.000000    0.000000 0 0    1.542000   0.000000          0 3.00 0
H5201   92 S5T1-S5T2           56    42       0   4                       Direct range
H5401   92       0.000000    0.000000 0 0    1.542000   0.000000          0 3.00 0
H5201   93 S5T1-S6T1           56    46       0   4                       Direct range
H5401   93       0.000000    0.000000 0 0    1.542000   0.000000          0 3.00 0
H5201   94 S5T1-S6T2           56    47       0   4                       Direct range
H5401   94       0.000000    0.000000 0 0    1.542000   0.000000          0 3.00 0
H5201   95 S3T2-S4T1           57    43       0   4                       Direct range
H5401   95       0.000000    0.000000 0 0    1.542000   0.000000          0 3.00 0
H5201   96 S3T2-S4T2           57    44       0   4                       Direct range
H5401   96       0.000000    0.000000 0 0    1.542000   0.000000          0 3.00 0
H5201   97 S3T2-S4T3           57    45       0   4                       Direct range
H5401   97       0.000000    0.000000 0 0    1.542000   0.000000          0 3.00 0
```

```
H5201    98 S3T4-S4T4         59   55      0   4                 Direct range
H5401    98      0.000000   0.000000 0 0    1.542000   0.000000     0 3.00 0
H5201    99 S3T4-S4T5         59   37      0   4                 Direct range
H5401    99      0.000000   0.000000 0 0    1.542000   0.000000     0 3.00 0
H5201   100 S6T1-S6T2         46   47      0   4                 Direct range
H5401   100      0.000000   0.000000 0 0    1.542000   0.000000     0 3.00 0
H5201   101 S5T3-S4T2         60   44      0   4                 Direct range
H5401   101      0.000000   0.000000 0 0    1.542000   0.000000     0 3.00 0
H5201   102 S5T3-S4T3         60   45      0   4                 Direct range
H5401   102      0.000000   0.000000 0 0    1.542000   0.000000     0 3.00 0
H5201   103 S5T3-S6T2         60   47      0   4                 Direct range
H5401   103      0.000000   0.000000 0 0    1.542000   0.000000     0 3.00 0
H5201   104 S5T3-S6T3         60   48      0   4                 Direct range
H5401   104      0.000000   0.000000 0 0    1.542000   0.000000     0 3.00 0
H5201   105 S5T4-S4T4         61   55      0   4                 Direct range
H5401   105      0.000000   0.000000 0 0    1.542000   0.000000     0 3.00 0
H5201   106 S5T4-S4T5         61   37      0   4                 Direct range
H5401   106      0.000000   0.000000 0 0    1.542000   0.000000     0 3.00 0
H5201   107 S5T4-S6T4         61   52      0   4                 Direct range
H5401   107      0.000000   0.000000 0 0    1.542000   0.000000     0 3.00 0
H5201   108 S5T4-S6T5         61   53      0   4                 Direct range
H5401   108      0.000000   0.000000 0 0    1.542000   0.000000     0 3.00 0
H5201   109 N1T1-S4T4         69   55      0   4                 Direct range
H5401   109      0.000000   0.000000 0 0    1.542000   0.000000     0 3.00 0
H5201   110 N1T1-S4T5         69   37      0   4                 Direct range
H5401   110      0.000000   0.000000 0 0    1.542000   0.000000     0 3.00 0
H5201   111 S4T1-S4T2         43   44      0   4                 Direct range
H5401   111      0.000000   0.000000 0 0    1.542000   0.000000     0 3.00 0
H5201   112 V2T1-S4T4         63   55      0   4                 Direct range
H5401   112      0.000000   0.000000 0 0    1.542000   0.000000     0 3.00 0
H5201   113 V2T1-S4T5         63   37      0   4                 Direct range
H5401   113      0.000000   0.000000 0 0    1.542000   0.000000     0 3.00 0
H5201   114 V2T1-S5T5         63   51      0   4                 Direct range
H5401   114      0.000000   0.000000 0 0    1.542000   0.000000     0 3.00 0
H5201   115 V2T1-S6T4         63   52      0   4                 Direct range
H5401   115      0.000000   0.000000 0 0    1.542000   0.000000     0 3.00 0
H5201   116 V2T1-S6T5         63   53      0   4                 Direct range
H5401   116      0.000000   0.000000 0 0    1.542000   0.000000     0 3.00 0
H5201   117 S3T1-S4T1         58   43      0   4                 Direct range
H5401   117      0.000000   0.000000 0 0    1.542000   0.000000     0 3.00 0
H5201   118 S3T1-S4T2         58   44      0   4                 Direct range
H5401   118      0.000000   0.000000 0 0    1.542000   0.000000     0 3.00 0
H5201   119 V1T1-V2T1         64   63      0   4                 Direct range
H5401   119      0.000000   0.000000 0 0    1.542000   0.000000     0 3.00 0
H5201   120 V1T1-S4T5         64   37      0   4                 Direct range
H5401   120      0.000000   0.000000 0 0    1.542000   0.000000     0 3.00 0
H5216   121 MAL               10    1      3   4                  Vessel gyro
```

```
H5416  121      0.000000       0.000000 0 0     1.000000   0.000000        0 2.00 0
H5216  122 GEX                    2     1       3    0                    Vessel gyro
H5416  122      0.000000       0.000000 0 0     1.000000   0.000000        0 0.10 0
H5213  123            gps0011     1     1       3    5                  Lat/lon pair
H5413  123      0.000000       0.000000 0 0     1.000000   0.000000        0 7.00 0
H5213  124            gps0012    70    70       3    6                  Lat/lon pair
H5413  124      0.000000       0.000000 0 0     1.000000   0.000000        0 2.00 0
H5200  125                        0     0       0    7                    Not in use
H5400  125      0.000000       0.000000 0 0     1.000000   0.000000        0 2.00 0
H5200  126                        0     0       0    0                    Not in use
H5400  126      0.000000       0.000000 0 0     1.000000   0.000000        0 2.00 0
H5200  127                        0     0       0    0                    Not in use
H5400  127      0.000000       0.000000 0 0     1.000000   0.000000        0 2.00 0
H5200  128                        0     0       0    0                    Not in use
H5400  128      0.000000       0.000000 0 0     1.000000   0.000000        0 2.00 0
H5200  129                        0     0       0    0                    Not in use
H5400  129      0.000000       0.000000 0 0     1.000000   0.000000        0 2.00 0
H5200  130                        0     0       0    0                    Not in use
H5400  130      0.000000       0.000000 0 0     1.000000   0.000000        0 2.00 0
H5200  131                        0     0       0    0                    Not in use
H5400  131      0.000000       0.000000 0 0     1.000000   0.000000        0 2.00 0
H5200  132                        0     0       0    0                    Not in use
H5400  132      0.000000       0.000000 0 0     1.000000   0.000000        0 2.00 0
H5200  133                        0     0       0    0                    Not in use
H5400  133      0.000000       0.000000 0 0     1.000000   0.000000        0 2.00 0
H5200  134                        0     0       0    0                    Not in use
H5400  134      0.000000       0.000000 0 0     1.000000   0.000000        0 2.00 0
H5210  135               1101    81     0       3    0              Streamer compass
H5410  135      0.000000       0.000000 0 0     0.100000   0.000000        0 2.00 0
H5210  136               1102    82     0       3    8              Streamer compass
H5410  136      0.000000       0.000000 0 0     0.100000   0.000000        0 2.00 0
H5210  137               1103    83     0       3    8              Streamer compass
H5410  137      0.000000       0.000000 0 0     0.100000   0.000000        0 2.00 0
H5210  138               1104    84     0       3    8              Streamer compass
H5410  138      0.000000       0.000000 0 0     0.100000   0.000000        0 2.00 0
H5210  139               1105    85     0       3    8              Streamer compass
H5410  139      0.000000       0.000000 0 0     0.100000   0.000000        0 2.00 0
H5210  140               1106    86     0       3    8              Streamer compass
H5410  140      0.000000       0.000000 0 0     0.100000   0.000000        0 2.00 0
H5210  141               1107    87     0       3    8              Streamer compass
H5410  141      0.000000       0.000000 0 0     0.100000   0.000000        0 2.00 0
H5210  142               1108    88     0       3    8              Streamer compass
H5410  142      0.000000       0.000000 0 0     0.100000   0.000000        0 2.00 0
H5210  143               1109    89     0       3    8              Streamer compass
H5410  143      0.000000       0.000000 0 0     0.100000   0.000000        0 2.00 0
H5210  144               1110    90     0       3    8              Streamer compass
H5410  144      0.000000       0.000000 0 0     0.100000   0.000000        0 2.00 0
```

```
H5210  145             1111   91    0      3   8            Streamer compass
H5410  145   0.000000      0.000000 0 0     0.100000   0.000000        0 2.00 0
H5210  146             1112   92    0      3   8            Streamer compass
H5410  146   0.000000      0.000000 0 0     0.100000   0.000000        0 2.00 0
H5210  147             1113   93    0      3   8            Streamer compass
H5410  147   0.000000      0.000000 0 0     0.100000   0.000000        0 2.00 0
H5210  148             1114   94    0      3   8            Streamer compass
H5410  148   0.000000      0.000000 0 0     0.100000   0.000000        0 2.00 0
H5210  149             1201   95    0      3   8            Streamer compass
H5410  149   0.000000      0.000000 0 0     0.100000   0.000000        0 2.00 0
H5210  150             1202   96    0      3   8            Streamer compass
H5410  150   0.000000      0.000000 0 0     0.100000   0.000000        0 2.00 0
H5210  151             1203   97    0      3   8            Streamer compass
H5410  151   0.000000      0.000000 0 0     0.100000   0.000000        0 2.00 0
H5210  152             1204   98    0      3   8            Streamer compass
H5410  152   0.000000      0.000000 0 0     0.100000   0.000000        0 2.00 0
H5210  153             1205   99    0      3   8            Streamer compass
H5410  153   0.000000      0.000000 0 0     0.100000   0.000000        0 2.00 0
H5210  154             1206  100    0      3   8            Streamer compass
H5410  154   0.000000      0.000000 0 0     0.100000   0.000000        0 2.00 0
H5210  155             1207  101    0      3   8            Streamer compass
H5410  155   0.000000      0.000000 0 0     0.100000   0.000000        0 2.00 0
H5210  156             1208  102    0      3   8            Streamer compass
H5410  156   0.000000      0.000000 0 0     0.100000   0.000000        0 2.00 0
H5210  157             1209  103    0      3   8            Streamer compass
H5410  157   0.000000      0.000000 0 0     0.100000   0.000000        0 2.00 0
H5210  158             1210  104    0      3   8            Streamer compass
H5410  158   0.000000      0.000000 0 0     0.100000   0.000000        0 2.00 0
H5210  159             1211  105    0      3   8            Streamer compass
H5410  159   0.000000      0.000000 0 0     0.100000   0.000000        0 2.00 0
H5210  160             1212  106    0      3   8            Streamer compass
H5410  160   0.000000      0.000000 0 0     0.100000   0.000000        0 2.00 0
H5210  161             1213  107    0      3   8            Streamer compass
H5410  161   0.000000      0.000000 0 0     0.100000   0.000000        0 2.00 0
H5210  162             1214  108    0      3   8            Streamer compass
H5410  162   0.000000      0.000000 0 0     0.100000   0.000000        0 2.00 0
H5210  163             1301  109    0      3   8            Streamer compass
H5410  163   0.000000      0.000000 0 0     0.100000   0.000000        0 2.00 0
H5210  164             1302  110    0      3   8            Streamer compass
H5410  164   0.000000      0.000000 0 0     0.100000   0.000000        0 2.00 0
H5210  165             1303  111    0      3   8            Streamer compass
H5410  165   0.000000      0.000000 0 0     0.100000   0.000000        0 2.00 0
H5210  166             1304  112    0      3   8            Streamer compass
H5410  166   0.000000      0.000000 0 0     0.100000   0.000000        0 2.00 0
H5210  167             1305  113    0      3   8            Streamer compass
H5410  167   0.000000      0.000000 0 0     0.100000   0.000000        0 2.00 0
H5210  168             1306  114    0      3   8            Streamer compass
```

```
H5410  168     0.000000     0.000000 0 0    0.100000   0.000000         0 2.00 0
H5210  169             1307  115     0       3   8              Streamer compass
H5410  169     0.000000     0.000000 0 0    0.100000   0.000000         0 2.00 0
H5210  170             1308  116     0       3   8              Streamer compass
H5410  170     0.000000     0.000000 0 0    0.100000   0.000000         0 2.00 0
H5210  171             1309  117     0       3   8              Streamer compass
H5410  171     0.000000     0.000000 0 0    0.100000   0.000000         0 2.00 0
H5210  172             1310  118     0       3   8              Streamer compass
H5410  172     0.000000     0.000000 0 0    0.100000   0.000000         0 2.00 0
H5210  173             1311  119     0       3   8              Streamer compass
H5410  173     0.000000     0.000000 0 0    0.100000   0.000000         0 2.00 0
H5210  174             1312  120     0       3   8              Streamer compass
H5410  174     0.000000     0.000000 0 0    0.100000   0.000000         0 2.00 0
H5210  175             1313  121     0       3   8              Streamer compass
H5410  175     0.000000     0.000000 0 0    0.100000   0.000000         0 2.00 0
H5210  176             1314  122     0       3   8              Streamer compass
H5410  176     0.000000     0.000000 0 0    0.100000   0.000000         0 2.00 0
H5210  177             1401  123     0       3   8              Streamer compass
H5410  177     0.000000     0.000000 0 0    0.100000   0.000000         0 2.00 0
H5210  178             1402  124     0       3   9              Streamer compass
H5410  178     0.000000     0.000000 0 0    0.100000   0.000000         0 2.00 0
H5210  179             1403  125     0       3   9              Streamer compass
H5410  179     0.000000     0.000000 0 0    0.100000   0.000000         0 2.00 0
H5210  180             1404  126     0       3   9              Streamer compass
H5410  180     0.000000     0.000000 0 0    0.100000   0.000000         0 2.00 0
H5210  181             1405  127     0       3   9              Streamer compass
H5410  181     0.000000     0.000000 0 0    0.100000   0.000000         0 2.00 0
H5210  182             1406  128     0       3   9              Streamer compass
H5410  182     0.000000     0.000000 0 0    0.100000   0.000000         0 2.00 0
H5210  183             1407  129     0       3   9              Streamer compass
H5410  183     0.000000     0.000000 0 0    0.100000   0.000000         0 2.00 0
H5210  184             1408  130     0       3   9              Streamer compass
H5410  184     0.000000     0.000000 0 0    0.100000   0.000000         0 2.00 0
H5210  185             1409  131     0       3   9              Streamer compass
H5410  185     0.000000     0.000000 0 0    0.100000   0.000000         0 2.00 0
H5210  186             1410  132     0       3   9              Streamer compass
H5410  186     0.000000     0.000000 0 0    0.100000   0.000000         0 2.00 0
H5210  187             1411  133     0       3   9              Streamer compass
H5410  187     0.000000     0.000000 0 0    0.100000   0.000000         0 2.00 0
H5210  188             1412  134     0       3   9              Streamer compass
H5410  188     0.000000     0.000000 0 0    0.100000   0.000000         0 2.00 0
H5210  189             1413  135     0       3   9              Streamer compass
H5410  189     0.000000     0.000000 0 0    0.100000   0.000000         0 2.00 0
H5210  190             1414  136     0       3   9              Streamer compass
H5410  190     0.000000     0.000000 0 0    0.100000   0.000000         0 2.00 0
H5210  191             1501  137     0       3   9              Streamer compass
H5410  191     0.000000     0.000000 0 0    0.100000   0.000000         0 2.00 0
```

```
H5210  192          1502  138    0        3   9            Streamer compass
H5410  192    0.000000     0.000000 0 0    0.100000   0.000000        0 2.00 0
H5210  193          1503  139    0        3   9            Streamer compass
H5410  193    0.000000     0.000000 0 0    0.100000   0.000000        0 2.00 0
H5210  194          1504  140    0        3   9            Streamer compass
H5410  194    0.000000     0.000000 0 0    0.100000   0.000000        0 2.00 0
H5210  195          1505  141    0        3   9            Streamer compass
H5410  195    0.000000     0.000000 0 0    0.100000   0.000000        0 2.00 0
H5210  196          1506  142    0        3   9            Streamer compass
H5410  196    0.000000     0.000000 0 0    0.100000   0.000000        0 2.00 0
H5210  197          1507  143    0        3   9            Streamer compass
H5410  197    0.000000     0.000000 0 0    0.100000   0.000000        0 2.00 0
H5210  198          1508  144    0        3   9            Streamer compass
H5410  198    0.000000     0.000000 0 0    0.100000   0.000000        0 2.00 0
H5210  199          1509  145    0        3   9            Streamer compass
H5410  199    0.000000     0.000000 0 0    0.100000   0.000000        0 2.00 0
H5210  200          1510  146    0        3   9            Streamer compass
H5410  200    0.000000     0.000000 0 0    0.100000   0.000000        0 2.00 0
H5210  201          1511  147    0        3   9            Streamer compass
H5410  201    0.000000     0.000000 0 0    0.100000   0.000000        0 2.00 0
H5210  202          1512  148    0        3   9            Streamer compass
H5410  202    0.000000     0.000000 0 0    0.100000   0.000000        0 2.00 0
H5210  203          1513  149    0        3   9            Streamer compass
H5410  203    0.000000     0.000000 0 0    0.100000   0.000000        0 2.00 0
H5210  204          1514  150    0        3   9            Streamer compass
H5410  204    0.000000     0.000000 0 0    0.100000   0.000000        0 2.00 0
H5210  205          1601  151    0        3   9            Streamer compass
H5410  205    0.000000     0.000000 0 0    0.100000   0.000000        0 2.00 0
H5210  206          1602  152    0        3   9            Streamer compass
H5410  206    0.000000     0.000000 0 0    0.100000   0.000000        0 2.00 0
H5210  207          1603  153    0        3   9            Streamer compass
H5410  207    0.000000     0.000000 0 0    0.100000   0.000000        0 2.00 0
H5210  208          1604  154    0        3   9            Streamer compass
H5410  208    0.000000     0.000000 0 0    0.100000   0.000000        0 2.00 0
H5210  209          1605  155    0        3   9            Streamer compass
H5410  209    0.000000     0.000000 0 0    0.100000   0.000000        0 2.00 0
H5210  210          1606  156    0        3   9            Streamer compass
H5410  210    0.000000     0.000000 0 0    0.100000   0.000000        0 2.00 0
H5210  211          1607  157    0        3   9            Streamer compass
H5410  211    0.000000     0.000000 0 0    0.100000   0.000000        0 2.00 0
H5210  212          1608  158    0        3   9            Streamer compass
H5410  212    0.000000     0.000000 0 0    0.100000   0.000000        0 2.00 0
H5210  213          1609  159    0        3   9            Streamer compass
H5410  213    0.000000     0.000000 0 0    0.100000   0.000000        0 2.00 0
H5210  214          1610  160    0        3   9            Streamer compass
H5410  214    0.000000     0.000000 0 0    0.100000   0.000000        0 2.00 0
H5210  215          1611  161    0        3   9            Streamer compass
```

H5410	215	0.000000	0.000000	0 0	0.100000	0.000000	0 2.00 0
H5210	216	1612	162	0	3 9		Streamer compass
H5410	216	0.000000	0.000000	0 0	0.100000	0.000000	0 2.00 0
H5210	217	1613	163	0	3 9		Streamer compass
H5410	217	0.000000	0.000000	0 0	0.100000	0.000000	0 2.00 0
H5210	218	1614	164	0	3 9		Streamer compass
H5410	218	0.000000	0.000000	0 0	0.100000	0.000000	0 2.00 0
H5200	219		0	0	0 9		Not in use
H5400	219	0.000000	0.000000	0 0	1.000000	0.000000	0 0.00 0
H5200	220		0	0	0 10		Not in use
H5400	220	0.000000	0.000000	0 0	1.000000	0.000000	0 0.00 0
H5200	221		0	0	0 0		Not in use
H5400	221	0.000000	0.000000	0 0	1.000000	0.000000	0 0.00 0
H5200	222		0	0	0 0		Not in use
H5400	222	0.000000	0.000000	0 0	1.000000	0.000000	0 0.00 0
H5200	223		0	0	0 0		Not in use
H5400	223	0.000000	0.000000	0 0	1.000000	0.000000	0 0.00 0
H5200	224		0	0	0 0		Not in use
H5400	224	0.000000	0.000000	0 0	1.000000	0.000000	0 0.00 0
H5200	225		0	0	0 0		Not in use
H5400	225	0.000000	0.000000	0 0	1.000000	0.000000	0 0.00 0
H5200	226		0	0	0 0		Not in use
H5400	226	0.000000	0.000000	0 0	1.000000	0.000000	0 0.00 0
H5200	227		0	0	0 0		Not in use
H5400	227	0.000000	0.000000	0 0	1.000000	0.000000	0 0.00 0
H5200	228		0	0	0 0		Not in use
H5400	228	0.000000	0.000000	0 0	1.000000	0.000000	0 0.00 0
H5200	229		0	0	0 0		Not in use
H5400	229	0.000000	0.000000	0 0	1.000000	0.000000	0 0.00 0
H5200	230		0	0	0 0		Not in use
H5400	230	0.000000	0.000000	0 0	1.000000	0.000000	0 0.00 0
H5200	231		0	0	0 0		Not in use
H5400	231	0.000000	0.000000	0 0	1.000000	0.000000	0 0.00 0
H5200	232		0	0	0 0		Not in use
H5400	232	0.000000	0.000000	0 0	1.000000	0.000000	0 0.00 0
H5200	233		0	0	0 0		Not in use
H5400	233	0.000000	0.000000	0 0	1.000000	0.000000	0 0.00 0
H5200	234		0	0	0 0		Not in use
H5400	234	0.000000	0.000000	0 0	1.000000	0.000000	0 0.00 0
H5200	235		0	0	0 0		Not in use
H5400	235	0.000000	0.000000	0 0	1.000000	0.000000	0 0.00 0
H5200	236		0	0	0 0		Not in use
H5400	236	0.000000	0.000000	0 0	1.000000	0.000000	0 0.00 0
H5200	237		0	0	0 0		Not in use
H5400	237	0.000000	0.000000	0 0	1.000000	0.000000	0 0.00 0
H5200	238		0	0	0 0		Not in use
H5400	238	0.000000	0.000000	0 0	1.000000	0.000000	0 0.00 0

```
H5200  239                          0     0       0    0                          Not in use
H5400  239      0.000000      0.000000 0 0      1.000000     0.000000              0 0.00 0
H5200  240                          0     0       0    0                          Not in use
H5400  240      0.000000      0.000000 0 0      1.000000     0.000000              0 0.00 0
H5200  241                          0     0       0    0                          Not in use
H5400  241      0.000000      0.000000 0 0      1.000000     0.000000              0 0.00 0
H5200  242                          0     0       0    0                          Not in use
H5400  242      0.000000      0.000000 0 0      1.000000     0.000000              0 0.00 0
H5200  243                          0     0       0    0                          Not in use
H5400  243      0.000000      0.000000 0 0      1.000000     0.000000              0 0.00 0
H5200  244                          0     0       0    0                          Not in use
H5400  244      0.000000      0.000000 0 0      1.000000     0.000000              0 0.00 0
H5200  245                          0     0       0    0                          Not in use
H5400  245      0.000000      0.000000 0 0      1.000000     0.000000              0 0.00 0
H5200  246                          0     0       0    0                          Not in use
H5400  246      0.000000      0.000000 0 0      1.000000     0.000000              0 0.00 0
H5200  247                          0     0       0    0                          Not in use
H5400  247      0.000000      0.000000 0 0      1.000000     0.000000              0 0.00 0
H5200  248                          0     0       0    0                          Not in use
H5400  248      0.000000      0.000000 0 0      1.000000     0.000000              0 0.00 0
H5200  249                          0     0       0    0                          Not in use
H5400  249      0.000000      0.000000 0 0      1.000000     0.000000              0 0.00 0
H5200  250                          0     0       0    0                          Not in use
H5400  250      0.000000      0.000000 0 0      1.000000     0.000000              0 0.00 0
H6001 DGPS      1
H6201     1 1 101     0.0     0.0   26.2                                                V1G1
H6201     2 2 101     0.0     0.0    0.0                                                  V1
H6201     3 3 101    -0.2    -0.1   25.9                                                V1G2
H6201     4 4 202     0.0 -3471.7   -6.0                                               TB2G1
H6201     5 5 202     0.0 -3471.7   -8.0                                                 TB2
H6201     7 6 301    37.5  -143.2    0.1                                                A2G1
H6201     9 7 102     0.7     0.0   19.9                                                V2G1
H6201    10 8 102     0.0     0.0    0.0                                                  V2
H6201    11 9 201   150.0 -3471.7   -6.0                                               TB1G1
H6201    1210 201   150.0 -3471.7   -8.0                                                 TB1
H6201    1411 203  -150.0 -3471.7    2.0                                               TB3G1
H6201    1512 203  -150.0 -3471.7    0.0                                                 TB3
H6201    1713 204   150.0 -3522.7    2.0                                               TB4G1
H6201    1814 204   150.0 -3522.7    0.0                                                 TB4
H6201    2015 205     0.0 -3522.7   -6.0                                               TB5G1
H6201    2116 205     0.0 -3522.7   -8.0                                                 TB5
H6201    2317 206  -150.0 -3522.7   -6.0                                               TB6G1
H6201    2418 206  -150.0 -3522.7   -8.0                                                 TB6
H6201    6619 401   -25.0  -332.0    2.0                                                N1G1
H6201    6720 401   -25.0  -332.0    0.0                                                  N1
H6201    7021 101     0.0     0.0    0.0                                               QMA_1
T5201  21      2.000    1741 54  22     0.200    1741 54  23     0.000    1741 54
```

```
T5209  24     0.000    1741 54  25     0.000    1741 54  26     0.000    1741 54
T5201  27     0.000    1741 54  28     0.000    1741 54
T6202  702192122.435N 72 245.164E   0.00                                1741 54
T5201   1  3458.100    1741 54   2   169.931    1741 54   3  3466.300    1741 54
T5211   4   172.783    1741 54   5  3482.500    1741 54   6   175.163    1741 54
T5201   7  3428.500    1741 54   8   177.067    1741 54   9  3451.700    1741 54
T5211  10   179.781    1741 54  11  3473.300    1741 54  12   182.467    1741 54
T5201  13   454.200    1741 54  14   274.865    1741 54  15   149.200    1741 54
T5211  16   158.547    1741 54  17     0.000    1741 54  18     0.000    1741 54
T5201  19   331.900    1741 54  20   174.845    1741 54
T6202   11192122.266N 72 245.156E  26.20                                1741 55
E1000                      735               735 1995  88 174156.3   1
T5216 122   350.700    1741 56
T6202  702192122.644N 72 245.186E   0.00                                1741 56
T5201  21     2.000    1741 56  22     0.200    1741 56  23     0.000    1741 56
T5209  24     0.000    1741 56  25     0.000    1741 56  26     0.000    1741 56
T5201  27     0.000    1741 56  28     0.000    1741 56
T5201   1  3458.200    1741 57   2   169.923    1741 57   3  3466.200    1741 57
T5211   4   172.773    1741 57   5  3482.200    1741 57   6   175.159    1741 57
T5201   7  3428.700    1741 57   8   177.064    1741 57   9  3451.700    1741 57
T5211  10   179.778    1741 57  11  3472.900    1741 57  12   182.463    1741 57
T5201  13   453.800    1741 57  14   274.888    1741 57  15   149.300    1741 57
T5211  16   158.443    1741 57  17     0.000    1741 57  18     0.000    1741 57
T5201  19   331.900    1741 57  20   174.798    1741 57
```

Glossary

Almanac — Data transmitted by a GPS satellite which include orbit information on all the satellites, clock correction and atmospheric delay parameters. These data are used to facilitate rapid SV acquisition. The orbit information is a subset of the ephemeris data with reduced accuracy.

Ambiguity — The unknown integer number of cycles of the reconstructed carrier phase contained in an unbroken set of measurements from a single satellite pass at a single receiver.

a posteriori — A value determined as a result of a positioning algorithm.

a priori — Pre-determined value used as a constant in a positioning algorithm.

Arc to chord correction, δ — The angular quantity to be added algebraically to a grid bearing to obtain a plane bearing:

$$\theta = \beta + \delta = \alpha + \gamma + \delta \tag{G.1}$$

The arc to chord corrections differ in amount and sign at the two ends of a line. For lines which do not cross the central meridian less than one-third of its length from one end, the bow is determined by the longer part. Note that:

$$\Delta\beta = \delta_1 - \delta_2 \tag{G.2}$$

The arc to chord correction is sometimes called the 't— T' correction.

Simplified formulae for arc to chord — on the zone boundary the arc to chord correction for a north-south line is about 0.85 second per kilometre. The formula for arc to chord correction can be written

$$\delta_1'' = -(N_2 - N_1)\left(2E_1 + E_2 - 1.5 \times 10^6\right) / \left(6r_m^{\ 2} \sin 1''\right) \tag{G.3}$$

This yields results correct to 0.01second for a line 30 kilometres long.

For programming applications, this formula may be written as follows:

$$\sin\delta_1 = -(N_2 - N_1)\left(2E_1 + E_2 - 1.5 \times 10^6\right) / 6r_m^{\ 2} \tag{G.4}$$

If the numerical difference between the arc to chord corrections at each end of the line can be neglected the formula can be simplidied to:

$$\delta_1'' = \delta_2'' - (N_2 - N_1)\left(E_2 + E_1 - 10^6\right) / \left(4r_m^{\ 2} \sin 1''\right) \tag{G.5}$$

The function $1/4rm^2 \sin 1''$ has a value of 0.127×10^{-8} anywhere on the Australian Map Grid and on the UTM grid (WGS 72 spheroid) between 8° and 57° South Latitude. Using this value, the formula can be reduced to:

$$\delta_1'' = -\delta'' = -(N_2 - N_1)\left(E_2 + E_1 - 10^6\right)\left(0.127 \times 10^{-8}\right) \tag{G.6}$$

The error in this simplified formula is 0".6 on the test line Buninyong— Flinders Peak.

The sign of the arc to chord correction will always follow strictly from the given formula but can be quickly checked by reference to a traverse diagram.However, some care is needed in the case where a line crosses the central meridian.

On an orthomorphic grid such as the AMG or the UTM Grid, the geodesic joining wto points will always plot as a curved line lying on that side of the straight line joining the two points where the projection scale factor is greater. Therefore, the curvature of this line will be reversed at the point where the line crosses the central meridian. However, in the case where the central meridian divides a line in such a way that one part of the line is less than one third of the total line length, the diagram approach for determination of the sign of the arc to chord correction will fail. In this case the sign is determined by the concavity of the longer part..

The digram in Figure G.1which is greatly exaggerated, illustrates this special case.

Argument of latitude — The sum of the true anomaly and the argument of perigee.

Argument of perigee — The angle or arc from the ascending node to the closest approach of the orbiting body to the focus or perigee point, as measured at the focus of an elliptical orbit, in the orbital plane in the direction of motion of the orbiting body.

Ascending node — The point at which an object's orbit crosses the reference plane (e.g., equatorial plane) from south to north.

Azimuth α (see Figure G.2) — A horizontal angle measured from the spheroidal meridian clockwise from north through 360°.

Bandwidth — A measure of the width of the spectrum of a signal (frequency domain representation of a signal) measured in Hertz. In a GPS receiver, the tracking bandwidth is dependent upon the receiver dynamics and affects the amount of noise included in the received signal and hence the position computation. In the differential link, bandwidth is one of the primary controls on data rate.

Baseline — The three-dimensional vector distance between a pair of stations for which simultaneous GPS data has been collected and processed with differential techniques. The most accurate GPS result.

Bias — See Integer bias terms.

Binary biphase modulation — Phase changes of either 0 or 180 degrees on a constant frequency carrier (representing a binary 0 or 1 respectively). GPS signals are biphase modulated.

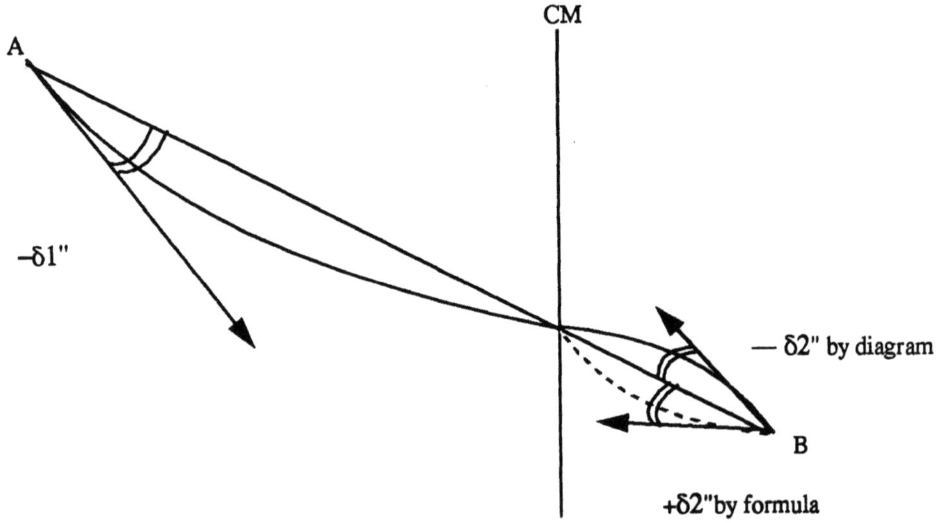

Figure G.1 Arc to chord.

Binary pulse code modulation — Pulse modulation using a string of binary numbers (codes). This coding is usually represented by ones and zeros with definite meanings assigned to them, such as changes in phase or direction of a wave.

Blue book — Slang term derived from a blue NGS reference book. The book contains information and formats required by NGS for survey data that is submitted to be considered for use in the national network.

C/A code — The coarse/acquisition (or clear/acquisition) code modulated onto the GPS L_1 signal. This code is a sequence of 1023 pseudo-random binary biphase modulations on the GPS carrier at a chipping rate of 1.023 MHz, thus having a code repetition period of 1 millisecond.

Carrier — A radio wave having at least one characteristic (such as frequency, amplitude, phase) which may be varied from a known reference value by modulation.

Carrier beat phase — The phase of the GPS signal remaining when the incoming Doppler— shifted carrier is differenced to (beat with) the nominally constant reference frequency generated by the receiver itself.

Carrier frequency — The frequency of the unmodulated fundamental output of a radio transmitter. the GPS L_1 carrier frequency is 1575.42MHz.

Celestial equator — The great circle that is the projection of the earth's geographical equator of rotation onto the celestial sphere. Its poles are the north and south celestial poles.

Celestial meridian — The vertical circle through the elevated celestial pole. It also passes through the other celestial pole, the astronomical zenith and the nadir.

Chip — The length of time to transmit either a zero or a one in a binary pulse code.

Chip rate — Number of chips per second (e.g., C/A code = 1.023 MHz).

Clock dither — A GPS space segment error introduced under Selective Availability whereby the transmission times of the pseudo— random noise (PRN) code from any Block II satellite are slightly advanced or retarded to corrupt the measurement of true signal travel time by the receiver, hence increasing the UERE and degrading the fix accuracy. See pseudo-range.

Clock offset — Constant time difference between two clocks.

Code division multiple access (CDMA) — A method of frequency reuse whereby many radios use the same frequency but with each one having separate and unique cross-correlation properties.

Control hub — A centralized onshore facility that allows the reception of pseudo-range data from several reference stations, normally via modem. Based on this facility to compare and contrast data, a network solution is then possible to improve QC provision and system accuracy.

Conventional International Origin (CIO) — Average position of earth's rotation axis during the years 1900 — 1905.

Convergence — The angle from true north (the direction of the meridian) to grid north. The convergence is positive for a point lying to the east of the central meridian.in the northern hemisphere and positive for a point lying to the west of the central meridian in the southern hemisphere.

Correlation-type channel — A GPS receiver channel which uses a delay lock loop to maintain an alignment (correlation peak) between the replica of the GPS code generated in the receiver and the received code.

Deflection of the vertical — The angle between the normal to the ellipsoid and the vertical (true plumbline). Since this angle has both a magnitude and a direction, it is usually resolved into two components: one in the meridian and the other perpendicular to it in the prime vertical.

Delay lock — the technique by which the received code (generated by the satellite clock) is compared with the internal code (generated by the receiver clock) and the latter is shifted in time until the two codes match. Delay lock loops can be implemented in several ways; tau dither and early-minus-late gating.

Delta pseudo-range — See Reconstructed carrier phase.

Differential processing — GPS measurements can be differenced between receivers, satellites and epochs. Note that this term is **not** the same as differential positioning, which is something completely different. The present convention for differential processing of GPS phase measurements is to take differences between receivers (single difference), then between satellites (double difference), then between measurement epochs (triple difference).

- A single difference measurement between receivers is the instantaneous difference in phase of the signal from the same satellite experienced by two receivers simultaneously.
- A double difference measurement is obtained by taking the difference between the single difference for one satellite and that of a chosen reference satellite.
- A triple difference measurement is the difference between a double difference at one epoch of time and the same double difference at the preceding epoch.

Differential (relative) positioning — The determination of relative positions of two or more navigation receivers which are simultaneously tracking the same radio-positioning signals. The technique relies on the principle that many of the error sources affecting the accuracy of a radio-positioning fix are systematic and relatively constant over a wide area. Hence the position error measured at one known location can be applied to stand alone fixes at other unknown locations in the form of correction to significantly improve their accuracy (dynamic differential positioning is a real-time calibration technique achieved by sending corrections to the mobile user from one or more monitor stations.)

DOP — dilution of precision. A description of the purely geometrical contribution to the uncertainty in a position fix. At the end of the least squares computation for position we end up with a covariance matrix. The trace of the covariance matrix is another name for its leading diagonal, and the matrix is always symmetrical about the leading diagonal. Different kinds of DOPs are simply the square roots of the traces of different sub-matrices of the overall covariance matrix of the parameters. These can be listed as follows:

- GDOP — geometric (three position coordinates plus the clock offset in the solution)
- PDOP — position (three coordinates)
- HDOP — horizontal (two horizontal coordinates)
- VDOP — vertical (height only)
- TDOP — time (clock only)
- RDOP — relative (normalized to 60 seconds).

Doppler aiding — The use of Doppler carrier-phase measurements to smooth code- phase position measurements.

Doppler shift — The apparent change in frequency of a received signal due to the rate of change of distance between the transmitter and receiver.

Double difference method — A method to determine that set of ambiguity values which minimizes the variance of the solution for a receiver pair baseline vector

Dynamic positioning — Determination of a timed series of sets of coordinates for a moving receiver, each set of coordinates being determined from a single data sample, and usually computed in real time.

Earth-centered earth-fixed (ECEF) — Cartesian coordinate system where the X direction is the intersection of the prime meridian (Greenwich) with the equator. The vectors rotate with the earth. Z is the direction of the spin axis.

Eccentricity anomaly E — The regularizing variable in the two-body problem. E is related to the mean anomaly M by Kepler's equation: $M = E - e\sin E$ (e stands for eccentricity).

Eccentricity — The ratio of the distance from the centre of an ellipse to its focus to the semimajor axis,

$$e = \left(1 - \frac{b^2}{a^2}\right)^{-1/2} \tag{G.7}$$

where a and b are the semimajor and semiminor axis of the ellipse.

Ecliptic — The earth — sun orbital plane.

Elevation mask — The angle below which satellites should not be tracked — usually 15° but sometimes relaxed to 10°. The angle is set to minimize reflection and multipath, and to lessen the effect of refraction through the atmosphere.

Ellipsoid — In geodesy, unless otherwise specified, a mathematical figure formed by revolving an ellipse about its minor axis, It is often used interchangeably with 'spheroid'. Two quantities define an ellipsoid; these are usually given as the length of the semimajor axis a, and the flattening, $f = (a - b)/a$, where b is the length of the semiminor axis. Prolate and triaxial ellipsoids are invariably described as such.

Ellipsoidal height — The height of the antenna above the chosen ellipsoid (spheroid). This is NOT the same as the height above mean sea level, which is approximately the geoidal height. The difference between the two can be tens of metres.

Ephemeris — A set of orbital parameters transmitted by a satellite in real time (predicted data, or broadcast ephemeris), or available later after observation of the actual orbit by the tracking station network (precise ephemeris).

Epoch — The measurement interval or data frequency — usually 15 seconds in static GPS. Any one particular time, used in relation with the time of a position fix.

Fast switching channel — A switching channel with a sequence time short enough to recover (through software prediction) the integer part of the carrier beat phase.

Flattening

$$f = (a - b)/a = l - (-e^2)^{1/2} \tag{G.8}$$

where: a = semimajor axis

b = semiminor axis

c = eccentricity

Frequency band — A range of frequencies in a particular region of the electromagnetic spectrum.

Frequency spectrum — The distribution amplitudes as a function of frequency of the constituent waves in a signal.

Fullwave — Term used to differentiate between measurements made with signal-squared (codeless) and code-tracking receivers. Specifically, a receiver tracking L_2 P-code can make measurements using the whole L_2 wavelength (23cm).

Fundamental frequency — the fundamental GPS frequency is 10.23MHz. All other frequencies are derived from this. The L_1 frequency is 154 times the fundamental at 1575.42MHz, and the L_2 frequency is 120 times the fundamental frequency at 1227.60MHz.

GDOP — Geometric dilution of precision. The relationship between errors in user position and time and in satellite range.

$$GDOP^2 = PDOP^2 + TDOP^2 \tag{G.9}$$

See PDOP.

Geocentre — The centre or the earth.

Geodetic datum — A mathematical model designed to best fit part or all of the geoid. It is defined by an ellipsoid and the relationship between the ellipsoid and a point on the topographic surface established as the origin of the datum. This relationship can be defined by six quantities, generally (but not necessarily) the geodetic latitude, longitude, and the height of the origin, the two components of the deflection of the vertical at the origin and the geodetic azimuth of a line from the origin to some other point. the GPS uses WGS84, which see.

Geodetic symbols — Here is a list of internationally recognized geodetic symbols. These are used with the Australian Map Grid, The Australian National Spheroid, WGS72 spheroid and the UTM grid. See the entry under Greek alphabet for aid with the symbols.

ϕ	Geodetic latitude, negative south of the equator
ϕ_1, ϕ_2	Latitude at points 1 and 2 respectively
ϕ_m	$(\phi_1+ \phi_2) /2$
$\Delta\phi$	$\phi_1 - \phi_2$
$\Delta\phi''$	$\Delta\phi$ expressed in seconds of arc
λ	Geodetic longitude measured from Greenwich, positive eastwards
$\lambda 0$	Geodetic longirude of a central meridian
ω	Geodetic longitude measured from a central meridian, positive eastwards.: $\omega = \lambda - \lambda 0$
E'	Easting measured from a Central Meridian, positive eastwards
N'	Northing measured from the equator, negative southwards
E	E' + 500 000 metres
N	N' in the northern hemisphere
	N' + 10 000 000 metres in the southern hemisphere
ρ,ν	Radii of curvature of the spheroid in meridian and prime vertical respectively
α	Azimuth, clockwise through 360° from true north
β	Grid bearing, clockwise through 360° from grid north
θ	Plane bearing, clockwise through 360° from grid north
γ	Grid convergence, positive when grid north is west of true north, negative when grid north is east of true north: $\beta = \alpha + \gamma$
δ	Arc to chord correction with sign defined by the equations: $\theta = \beta + \delta = \alpha + \gamma + \delta$
$\Delta\alpha$	Meridian convergence
$\Delta\beta$	Line curvature
s	Spheroidal distance
s'	Sea level, or geoidal, distance
S	Grid distance.
L	Plane distance
m	Meridian distance, true distance from the equator, negative southwards
G	Mean length of an arc of one degree of the meridian
σ	m/G
a, b	Major and minor semi— axis of the spheroid
f	$(a - b) /a$ = flattening
e^2	$(a^2 - b^2) /a^2$ = (eccentricity) ^{2}

e'^2	$(a^2 - b^2)/b^2$ = (second eccentricity) ^{2}
n	$(a - b)/(a + b) = f/(2 - f)$
k_0	Central scale factor = 0.999 6
k	Point scale factor
K	Line scale factor
t	tan ϕ
ψ	ν/ρ
ϕ'	Latitude for which $m = N'/k_0$ = Foot — point latitude
	t', ψ', ρ', ν' are functions of the latitude ϕ'
R^2	$\rho\nu$.
$R\alpha$	Radius of curvature in a given azimuth.
r^2	$R^2 k_0^2 = \rho\nu k_0^2$
rm^2	$\rho\nu k_0^2$ at ϕm
h	Spheroidal height
H	Height of a point above the geoid, or orthometric height
N	h — H = geoid— [heroid separation.

Note:

E', N', E, N, S, L, r, k and K include the central scale factor, k_0.

s, ρ, ν, R and m are true distances, which must be specifically multiplied by k_0 when necessary.

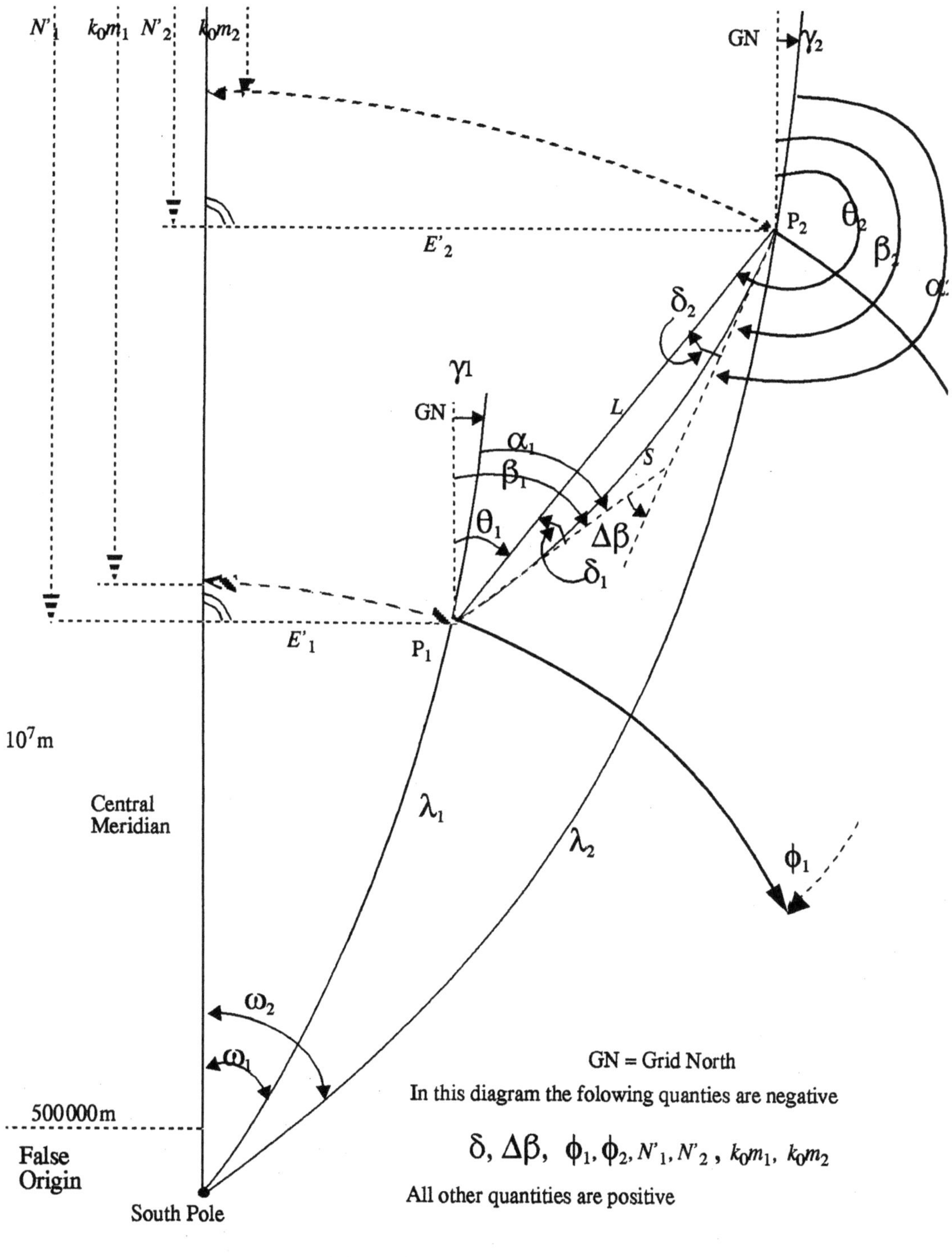

Figure G.2 Geodetic symbols.

Geoid — The particular equipotential surface which coincides with mean sea level, and which may be imagined to extend through the continents. This surface is everywhere perpendicular to the force of gravity.

Geoid height — The height above the geoid is often called elevation above mean sea level.

GPS — Global positioning system consisting of:

- a space segment (up to 24 NAVSTAR satellites in 6 different orbits)
- the control segment (5 monitor station, 1 master control station and 3 upload stations)
- the user segment (GPS receivers).

NAVSTAR satellites carry extremely accurate atomic clocks and broadcast coherent simultaneous signals.

GPS ICD— 200 — The GPS Interface Control Document is a government document that contains the full technical description of the interface between the satellites and the user. GPS receivers must comply with this specification if they are to receive and process GPS signals properly.

Gravitational Constant — The proportionality constant in Newton's Law of Gravitation:

$$G = 6.672 \times 10^{-11} Nm \quad (G.10)$$

Greek alphabet

Alpha	A	α
Beta	B	β
Gamma	Γ	γ
Delta	Δ	δ
Epsilon	E	ε
Zeta	Z	ζ
Eta	H	η
Theta	Θ	θ
Iota	I	ι
Kappa	K	κ
Lambda	Λ	λ
Mu	M	μ
Nu	N	ν
Xi	Ξ	ξ

Omicron	O	o
Pi	Π	π
Rho	P	ρ
Sigma	Σ	σ
Tau	T	τ
Upsilon	Y	υ
Phi	Φ	φ
Chi	X	χ
Psi	Ψ	ψ
Omega	Ω	ω

Greenwich Mean Time (GMT) — GMT and UT are often used interchangeably.

Grid bearing, line curvature and grid distance (see Figure G.2) — A line on the spheroid of length *s* is projected on the grid as an arc.

Grid bearing, β at a point on the arc, is the angle between grid north and the tangent to the arc at the point. It is measured from grid north clockwise through 360°.

Line curvature, $\Delta\beta$, is the change in grid bearing between two points on the arc.

$$\beta_2 = \beta_1 + \Delta\beta \pm 180° \tag{G.11}$$

Grid distance, *S* is the length measured along the arc of the projected line whose spheroidal distance is s.

Grid convergence γ — The angular quantity to be added algebraically to an azimuth to obtain agrid bearing:

$$GridBearing = Azimuth + GridConvergence \tag{G.12}$$

$$\beta = \alpha + \gamma \tag{G.13}$$

In the southern hemisphere the sign of the grid convergence is positive for points east of a central meridan and negative for points west of it.

Simplified formulae for grid convergence — When performing simplified computations on the AMG or UTM Grid it may occasionally be required to convert an azimuth to a grid bearing. A grid bearing, β, is derived from an azimuth, α, algebraically adding the grid convervence, γ, to the azimuth. In the southern hemisphere, grid convergence is positive east of the central meridian.

$$\beta = \alpha + \gamma \tag{G.14}$$

Grid convergence is most easily determined from geographical coordinates. The rigorous formula for obtaining frid convervence from geographical coordinates may be reduced to:

$$\gamma'' = \omega'' \sin\phi + \frac{1}{3\,(206265)^2}\,(\omega'')^3 \sin\phi\,(\cos\phi)^2\left(2\psi^2 - \psi\right) \quad \text{(G.15)}$$

This formula may be further reduced to:

$$\tan\gamma = -\sin\phi\tan\omega \quad \text{(G.16)}$$

which is accurate to less than 0.1" on both of the test lines.

Depending upon the accuracy required, values for ϕ and ω (= λ — $\lambda 0$) may be taken from any good quality map.

Alternatively, the rigorous formula for obtaining grid convergence from AMG or UTM grid coordinates may be reduced to :

for angles in seconds of arc

$$\gamma'' = \left[-t'x + t'\frac{\left(x^3\right)}{3}\left(-2\psi'^2 + 3\psi' + t'^2\right)\right]206265 \quad \text{(G.17)}$$

where $x = E'/k_0 \nu'$

Halfwave — Measurements made using L_2-squared measurements. The squaring process results in only half of the original L_2 wavelength being available.

HDOP — Horizontal dilution of precision.

Heights, spheroidal height, *h* — The distance of a point above the spheroid measured along the normal from that point to the surface of the spheroid.

Geoidal height, *H* — The distance of a point above the geoid measured along the normal from that pont to the surface of the geoid. It is also referred to as the orthometric height.

Geoid— Spheroid separation, *N* — The distance from the surface of the spheroid to the surface of the geoid measured along the normal to the spheroid. This distance can be either positive or negative depending upon whether the deoid is respectively above or below the spheroid.

HOW — Handover word. The word in the GPS message that contains time synchronization information for the transfer from the C/A code to the P code.

Inclination — The angle between the orbital plane of a body and some reference plane (e.g. the equatorial plane).

INS — Inertial navigation system, which contains an inertial measurement unit (IMU).

Integer bias term — The receiver counts the radio waves from the satellite, as they pass the antenna, to a high degree of accuracy. However, it has no information on the number of waves to the satellite at the time it started counting. This unknown number of wavelengths between the satellite and the antenna is the integer bias term.

Integrated Doppler — A measurement of Doppler shift frequency or phase over time.

Ionospheric delay — A wave propagating through the ionosphere (which is a nonhomogeneous, in space and time, and dispersive medium) experiences delay. Phase delay depends on electron content and affects carrier signals. Group delay depends on dispersion in the ionosphere as well, and affects signal modulation (codes). the phase and group delay are of the same magnitude but opposite sign.

JPO — the Joint Program Office for GPS located at the USAF Space Division in El Segundo, California. THE JPO consists of the USAF Program Manager and Deputy Program Managers representing the Army, Navy, Marine Corps, Coast Guard, Defence Mapping Agency and NATO.

Kalman filter — A numerical method used to track a time— varying signal in the presence of noise. If the signal can be characterized by some number of parameters that vary slowly with time, then Kalman filtering can be used to tell how incoming raw measurements should be processed to best estimate those parameters as a function of time.

Kinematic surveying — A form of continuous differential carrier— phase surveying requiring only short periods of data observations. Operational constraints include starting from or determining a known baseline, and tracking a minimum of four satellites. One receiver is statically located at a control point, while others are moved between points to be measured.

Keplerian orbital elements — Elements allowing description of any astronomical orbit:

a: semimajor axis

e: eccentricity

o: argument of perigee

l: right ascension of ascending node

i: inclination

t: true anomaly

L_1 — The primary L-band signal radiated by each NAVSTAR satellite at 1575.42MHz. The L_1 beacon is modulated with the C/A and P codes, and with the NAV message. L_2 is centered at 1227.60MHz and is modulated with the P-code and the NAV message.

Lane — The area (or volume) enclosed by adjacent lines (or surfaces) of zero phase of either the carrier beat phase signal or the difference between two carrier beat phase signals. On the earth's surface a line of zero phase is the locus of all points for which the observed value would have an exact integer value for the complete instantaneous phase measurement. In three dimensions, this locus becomes a surface.

L band — the radio-frequency band extending from 390MHz to (nominally) 1550MHz.

Mean anomaly

$$M = n(t-T) \tag{G.18}$$

where n is the mean motion,

t is the time

T is the instant of perigee passage.

Mean motion

$$n = 2/P \tag{G.19}$$

where P is the period of revolution.

Meridian convergence (see Figure G.2) — $\Delta\alpha$ — is the change in the azimuth of a geodesic between two points on the spheroid:

$$ReverseAzimuth = ForwardAzimuth + MeridianConvergence \pm 180°$$
$$\alpha_2 = \alpha_1 + \Delta\alpha \pm 180° \tag{G.20}$$

Microstrip antenna — GPS antenna made from a precisely cut two-dimensional flat piece of metal foil glued to a substrate.

Monitor station — Equipment installed at a known location on- or offshore, intended to measure one or more aspects of the performance of a differential system, including reception of the differential corrections and accuracy of the differential position or pseudo— range corrections. A worldwide group of stations is used in the GPS control segment to monitor satellite clock and orbital parameters. Data collected here is linked to a master station where corrections are calculated and controlled. These data are uploaded to each satellite at least once per day from an upload station.

Mobile station — Any other set of user equipment within a differential system that is not a reference or monitor station.

Multi-baseline (multiple reference station) solution — Computation of a position fix using differential corrections from a number of errors inherent in any one set of data.

Multichannel receiver — A receiver containing many independent channels. Such a receiver offers highest SNR because each channel tracks one satellite continuously.

Multipath — Interference similar to 'ghosting' on a TV screen which occurs when the satellite signal arrives at the antenna via different paths, causing a phase change. An error resulting from interference between radio waves which have travelled between the transmitter and the receiver by two paths of different electrical lengths. Affects the GPS receiver as noise superimposed on to the position fix.

Multipath error — A positioning error resulting from interference between radio waves which have travelled between the transmitter and the receiver by two paths of different electrical lengths.

Multiplexing channel — A receiver channel which is sequenced through several satellite signals (each from a specific satellite and at a specific frequency) at a rate which is synchronous with the satellite message bit— rate (50 bits) per second, or 20 milliseconds per bit). Thus one complete sequence is completed in a multiple of 20 milliseconds.

NAD83 — North American Datum, 1983.

NAVDATA — The 1500-bit navigation message broadcast by each satellite at 50bps on both L_1 or L_2 beacons. This message contains system time, clock correction parameters, ionospheric delay model parameters, and the vehicle's ephemeris and health. this information is used to process GPS signals to obtain user position and velocity.

NAVSTAR — The name given to GPS satellites, built by Rockwell International which is an acronym formed from NAVigation System with Time and Ranging.

Observing session — The period of time over which GPS data are collected simultaneously by two or more receivers.

Outage — The occurrence in time and space of a GPS dilution of precision value exceeding a specified maximum.

Over determination — The use of redundant measurements in a position computation in order to provide an independent check on the accuracy; one of the fundamentals of surveying. In GPS terms this means, for example, using more than four pseudo-ranges to solve for the 4 unknowns (X, Y, Z and time)) in a 3D fix and hence obtaining residuals from the least squares computation that indicate the quality of each input measurement.

P— code — The protected or precise code used on both L1 and L2 GPS beacons. This code will be made available by the DOD only to authorized users.

PDOP position dilution of precision — A unitless figure expressing the confidence level of the user's position. A figure of about 3 or less is considered good for positioning, while a figure of over 7 would be considered poor. PDOP is directly related to satellite geometry, and small figures are obtained when the satellites in view are widely separated.

Parity error — A digital message is composed of 1s and 0s. Parity can be defined as the sum of these bits within a word unit. A parity error results when one of the bits is changed so that the parity calculated at reception is not the same as it was at transmission of the message.

Perigee — That point in a geocentric orbit where the geometric distance is a minimum. The closest approach of a body.

Phase lock — The technique whereby the phase of an oscillator signal is made to follow exactly the phase of a reference signal by first comparing the phases of the two signals, and then using the resulting phase difference signal to adjust the reference oscillator frequency to eliminate phase difference when the two signals are next compared.

Phase observable — See Reconstructed carrier phase.

Plane bearing and plane distance — A straight line can be drawn on the grid between the ends of the arc defined in the entry under Grid Bearing

Plane bearing θ, is the angle between grid north and this straight line.

Plane distance L is the length of this straight line. The difference in length between the plane distance, L, and the grid distance, S, is nearly always negligible.

Using plane bearings and plane distances, the formulae of plane trigonometry hold with complete rigour:

$$\tan\theta = \frac{\Delta E}{\Delta N} \tag{G.21}$$

$$\Delta E = L\sin\theta \tag{G.22}$$

$$\Delta N = L\cos\theta \tag{G.23}$$

Point positioning — A geographic position produced from one receiver in a standalone mode. At best, position accuracy obtained from a standalone receiver is 15 — 25m, depending on satellite geometry. With the SA switched on it is considerably worse.

Polar motion — Motion of the instantaneous axis of the rotation of the earth with respect to the solid body of the Earth. Irregular but more or less circular motion with and amplitude of about 15m and a main period of about 430 days (called Chandler wobble).

Precise positioning service (PPS) — The highest level of military dynamic positioning accuracy that will be provided by GPS, based on the dual-frequency P-code and having high anti-jam and anti-spoof qualities.

PRN — Pseudo-random noise — A sequence of digital 1s and 0s which appears to be randomly distributed like noise, but which in fact can be exactly reproduced.

Pseudolite — A ground-based GPS transmitter station which broadcasts a signal with a structure similar to that of an actual GPS satellite.

Pseudo-range — A measure of the apparent propagation time from the satellite to the receiver antenna, expressed as a distance. Pseudo-range is obtained by multiplying the apparent propagation time by the speed of light. It differs from the actual range by the amount that the satellite and user clocks are offset, by propagation delays and by other errors.

Range rate — The rate of change of range between the satellite and receiver. The range to a satellite changes due to satellite and observer motions. Range rate is determined by measuring the Doppler shift of the satellite beacon carrier.

RDOP relative dilution of precision

$$\frac{(\sigma'_{DX^2} + \sigma_{DY'^2} + \sigma_{DZ^2})^{1/2}}{\sigma_{DD}} \tag{G.24}$$

usually in units of m/cycle. Multiplying RDOP by of a double-difference measurement yields the spherical relative-position error.

Reconstructed carrier phase — The difference between the phase of the incoming Coppler— shifted GPS carrier and the phase of a nominally constant reference frequency generated in the receiver. For static positioning, the reconstructed carrier phase is sampled at epochs determined by a clock in the receiver.

The reconstructed carrier phase changes according to the continously integrated Doppler shift of the incoming signal, biased by the integral of the frequency offset between the satellite and receiver reference oscillators.

The reconstructed carrier phase can be related to the satellite— to— receiver range, once the initial range (or phase ambiguity) has been determined. A change in the satellite— to— receiver range of one wavelenfth of the GPS carrier (19cm for L_1) will result in a one cycle change in the phase of the reconstructed carrier.

Reduction of measured distances to the spheroid — Modern geodetic measurments are undertaken eith electronic distance meaurement equipment (EDM) between points on the earth's surface at different elevations above the surface of the spheroid. Due to the effects of atmospheric refraction, the light waves or microwaves used by EDM follow a curved path. Before this curved wave path distance can be used for any geodetic computations, it should be reduced to the surface of the spheroid by the application of both physical and geometric corrections

The physical corrections, which involve the application of certain velocity corrections to the measured wave path distance,are not dealt with here. The geometric corrections to the measured wave path distance, d_1 firstly to the wave path chord d_2, thence to the spheroidal chord distance d_3, and finally to the spheroidal distance s_0. All geometric corrections can be combined in the one rigorous formula to give the length of the spheroidal normal section(s) as follows:

$$s = 2R_{\alpha} arc\sin\frac{d_3}{2R_{\alpha}} \tag{G.25}$$

where

$$d_3 = \left|\frac{d_2^2 - (h_A - h_B)^2}{(1 + h_A/R_{\alpha})(1 + h_B/R_{\alpha})}\right|^{\frac{1}{2}} \tag{G.26}$$

The difference between the wave path length d_1 and the wave path chord $d_2$is a funtions of the EDM measuring medium used and also of the meteorological conditions prevailing alon the wave path at the time of measurement. This difference can usually be ignored for disatnce measurments of up to 15 kilometres in length using either light waves or microwaves.

The formulae given in equation (G.25) and equation (G.26), which enable the wave path chord d_2 to be directly reduced to the spheroidal distance(s), include the slope, spheroid level and chord to arc corrections. Thes slope correction reduces the wave path chord to a horizontal distance at the mean elevation of the terminals of the line. The spheroid level correction reduces the horizontal distance to the spheroidal chord distance d_3 and the chord to arc correction when applied to the spheroidal chord distance gives the spheroidal distance(s).

$$SlopeCorrection = \left(d_2^2 - \Delta h^2\right)^{\frac{1}{2}} - d_2 \tag{G.27}$$

$$SpheroidLevelCorrection = \frac{h_m}{R}\left(d_2^2 - \Delta h^2\right)^{\frac{1}{2}} \tag{G.28}$$

$$ChordToArcCorrection = \frac{d_3^3}{24R^2} \tag{G.29}$$

The chord to arc correction can usually be ignored for all but the most precise geodetic surveys. For a wave path distance d_3 of 30 kilometres, the correctin is 0.028 metre anywhere on the Australian National Spheroid. For a distance of 50 kilometres, the correction reaches approximately 0.13 metre.

The spheroid level correction can be re-written to bive a locality or line height factor as follows:

$$HeightFactor = 1 - \frac{h_m}{R + h_m} \tag{G.30}$$

If an accuracy of $1{:}10^5$ is required for the reduced distance, the mean height should be correct to 60 metres.

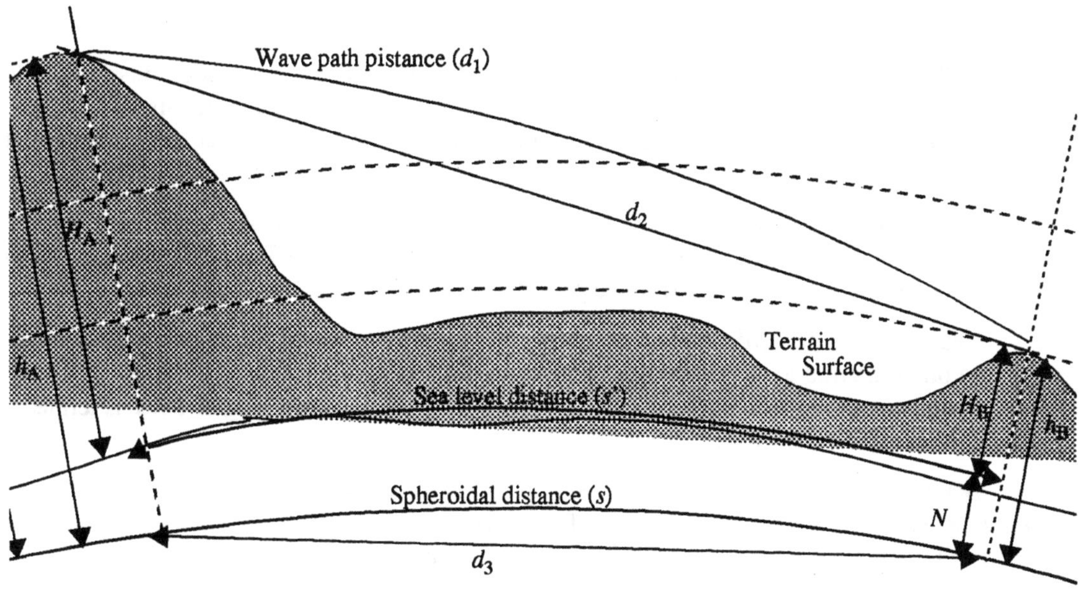

N values ar shoenin terms of N = + 4.9 metres at Johnston Geodetic Station

Figure G.3 Reduction of measured distance to the spheroid.

Relative navigation — A technique similar to relative positioning except that one or both, of the points may be moving. The pilot of a ship or aircraft may need to know his position relative to a harbour or runway. A datalink is used to relay the error terms to the moving vessel to allow real-time navigation.

Relative positioning — The process of determining the relative difference in position between two marks with greater precision than that to which the position of a single point can be determined. Here, a receiver (antenna) is placed over each spot and measurements are made by observing the same satellite at the same time. This technique allows cancellation (during computations) of all errors which are common to both observers, such as satellite clock errors, propagation delays, etc.

Right ascension of ascending node — The angular distance measured from the vernal equinox, positive to the east, along the celestial equator to the ascending node. Typically denoted by a capital omega. Used to discriminate between orbital planes.

RINEX — Receiver INdependent EXchange format — A set of standard definitions and formats to provide the free exchange of GPS data and facilitate the use of data from any GPS receiver with any software package. The format includes definitions for three fundamental GPS observables: time, phase and range. A complete description of the format is in the Commission VIII International Coordination of Space Techniques for Geodesy and Geodynamics GPS Bulletin, May— June 1989.

RTCM — Radio Technical Commission for Maritime Services — Commission set up to define a differential data link to relay GPS correction messages from a monitor station to a filed user. RTCM SC-104 recommendations define the correction message format and 16 different message types.

SATNAV — A local term referring to use of the older TRANSIT system for satellite nabigation. One major difference between TRANSIT and GPS is that the TRANSIT satellites are in low— altitude polar orbits with a 90 minute period.

Sea Level or geoidal distance *s'* — (see Figure G.2) When heights above the geoid, which are often referred to as orthometric heights, are used in the reduction of distances measured between points on the earth's surface to the surface of the geoid, the resulting distance is termed a sea level, or geoidal, distance *s'*.

Selective availability (SA) — A DOD program to control the accuracy of pseudo-range measurements, whereby the user receives a false pseudo-range which is in error by a controlled amount. Differential GPS techniqes can reduce these effects for local applications.

Semimajor axis — One half of the major axis of an ellipse.

SEP — Spherical error probable, a statistical measure of precision defined as the 50th percentile value of the three-dimensional position error statistics. Thus, half of the results are within a 3-SEP value.

Scale Factors, point pcale factor, *k* — is the ratio of an infinitesimal distance at a point on the grid to the coirresponding distance on the speroid:

$$k = \frac{dL}{ds} = \frac{dS}{ds} \qquad \text{(G.31)}$$

It is the distinguishing feature of conformal projections, such as the Universal Transverse Mercator, that this ratio is independent of the azimuth of the infinitesimal distance.

Line scale factor, *K* — From point to point along a line on the grid the point scale factor wil in general vary. The line scale factor is the ratio of a plane distance, L, on the grid to the corresponding speroidal distance, *s*:

Height Factor — is the ratio of an infinitesimal horizontal distance at a point on the speroid to the corresponding distance at a particular height above)or below) the speroid.

Combined Point Scale and Height Factor — is the product of the point scale factor and the height factor. It may be used to convert measured horizontal distances to grid distances. Its use is not rigorous, because the point scale factor is an approximation for the line or locality being computed.

Scale Factors simplified formulae — In order to convert a spheroidal distance, *s*, to a plane distance, *L*, or to obtain the spheroidal distance from a plane or grid distance, it is necessary to calculate the line scale factor. Since a Transverse Mercator projection is conformal, the scale at any point is the same in all directions, but varies with the distance from the central meridian.

To minimise the scale factor over the whole of the zone width, a central scale factor k0 = 0.999 6 is superimposed over all projected distances. After the application of this central scale factor, the point scale factor will vary from 0.999 6 on a central meridian to almost 1.001 0 on the zone boundary at latitude 8° South and to 1.000 3 on a zone boundary at latitude 45° South for the Australian Map Grid. For the UTM Grid (WGS 72 spheroid) the point scale factors will vary between the same values between the equator and latitude 56° South.

Point Scale Factor — The rigorous equation for point scale factor can be reduced to:

$$k = k_0\left[1 + \left(\frac{E'^2}{2r_m^{\ 2}}\right) + \left(\frac{E'^4}{24r_m^{\ 4}}\right)\right] \qquad \text{(G.32)}$$

This equation is accurate to 1 part in 10 million. Omission of the last term results in an accuracy of 2 parts in 10 million.

The equation can be simplified further using the following approximation:

$$k = 0.9996 + 1.23E'^2 10^{-14} \tag{G.33}$$

which is easy to remember. The formula is correct at latitude 34° and is accurate to 8ppm at the equator.

Examples

AMG	UTM
Station: BUNINYONG, Zone 55	*Station*: M, Zone 58
E' = −271300	E' = 287400
$k = 0.9996 + 1.23\,(-271300)^2 \times 10^{-14}$	$k = 0.9996 + 1.23\,(287400)^2 \times 10^{-14}$
= 1.000505	= 1.000616
which is accurate to 2 ppm	which is accurate to 4 ppm

For surveys of low acuracy and limited extent, the point scale factor can be substituted for the line scale factor and combined with the height factor to reduce slope distances for use in AMG or UTM Grid Computations.

Line scale factor — The scale factor will in general vary from one end of a line to the other, and the method used to determine ascale factor for the whole line will depend upon the accuracy of the survey and the length of the line.

The equation for line scale factor can be reduced to:

$$K = k_0\left[1 + \left(E'^2_1 + E'_1 E'_2 + E'^2_2\right)/6r_m^2\right]$$

For single lines or traverses, line scale factors can be obtained anywhere on the AMG (or UTM grid) by using either:

1. the point scale factor for the mean easting of the the line or traverse. This procedure is accurate to 1ppm in any line or traverse extending 33 kilometres in easting:

or

2. the mean of the point scale factors at the extremities of the line or traverse. This procedure is accurate to 1ppm in any line or traverse extending 16 kilometres in easting.

The accuracies stated above are independent of the location in the zone

Combined point scale and height factor — For surveys of lower accuracy and of limited extent, it may be quite adequate to obtain the line scale factor by using the methods given above or by simply deriving a locality scale factor based on the mean easting of the centre of the survey area. This factor can then be combined with a line or locality height factor, to obtain grid (plane) distances for computation of coordinates on the AMG or UTM grid. The procedure for the rigorous reduction of measured distances to the surface of the spheroid has already been covered in **Reduction of measured distances to the spheroid.**

The combined point scale and height factor is simply obtained by multiplying the point or locality scale factor by the line or locality height factor.

Numerical Example

s = 3 264.16

h_m = 400metres

$\phi_m = 35°$

E_m = 228 700

R_m = 6 370 800

E'_m = 271 300

Pointscale factor, $k = 0.9996 + 1.23 \times E'^2 x 10^{-14}$

$= 1.000505$

Height factor $= 1 - 400 / 6370800 + 400$

$= 0.999937$

Combined point scale and height factor = 1.000505 (0.999937)

Grid (plane distance = 3264.16 (1.000442)

= 3265.60 metres

A nomogram can be prepared to give the combined point scale and height factor for any given region.

Sidereal day — Time between two successive upper transits of the vernal equinox.

Simultaneous measurements — Measurements referenced to time— frame epochs which are either exactly equal, or else so closely spaced in time that the time misalignment can be accommodated by correction terms in the observation equation, rather than by parameter estimation.

Slope distance — The three dimensional vector distance from station one to station two. The shortest distance (a chord) between two points.

Slow switching channel — A switching channel with a sequencing period which is too long to allow recovery of the integer part of the carrier beat phase.

Solar day — Time between two successive upper transits of the sun.

Spheroid — See Ellipsoid.

Spheroidal distance *s* — The distance on the spheroid along either a normal section or a geodesic. The difference between the two is usually negligible, amounting to less than 20 millimetres in 3000 kilometres.

Spread spectrum — The received GPS signal is a wide bandwidth, low— power signal (— 160dBW). This property results from modulating the *L*- band signal with a PRN code in order to spread the signal energy over a bandwidth which is much greater than the signal information bandwidth. This is done to provide the ablility to receive all satellites unambiguously and to provide some resistance to noise and multipath.

Spread spectrum systems — A system in which the transmitted signal is spread over a frequency band much wider than the minimum bandwidth needed to transmit the information being sent.

Squaring-type channel — A GPS receiver channel which multiplies the received signal by itself to obtain a second harmonic of the carrier which does not contain the code modulation. Used in so— called codeless receiver channels.

SPS standard positioning service — The level of static and/or dynamic positioning capability that will be provided by GPS, based on the single-frequency C/A code. The accuracy of this service is to be set at a level consistent with national security.

Static positioning — Positioning applications in which the positions of static or neat static points are determined.

SV — Satellite vehicle or space vehicle.

Vertical — The line perpendicular to the geoid at any point. the direction of the force of gravity at that point. Plumb line.

WGS72 World Geodetic System (1972) — The mathematical reference ellipsoid previously used by GPS, having a semimajor axis of 6378.135km and a flattening of 1/298.26.

WGS84 World Geodetic System (1984) — The mathematical ellipsoid used by GPS since January 1987. The shift from WGS72 to WGS84 at our latitude (37degrees is about 13.6m east, 45m north and 2.7m up.

Widelane — A linear combination of L_1 andL_2 observations (L_1–L_2) used to partially remove ionospheric errors. This combination yields a solution in about one— third the time of a complete ionosphere-free solution.

Z-count — The GPS satellite clock time at the leading edge of the next data subframe of the transmitted GPS message (usually expressed as an integer number of 6 seconds).

Bibliography

Bomford, G.(1985) *Geodesy*, 4th edn.

Bowditch, N.(1977) *American Practical Navigator.*

Burnsid,e C.D.(1982) *Electronic Distance Measurement*, 2nd edn.

Cooper, M.A.R.(1982) *Fundamentals of Survey Measurement and Analysis.*

Cross, P.A.(1981) *The computation of position at sea*, Hydrographic Journal No. 20 April.

Cross, P.A.(1982) *Advanced Least Squares Applied to Position-Fixing*, North East London Polytechnic.

Cross, P.A., Hollwey, J.R., Small L.G. (1981) *Geodetic Appreciation, Working Papers*, North East London Polytechnic.

Cross, P.A., Pritchett, C.H. (1986) *A Kalman Filter for Real-Time Positioning during Geophysical Surveys at Sea*, FIG XVIII International Congress, Toronto. Canada.

Curtin University, School of Surveying and Land Information. *Geodesy* 384, 481 course notes.

Houtenbos, A.P.E.M.(1989) *Integrated Processing of 3D Marine Seismic Positioning Data*, The Hydrographic Journal. No. 52 April.

Hydrographer of the Navy (1965, 1973) *Admiralty Manual of Hydrographic Surveying*, Vol 1 and Vol 2.

Ingham, A.E. (1984) *Hydrography for the Surveyor and Engineer.*

Institution of Surveyors, Australia (1980) *AMG Without Tears.*

International Hydrographic Bureau, Monaco (1990) *Precise Positioning for Hydrographic Surveying*. IHB Special Publication No. 39.

Kemp, J.F. Young P. (1972) *Notes on Compass Work.*

Maloney, E.S. (1978) *Dutton's Navigation and Piloting.*

Milne, P.H.(1983) *Underwater Acoustic Positioning Systems.*

National Mapping Council of Australia (1986) *The Australian Geodetic Manual.* Special Publication 10.

Racal Survey, *Product Notes.*

Rainsford, H.F. (1968) *Survey Adjustments and Least Squares.*

Sun Microsystems Inc. (1988) *A RISC Tutorial.*

UKOOA Survey and Positioning Committee (1992) *Guidelines for the Measurement and Basic Quality Control of Offshore Positioning Data.*

UKOOA Survey and Positioning Committee (1992) *GPS in Oil and Gas Exploration.*

UKOOA Survey and Positioning Committee, *P1/84 Seismic Post Plot Data Format.*

UKOOA Survey and Positioning Committee, *P1/90 Post Plot Data Exchange Tape 1990 Format.*

UKOOA Survey and Positioning Committee (1986) *P2/86 Raw Marine Positioning Data Exchange Tape Format Version 1.1.*

UKOOA Survey and Positioning Committee (1992) *P2/91 Exchange Format for Raw Marine Positioning Data Version 1.1.*

Weltner, Grosjean, Schuster, Weber (1986) *Mathematics for Engineers and Scientists.*

Index

MIX
Papier aus verantwortungsvollen Quellen
Paper from responsible sources
FSC® C105338

If you have any concerns about our products,
you can contact us on
ProductSafety@springernature.com

In case Publisher is established outside the EU,
the EU authorized representative is:
Springer Nature Customer Service Center GmbH
Europaplatz 3, 69115 Heidelberg, Germany

Printed by Libri Plureos GmbH
in Hamburg, Germany